Astrophysics and Space Science Proceedings

Volume 41

More information about this series at
http://www.springer.com/series/7395

Jim Dungey with the first Royal Astronomical Society Dungey lecturer, Peter Cargill (Imperial College, London and St. Andrews University), on the occasion of the inaugural lecture, at Burlington House, London, 11 Jan 2013 (Photo credit/copyright: Q. Stanley, RAS)

David Southwood • Stanley W.H. Cowley FRS •
Simon Mitton
Editors

Magnetospheric Plasma Physics: The Impact of Jim Dungey's Research

Springer

Advancing Astronomy and Geophysics

Editors
David Southwood
Imperial College London
London
United Kingdom

Stanley W.H. Cowley FRS
Department of Physics & Astronomy
University of Leicester
Leicester
United Kingdom

Simon Mitton
St. Edmund's College
Cambridge
United Kingdom

ISSN 1570-6591 ISSN 1570-6605 (electronic)
Astrophysics and Space Science Proceedings
ISBN 978-3-319-18358-9 ISBN 978-3-319-18359-6 (eBook)
DOI 10.1007/978-3-319-18359-6

Library of Congress Control Number: 2015945381

Springer Cham Heidelberg New York Dordrecht London

Printed on acid-free paper

Springer International Publishing AG Switzerland is part of Springer Science+Business Media (www.springer.com)

Contents

Introduction: Jim Dungey and Magnetospheric Plasma Physics

This book's results make good background reading for much of modern magnetospheric physics. Its origin was a Festspiel for Professor Jim Dungey, former professor in the Physics Department at Imperial College on the occasion of his 90th birthday, 30 January 2013. Sadly, its appearance follows his death on May 9th this year (2015). Remarkably, although he retired 30 years ago, his pioneering and, often, maverick work in the 1950s through to the 1970s on solar terrestrial physics is probably more widely appreciated today than when he retired. Happily, the 90th birthday celebrations, which he enjoyed hugely, made his present high standing very clear.

Jim was a theoretical plasma physicist. Plasma physics itself was a new field when he did his Ph.D. at Cambridge in the late forties. The magnetosphere was a new, even unnamed regime when he began his career. At the time, ideas like frozen-in magnetic field and magnetohydrodynamic waves were opening up astrophysics as well as solar terrestrial physics. At the same time, it was becoming clear that plasmas, being perfect conductors, could lead to effects that were counter-intuitive to classical electromagnetic theory. I always suspected that the rather special properties of plasma electromagnetics and the departures from the behaviour of classical gases or the expectations of standard electromagnetism were part of Jim's fascination.

Stan Cowley's review here covers how Jim's reconnection model of the magnetosphere (see Fig. 1) evolved to become the standard model of solar terrestrial coupling. My own paper at the end of the collection argues that the paper resolved the confrontation between Scandinavian and UK schools that had begun many years earlier.

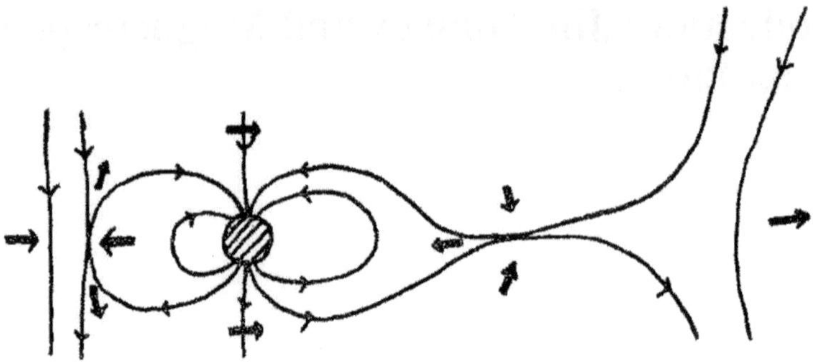

Fig. 1 A freehand sketch by Jim of his basic open magnetosphere idea. The sketch was originally published in J. Phys. Soc. Japan, 17, Suppl. A-ll, 15, 1962

Acceptance of Jim's open magnetosphere idea was long in coming. When Stan and I became students of Jim's in the 1960s we realised we were joining the "radical" school. I have always felt my career was kick-started by adopting and applying Jim's mode of thinking and thereby knowing many answers before others did.

Today's assessment of Jim's work is dominated by his epoch-making paper from 1961. In the 1980s, the sheer volume of simultaneous data from spacecraft, ground-based radars, auroral imagers and other sources provided incontrovertible evidence for a time-dependent model of how reconnection on the magnetopause and in the mid-tail modulates the exchange of magnetic flux between the magnetosphere and the solar wind and thus geomagnetic activity overall, the Dungey cycle. Here, Steve Milan follows up on Stan Cowley's work. One can see in his review how the many facilities built up in the last 30 years worldwide allow detailed, global monitoring of the solar terrestrial environment. Many, originally unsuspected, magnetospheric phenomena have now been found to fit naturally into the open magnetosphere picture. Accordingly, Dungey's open magnetosphere model now underpins a holistic picture explaining not only the magnetic and plasma structure of the magnetosphere but also its dynamics which can be monitored in real time.

Jim was always interested in computer simulation. This interest is represented here by the papers by Maha Abdalla and Jim Eastwood. In particular, Maha shows how modern-day simulation of solar terrestrial coupling can reproduce the real-time evolution of the solar terrestrial system in ways undreamt of in 1961.

However, if problems close to those worked on by Jim are radically changed by changes in computer power so have some of Jim's students gone off to apply ideas learnt from Jim in wider fields. Jim Eastwood takes as his starting point the many-body computation problem that he first encountered during his Ph.D. studies whilst undertaking simulations of non-adiabatic plasma flow in the current sheet of the geomagnetic tail. Jim Dungey's "light touch" technique for supervision provided key nuggets that the student could then exploit creatively. One such nugget stimulated Jim Eastwood to extend simulations in an array of situations beyond space plasma physics. Generalised fast computation methods made possible hitherto impracticable many-body calculations of galaxy formation, condensed matter, galaxy clustering, vacuum electronic devices and electromagnetic scattering in aerospace engineering applications.

One of the difficulties of grasping the depth of Jim's contributions to space plasma physics is the fact that his solution of the overall grand solar terrestrial coupling problem can eclipse the significance of his work on the many other fundamental physical processes that occur in near Earth space. In the 1960s, the field of solar terrestrial physics was rich with exotic physical phenomena newly revealed by the spacecraft data that were starting to appear in large quantities. Moreover, the legacy of International Geophysical Year (which had launched the space age) had left a heritage of data from the globally distributed ground-based networks set up in for the IGY. The new data from both sources suddenly meant that the magnetosphere of Earth made an excellent laboratory for collision-free plasma physics. The laboratory contained a menagerie of plasma waves in which charged particles can interact with wave electric and magnetic fields without collisional dissipation or damping.

The physics of the phenomenon of triggered very low frequency (VLF) emissions was picked out early on by Jim as a process that had to be both collision-free and also fundamentally non-linear. The first observed triggered VLF emissions were seen when rising or falling tone natural whistler mode signals appeared to be excited by the passage through the magnetosphere of terrestrial military Morse code transmissions. The non-linearity of the process was guaranteed by the fact that the triggered signals are initiated at a frequency shifted from the trigger frequency. Dave Nunn's paper here points out how Jim's basic ideas have remained explanative 50 years on. The experimental background has been transformed by controlled triggering experiments from the ground as well as the increased resolution of space sensors; theoretical work has been transformed by the enormous changes in computational capacity. Moreover, the triggering process has been found closely linked to the naturally occurring "chorus" emission that pervades the outer magnetosphere. The advent of advanced plasma wave particle measurements from instruments on space missions such as Cluster or the new Radiation Belt Storm Probes (now called the Van Allen Probes) is continuing to provide information showing nature capable of providing strings of rising and falling tones independent of the man-made stimulation that triggered the original interest.

At higher frequencies, an almost ubiquitous feature of magnetospheres is the emission of radio signals from somewhere above the auroral zones. In the case of the Earth, the discovery of Auroral Kilometric Radiation (AKR) came in the early 1970s. Finding the plasma mechanism behind the phenomenon was a great challenge. Robert Strangeway picks up the story here and the background to Wu and Lee providing the final explanation of the emissions not long after Professor Wu had spent some sabbatical time at Imperial College, London. Jim's contribution to the overall background to the problem was fundamental in explaining the phase space structure of auroral primary electron distributions above the aurora. Jim had given a Master's student, Steve Knight, a problem to find the relation between current and voltage along the magnetic field in the tenuous plasma above the aurora.

The "Knight relation" is now part of the standard model of ionospheric–magnetospheric coupling. It was the combination of Steve Knight's formalism with the insight provided by Wu and Lee's theory of relativistic electron behaviour that led to the solution of the AKR problem. Moreover, it was Jim's open magnetosphere model that gave the first explanation of why there should be intense electrical currents above the auroral zone which cause not only the light displays where electrons hit the atmosphere but also the radio emissions (which render Earth almost as bright as Saturn in the kilometric band).

One of the first surprises of the space age was the discovery that the Earth was surrounded by very energetic radiation belts. Jim, in a paper with Nakada and Hess, in 1965 applied the Liouville theorem and adiabatic theory to reorganise radiation belt data in the frame of the invariants expected to govern collision-free plasma in a closed system. The paper gave the first clear evidence that the bulk of the radiation belts had a source external to the magnetosphere and eliminated the original CRAND (Cosmic Ray Albedo Neutron Decay) theory. At the same time, Jim was interested in how radiation belt particles might be lost through scattering into the atmospheric loss cone by resonant interaction with plasma waves. He took some of the first steps towards what is now known as the Kennel–Petschek theory. Then he moved to look at longer time scale behaviour involving the charged particles' bounce along the field and guiding centre drift across the field. Understanding radiation belt variability remains a key challenge in space physics and is the principal goal of the recently launched NASA Van Allen probes mission. Relativistic electrons and ions in the Earth's radiation belts encounter a variety of wave modes including whistler mode chorus, plasmaspheric hiss, lightning generated whistlers, ultralow frequency (ULF) magnetohydrodynamic waves and electromagnetic ion cyclotron (EMIC) waves, all of which can interact efficiently with radiation belt particles. Jim took an interest in the significance of ULF waves very early. In 1953, Owen Storey suggested that the Earth must have an extended plasma environment on the basis of whistler mode propagation between hemispheres. In the next year, Jim followed up the idea by pointing out that if this was so, the extended plasma region (now, of course, called the magnetosphere) would provide cavities for magnetohydrodynamic waves which would resonate in the ULF (ultralow frequency) bands. It was a few years before IGY data showed that signals were seen in resonance (field line resonance) at magnetically conjugate points on the Earth. In this book, Jeff Hughes (Boston) takes up the story from there. He recounted the manner in which he had followed Jim's intuitions to compute the full wave solutions for ULF waves from the magnetosphere through the ionosphere, atmosphere and lithosphere. The Hughes rotation of polarisation, controversial at first but now accepted as standard, emerged from that work which then led on to the analysis of how resonant signals would damp and a prediction of the large-scale phase structure observed on the large-scale terrestrial magnetometer networks that would be deployed in the succeeding year. Analysis of space data revealed large amplitude signals with short transverse wavelengths which would not be seen on Earth due to the shielding effect of the ionosphere. Then the advent of high time resolution radar measurements from coherent scatter radars like STARE and

SABRE in Europe provided knowledge of the actual ionospheric behaviour and gave almost direct measurements of the electric field distribution there. Today the SuperDARN global network and multiple spacecraft measurements allow the use of ULF waves for detailed diagnostics of the magnetosphere.

In 1966, Jim submitted an idea for a multi-spacecraft magnetospheric mission to the European Space Research Organisation, predecessor of the European Space Agency (ESA). He called it TOPS, Tetrahedral Observatory Probe System. It was several decades before the science community as a whole caught up with his ideas. After a severe launch failure in 1996, in 2000 ESA launched the four Cluster spacecraft. Now Jim's original proposals for measuring plasma current structure through the "curlometer", as well as large-scale wave phase structure and numerous other techniques for separating space and time variation from moving sensors, could then be implemented. Matt Taylor (ESA) reviews here the science achievements of the Cluster mission concentrating particularly on the way in which the mission had revealed the detailed structure of some of the phenomena, large-scale currents, Kelvin–Helmholtz boundary waves, the auroral acceleration region and the structure of the dayside and nightside reconnection regions.

Cluster also means we need to return to magnetic reconnection, the process at the core of Jim's solution of the solar terrestrial coupling problem. Jonathan Eastwood (Imperial College) starts from Jim's 1953 paper in Philosophical Magazine, pointing out how much of the qualitative theory of reconnection was laid out then. He then turns to the last two decades in spacecraft observations of reconnection, driven very much by not only the availability of multipoint measurements from missions like Cluster but also large-scale computer simulations that Jim was always anticipating. As Jonathan points out, Jim, in the last paper that he published "Memories, Maxims and Motives" (Journal of Geophysical Research, vol. 99, pp 19189–19197, 1994), predicts the importance of the Hall effect in the central region.

The Festspiel introduced some advances that possibly Jim had not foreseen, although his ideas were fundamental. Margaret Kivelson (UCLA/University of Michigan) gave a review entitled, "Adventures in parameter space: Reconnection and the Magnetospheres of the Solar System", whose title played with one of Jim's seminal review papers from the 1960s. She took us through the variety of magnetospheres which have been seen directly during the space age. At one extreme, there is the magnetosphere of Ganymede, embedded in a sub-Alfvenic flow deep inside the Jovian magnetosphere, and at the other, there is the magnetosphere constituted by the heliosphere itself. The Voyager spacecraft, although detecting evidence of its presence ahead, has yet to definitively cross the heliopause. However, the IBEX (Interplanetary Boundary Explorer) spacecraft measures energetic neutral atoms (ENA) from the heliosphere and beyond and these can be used to image the regions where the ENA are formed. The discovery of the IBEX "ribbon" which she pointed out could be interpreted as the site of reconnection between galactic and heliospheric field at the heliopause. As IBEX has determined that there is no heliospheric shock and so is in sub-Alfvenic conditions, the largest magnetosphere associated with the solar system has much in common with the smallest. In between, the variations in spatial scale and reconnection time scale were discussed and compared in respect of Earth, Mercury, the giant planets, as well as Ganymede.

Jim's doctoral thesis work was undertaken with Fred Hoyle at Cambridge and was a study of magnetic reconnection. The penultimate paper of the book is by Peter Cargill and takes us back to this topic. It is based on the content of the first annual James Dungey lecture of the Royal Astronomical Society instituted immediately following his 90th birthday Festspiel. Peter starts by looking back to the 1950s examining the context and consequences of Jim's early work. He then moves to today and concludes with his personal views of how magnetic reconnection is likely to play a central role in energy release in the solar atmosphere. In particular, he shows how a range of reconnection events on different scales (flares, microflares and nanoflares) can account for coronal behaviour.

The collection concludes with a historical paper by me. As well as allowing me a shot at explaining why some ideas seen as obvious today took up to a century to emerge, it also permits me to put down on paper some of my own conversations with some of Jim's contemporaries, notably Hannes Alfvén, often seen as his major scientific antagonist. I look at the development of understanding of solar terrestrial coupling from the first identification of a possible material connection between Earth and Sun by Carrington and Balfour-Stewart during the great solar and geomagnetic disturbances of 1859 through to Jim's classic paper of 1961. From early days, there were competing views of how Sun and Earth might couple directly. By the early twentieth century, Scandinavian scientists and British scientists disagreed fundamentally. Controversy raged between British and Scandinavian schools through to the 1940s and 1950s. There were scientific difficulties at the heart of this. Nonetheless, Jim's 1961 paper effectively resolved the scientific issues between schools. Such is the human aspect of science, or rather scientists, that the divisions between the schools remained stoking intense arguments for another 20 years. At that point, not only had the experimental evidence become overwhelming for his model, but also Jim had retired. As shown in Fig. 2, a photograph taken at the time of his retirement, he was happily aware then that his open model ruled all. Citations of his 1961 paper increased in the years following his retirement. The citation rate then further doubled in the years following the launch of ESA's four Cluster spacecraft in 2000.

Fig. 2 Jim upholding the open magnetosphere—a weather vane presented by his old students to him on his retirement from Imperial College in 1984. Image credit (Jeff Hughes)

Jim did not receive as much recognition during his career as he was due. Nonetheless, he was honoured with the Fleming Medal of the American Geophysical Union, with Honorary Membership of the European Geophysical Society, its highest honour, and with the Gold Medal of the Royal Astronomical Society. Furthermore, the Council of the latter Society decided in 2012 to introduce an annual James Dungey Lecture in solar, planetary and solar terrestrial physics. Long may his ideas be explored and promulgated.

David Southwood
Space and Atmospheric Physics Group,
Blackett Laboratory,
Imperial College,
London, UK

Chapter 1
Dungey's Reconnection Model of the Earth's Magnetosphere: The First 40 Years

Stanley W.H. Cowley FRS

Abstract The proposal of the 'reconnection' or 'open' model of the Earth's magnetosphere was undoubtedly the most important contribution of Jim Dungey's remarkable scientific career, forming the theoretical basis which continues to underpin our understanding of the terrestrial outer plasma environment. In this paper we first consider the development of the ideas which led to this proposal, starting from PhD studies with Fred Hoyle in Cambridge in 1947. Work stimulated by Hoyle's theory of the auroras centred on the formation of current sheets in the vicinity of magnetic neutral points and the occurrence of field line reconnection, leading directly to a first description of the 'open' model in Jim's thesis submitted in September 1950. While the ideas concerned with magnetic reconnection were subsequently published in 1953, the 'open' model itself was not published until 1961, under circumstances in which in situ spacecraft data relevant to the magnetosphere and interplanetary medium were becoming available for the first time. Initial clear observational confirmation of the model, relating magnetic disturbances at the Earth to the direction of the interplanetary magnetic field, came 5 years later. We chart the development of the model and its many ramifications over the following three decades, by Jim Dungey until his retirement in 1984, and by colleagues at Imperial College, under circumstances in which ideas could increasingly be tested in detail against space- and ground-based data of greatly expanding volume and sophistication.

1.1 Overview

As made clear from the variety of topics discussed in this volume, during his ~40-year scientific career Jim Dungey made fundamental theoretical contributions to the understanding of a wide range of physical processes occurring in the Earth's plasma

S.W.H. Cowley FRS (✉)
Department of Physics and Astronomy, University of Leicester, Leicester LE1 7RH, UK
e-mail: swhc1@ion.le.ac.uk

D. Southwood et al. (eds.), *Magnetospheric Plasma Physics: The Impact of Jim Dungey's Research*, Astrophysics and Space Science Proceedings 41,
DOI 10.1007/978-3-319-18359-6_1

environment. These include, for example, first discussions of the Kelvin-Helmholtz instability (KHI) at the magnetopause boundary and mass and momentum transfer due to boundary waves, standing Alfvén waves on magnetospheric field lines and their connection with ultra-low frequency (ULF) micropulsations, and resonant interactions of various kinds between waves and particles in collisionless plasmas and their effect on wave growth and particle diffusion. Early discussion of the KHI and ULF waves in the magnetospheric context may be found in the papers by Dungey (1954, 1955, 1958a (§§8.5–8.6), 1963c, 1967), and of wave-particle interactions in space plasmas in Dungey (1962a, 1963a, b, c, 1965a, 1966a). Indeed, in his professorial inaugural lecture at Imperial College in May 1966 (Dungey 1966b), Dungey indicates that wave-particle interaction phenomena are his 'major interest', an assertion that is at least somewhat borne out by his published output in the ~20 years up to his retirement. Nevertheless, despite the significance of these achievements, there is no doubt that Jim Dungey's widest impact on scientific thought has been through his proposal of the 'reconnection' or 'open' model of the Earth's magnetosphere (Dungey 1961), which today is the essentially universally-accepted paradigm that underpins understanding of the terrestrial plasma environment. In this contribution we outline the development of these ideas, in three consecutive temporal phases.

The first phase corresponds to the pre-space era characterised by major deficiencies in relevant observations, starting from PhD studies supervised by Fred Hoyle in Cambridge beginning in 1947, through post-doctoral positions at the University of Sydney, Pennsylvania State University, and back to Cambridge, followed by a mathematics lectureship at King's College, Newcastle upon Tyne. In this first interval theoretical ideas were stimulated and simmered for over a decade. The transition from the first to the second phase occurred in 1960 with the writing of the brief but critical paper cited above which outlines the reconnection model, encouraged by a modicum of in situ space data then newly becoming available, followed by initial working through during the 1960s of many of the physical ramifications that follow. These developments took place initially during employment at the Atomic Weapons Research Establishment, Aldermaston, where wave-particle interaction studies cited above were also undertaken, the latter concerned with gyro-resonant diffusion and precipitation of charged particles trapped in the Earth's radiation belts. In 1963 Jim Dungey moved to Imperial College, London, being appointed (according to his inaugural address) to Patrick Blackett's 'second hand' professorial chair in 1965 (Dungey 1966b).[1] The first clear observational confirmation of a central prediction of the reconnection model, concerning the dependence of magnetic disturbance at Earth on the direction of the magnetic field in the interplanetary medium, was published shortly thereafter (Fairfield and Cahill 1966).

[1] This opening joke from Dungey's address refers to Blackett's retirement from the Department of Physics (subsequently the 'Blackett Laboratory') at Imperial College in July 1963, thus rendering available his 'used' professorial chair. Blackett was awarded the 1948 Nobel Prize for Physics, for particle physics research using high-energy cosmic rays.

The third phase of studies, from the 1970s to the early 1990s, saw considerable elaboration of both theoretical and observational themes, under circumstances in which massively expanding volumes of detailed data from both ground and space allowed detailed testing and verification of the predictions of the model. In this contribution we focus on the studies led by Jim Dungey with his PhD students and post docs up to his retirement in 1984, and by his colleagues at Imperial College thereafter.

The main research developments in each of these three phases are outlined in the sections below. Our discussion interfaces in this volume with the contribution by David Southwood on the one hand, who sets the prior scene in particular relation to the ideas of Kristian Birkeland, Sydney Chapman, and Hannes Alfvén, and with the contribution by Steve Milan on the other, which brings the story up to the present time.

1.2 Growth Phase

1.2.1 Giovanelli and Hoyle

The origins of the reconnection model of the magnetosphere can be traced back to the solar studies of Giovanelli (1946, 1947, 1948), who proposed a theory of solar flares involving neutral points in the magnetic field. Giovanelli had observed numerous solar flares during his graduate studies at Mount Stromlo Observatory in Canberra, Australia, in the late 1930s, leading to the conviction that flares occurred more frequently in complex sunspot groups where magnetic neutral points might be expected to form in the solar corona above the group. He suggested that electrons could run away in the vicinity of a neutral point in the presence of a locally quasi-uniform electric field associated with the evolving sunspot field, leading to large local electric current densities and acceleration of the electrons to high energies in the electric field.

In 1947 Giovanelli's research papers were submitted to the Australian National University in Canberra in part fulfilment of requirements for the Doctor of Science degree, for which Fred Hoyle at Cambridge was appointed external examiner. Hoyle was writing his own book on solar physics in the summer of 1947, in which discussion of electromagnetic effects in the solar atmosphere were 'mainly stimulated by the investigations of R.G. Giovanelli' [Hoyle 1949 (Author's Preface and §24)]. One outgrowth of this stimulation, however, was the suggestion by Hoyle that a similar idea may also apply to Earth and the auroras. Contrary to the interplanetary field-free paradigm of Chapman and Ferraro (1931, 1932), Hoyle considered the interaction of a magnetised solar stream with the Earth, discussed (Alfvén's recent 'frozen-in' theorem not withstanding) in terms of a 'dipole plus uniform field' magnetic field structure [Hoyle 1949 (§27)]. The dipole field was, of course, that of the Earth, while the uniform field was that carried outward by the 'corpuscular stream' from the Sun, the existence of which would not be proven by direct measurement for more than a decade. Hoyle noted the presence of two neutral points in such a field whose position depends on the strength and orientation of the uniform field, and suggested following Giovanelli that these are seats of strong particle acceleration through the presence of a quasi-uniform electric field, assumed similar in strength to that in the corpuscular stream.

Using modern units, the magnitude of the electric field was thus estimated by Hoyle as $E \approx uB$, where $u \approx 1{,}000$ km s^{-1} is the speed of the stream determined from the delay between solar flares and storm sudden commencements, and B is the interplanetary magnetic field strength near the Earth, estimated to be ~100 nT, such that $E \approx 0.1$ V m^{-1}. Such an interplanetary magnetic field gives neutral points at radial distances of ~7 R_E, depending modestly on the orientation of the external field. The acceleration region was guessed to have a length of ~1 % of the distance of the neutral point from Earth (~400 km), giving an accelerating potential of ~40 kV, sufficient, it was noted, to result in atmospheric penetration by the electrons to heights of ~100 km, in rough agreement with auroral observations. Assuming the accelerated particles follow magnetic field lines down to the Earth, auroras will then occur at the feet of the field lines mapping to the neutral points. Hoyle indicated that these are in general two in number for each neutral point, lying in the meridian of the neutral points and hence of the interplanetary field, leading to two auroral spots in the north and two in the south, on opposite sides of the pole. Their positions thus depend on the orientation and strength of the external field, thus providing an explanation of auroral motion. In the 'Supplementary Notes' on p. 129 of his treaties, Hoyle reports the results of calculations of the co-latitude of these spots made by his student J.W. Dungey, to whom development of these ideas was passed as a PhD project in 1947. The results, expanded upon somewhat in Dungey's thesis (Dungey 1950, Chapter VI), show that for a near-southward-directed field the co-latitude varies from ~12° for an external field of 1 nT, to ~43° for a field of ~1,000 nT, thus suggesting a possible reason for the equatorward expansion of auroras during magnetic storm intervals. Dungey (1950) also noted that a maximum auroral frequency at ~23° latitude implied an intermediate external field strength of ~30 nT. Published appreciation of the fact that the field lines from such magnetic neutral points generally form a fan that intersects the Earth in a closed ring around each pole at such latitudes only appeared rather later, as will be mentioned below.

1.2.2 Jim's Thesis and the Concept of Magnetic Reconnection

In addition to discussions of solar topics, Jim Dungey's thesis contains two chapters directly relevant to the present theme. Chapter V entitled 'Accelerating Processes' concerns the behaviour of conducting fluids in the vicinity of X-type magnetic neutral points such as those mentioned above, which introduced the concept of magnetic reconnection, though the term itself was not employed until several years later. Chapter VI, entitled 'Hoyle's Theory of the Aurora Polaris', then applied these ideas to the Earth's environment, and contained the first description of the reconnection model, including flux transport through the system driven by the flow of the solar stream, as well as acceleration of particles near the neutral points discussed along similar lines to those outlined by Hoyle above. However, only the first of these topics was subsequently 'written up' for publication (Dungey 1953, 1958a, b), which we now discuss here, while the reconnection model itself did not re-emerge for almost a decade, discussion of which is thus deferred to the following section.

The magnetic structure considered is shown in Fig. 1.1, taken from Dungey (1958b), in which the heavy curved lines are magnetic field lines, such that the current implied by Ampère's law is directed into the diagram. The lighter lines marked '*f*' then show the directions of the $\boldsymbol{j} \times \boldsymbol{B}$ force on the fluid, which, it was argued, will then flow in the same general direction as this force, causing the frozen-

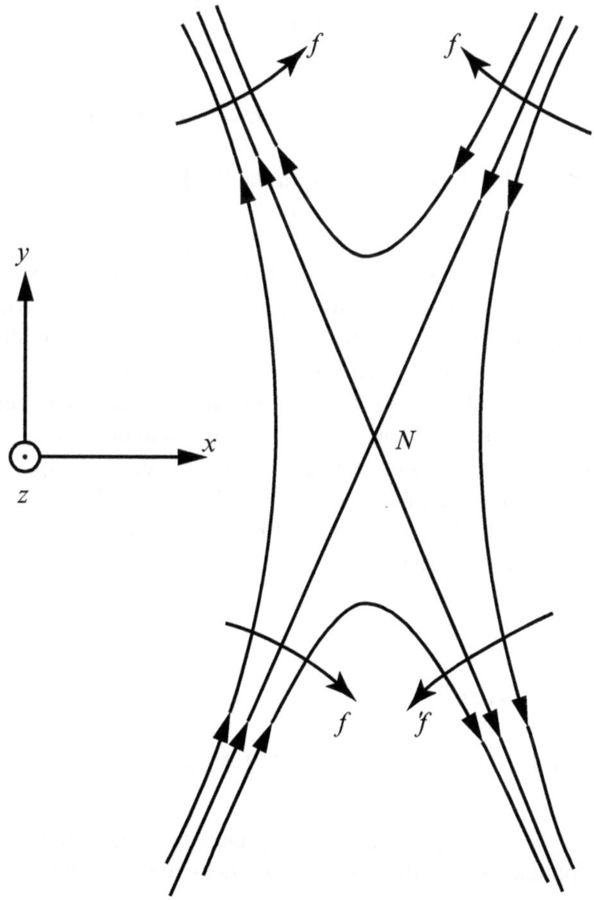

Fig. 1.1 Sketch of magnetic field lines, shown by the *heavy arrowed lines*, in the vicinity of a magnetic neutral point, *N*, with an electric current directed into the plane of the diagram. The *curved arrows* marked '*f*' show the direction of the $\boldsymbol{j} \times \boldsymbol{B}$ force on the plasma, which tends to make the plasma move in the same direction. From Dungey (1958b)

in field lines to move with it in such a way as to increase the current density. If such a field configuration with no initial current is thus perturbed, the principal axes of the field initially being orthogonal, the current will then grow to large values, thus reversing usual expectation based on 'Lenz's law'. Such behaviour was confirmed by numerical solution of the relevant equations on the EDSAC computer (Dungey 1950, 1953), which began operation in the Mathematical Laboratory at Cambridge in May 1949. According to Dungey's (1994) later account, an initial paper describing these researches submitted to *Monthly Notices* was rejected on the basis of the neglect of plasma pressure forces, but considerations included in the revised paper published by Dungey (1953) in *Philosophical Magazine* showed that while the build-up of pressure may slow the field collapse, it should not prevent it.

It will be noted that the above discussion of the current sheet instability assumed 'frozen-in flow' in the presence of a perfectly conducting plasma. In this limit the electric field remains zero at the neutral line, and there is no transfer of magnetic flux between the quadrants of the field structure. However, as the field collapse proceeds the current density near the neutral line increases to the point where the finite conductivity of the plasma becomes important, no matter how high the latter, such that the system may then approach a steady state. In this case the simplest form of Ohm's law can be written as

$$\boldsymbol{E} = -\boldsymbol{u} \times \boldsymbol{B} + \boldsymbol{j}/\sigma, \tag{1.1}$$

where $\boldsymbol{E}$ is the electric field, $\boldsymbol{u}$ the fluid velocity, $\boldsymbol{B}$ the magnetic field, $\boldsymbol{j}$ the current density, and σ the plasma conductivity. In the steady-state, Faraday's law shows that the electric field, directed into the plane of Fig. 1.1, becomes spatially uniform, with the first term on the right hand side of Eq. (1.1) being dominant away from the neutral line, associated with near 'frozen-in' transport of the field toward and away from the latter, while the second term, associated with diffusion of the field through the plasma, is dominant near the neutral line. As noted by Dungey (1950, 1953), 'the lines of force in [the inflow] can be regarded as being broken and rejoined to form those in [the outflow]'. The term 'magnetic reconnection' used by Dungey to describe this process appeared a little later in a paper arising from a symposium held in Stockholm in August 1956 (Dungey 1958b). The rate of such 'reconnection' of magnetic flux, per unit length along the neutral line, is equal to the electric field strength along the line.

1.3 Onset

1.3.1 Genesis of the 1961 Paper

Of the two papers published by Dungey (1953) and (1958b), the earlier discusses applications of the above theoretical ideas to the auroral problem as originally outlined by Hoyle, as well as to solar flares and interplanetary space, while the later is centred in the solar context. The latter is true of much of related work internationally during the 1950s, due to the availability of 'global' remote-sensing observations of the solar atmosphere, compared with the paucity of observations relevant to the region between the upper ionosphere at Earth and the Chapman-Ferraro magnetic field boundary. In the 1950s, Dungey called this region the 'outer atmosphere', before the term 'magnetosphere' was coined by Tommy Gold in 1959. Information on the plasma density in this region had been gleaned from observations of lightning-induced whistler waves by Owen Storey (Storey 1953), and while the auroras remained decidedly enigmatic, clues were available from the study of magnetic disturbances observed on the ground. However, while polar magnetic disturbances were already believed to be due to Hall currents driven in the ionosphere by electric fields [Dungey 1958a (§8.4)], and thus associated with ionospheric flow, the connection with magnetised solar streams at Earth and the auroral problem had yet to be made. Indeed, in his monograph *Cosmic Electrodynamics*, Dungey [1958a (§8.3)] notes that while both Hoyle [1949 (§27)] and Alfvén [1950 (§6.2)] had, unlike Chapman and Ferraro (1931, 1932), proposed models containing a magnetic field in the interplanetary medium, 'it is not yet certain whether this modification enables their models to account for more observations'. According to his later account, this represented one significant factor that delayed submission of the reconnection model for publication (Dungey 1994). Another was the somewhat sticky reception given to his ideas on reconnection itself.

The critical insight came in late 1960, in 'a flash while I was sitting in a sidewalk café in Montparnasse', preparing a seminar to be given at the Observatoire de Paris-Meudon, in which the three-dimensional magnetic and electric fields were envisaged, together with their connection with the related polar ionospheric Hall currents (Dungey 1983, 1994). The paper was immediately written 'when I got home', submitted to *Phys. Rev. Letters* in November 1960, revised in December, and published the following January (Dungey 1961). Its opening sentence refers to interest in Hoyle's theory of the aurora being 'reawakened' by the first measurements of the magnetic field in interplanetary space by the Pioneer 5 probe in March and April 1960 (Coleman et al. 1960). The model is illustrated in Fig. 1.2, which we have chosen to take from Jim Dungey's thesis (Dungey 1950). Panel (a) shows the noon-midnight meridian plane with the flow of the solar wind from left to right and the principal motions associated with reconnection near the neutral points, while panel (b) shows the time history of a given interplanetary field line. Similar to the classic published paper, the electromagnetic field essentially consists of a 'dipole plus uniform near-southward interplanetary magnetic field' as originally introduced conceptually by Hoyle, somewhat distorted by the flow of the solar wind, together with an electric field that points out of the plane of the diagram from dawn to dusk. We note that in Alfvén's electromagnetic model mentioned above [Alfvén 1950 (§6.2)], the interplanetary field was instead taken to point strictly northward along the dipole axis, with charged particles drifting essentially continuously antisunward throughout, via the agency of a uniform electric field directed everywhere

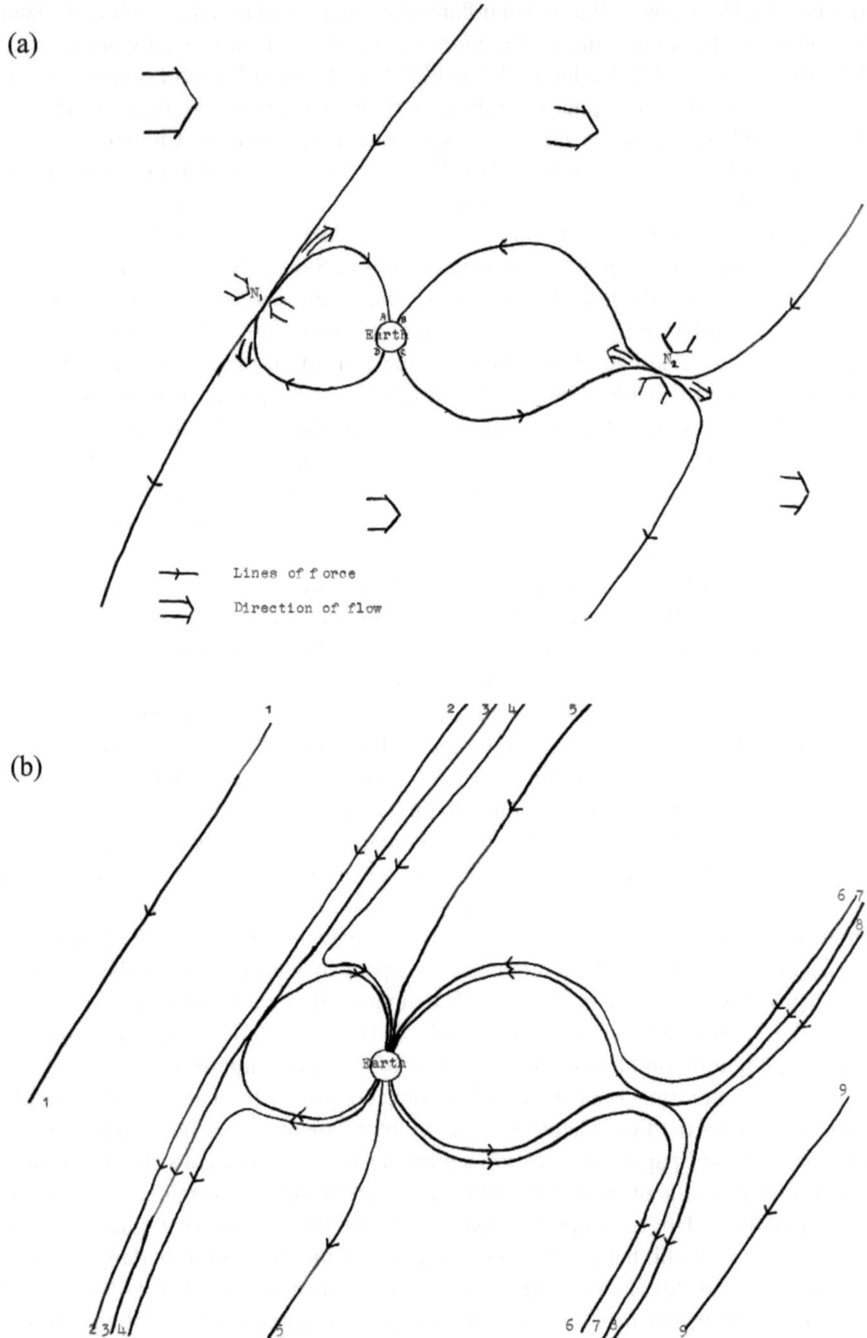

Fig. 1.2 Sketches of the field and flow in the reconnection model of the magnetosphere, showing the noon-midnight meridian plane with the solar wind blowing from left to right. Panel (**a**) shows the flow in the vicinity of the neutral points N_1 and N_2 resulting from reconnection in their vicinity, while panel (**b**) shows the time history of a given interplanetary field line. From Dungey (1950)

from dusk to dawn, both added fields thus being directed opposite to those in Dungey's picture.

As noted by Dungey (1961), the magnetic field geometry of the reconnection model in Fig. 1.2 contains a region of 'closed' field lines that map to the Earth at both ends, topologically a doughnut, two regions of 'open' field lines that map to the Earth at one end and into the interplanetary medium at the other, topologically cylinders, and an exterior region of 'interplanetary' field lines. The electric field directed out of the plane of the diagrams is associated with the $\boldsymbol{E} \times \boldsymbol{B}$ flow of field and plasma indicated by the large arrows in the upper panel, the field and plasma being 'frozen' together except in the immediate vicinity of the neutral lines on the day and night sides as indicated in Sect. 1.2.2 above. Anti-sunward flow of 'open' field lines takes place over the poles, reversing to a sunward return flow of 'closed' field lines at lower latitudes. Assuming the dominance of Hall currents flowing opposite to the direction of the plasma flow in the ionosphere (carried by E-region electrons), crucially this pattern of flow is consistent with the S_D current system deduced from magnetic disturbances observed at high latitudes. An idealised picture of these currents taken from Chapman and Bartels [1940 (§9.14)] is shown in Fig. 1.3, in a view looking down on the north pole with the Sun to the left. The current flows sunward in the polar region, and closes principally through intense currents in the immediate equatorward region via dawn and dusk that Chapman and Bartels note corresponds to the auroral zone. We note that this S_D field corresponds to the usual non-quiet disturbance field at high latitudes, not restricted to magnetic storm intervals. The distinction between 'substorm' intervals and other more general magnetic disturbances at high latitudes had yet to re-emerge

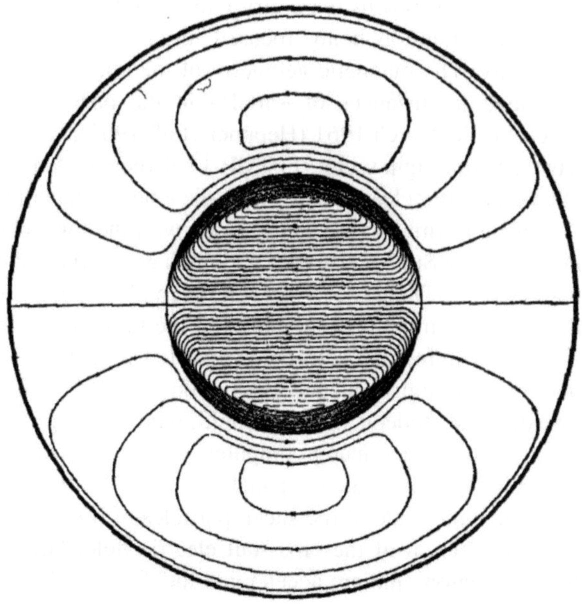

Fig. 1.3 Idealised diagram of the S_D overhead ionospheric current system, in a view looking down onto the northern pole, with the Sun to the left. The principal currents flow sunward at highest latitudes, and close in intense anti-sunward currents via dawn and dusk that map to the auroral zone. From Chapman and Bartels [1940 (§9.14)]

from IGY (1957–1958) data, following Birkeland's original discovery of magnetic 'polar elementary storms' more than 50 years before [Birkeland 1908; see also Egeland and Burke 2005 (§4.3)]. According to his account (Dungey 1994), Dungey was aware that a fully 'closed' model driven e.g. by waves at the boundary (discussed briefly in Dungey [1958a (§8.6)]), would produce similar flows and currents in the ionosphere, as first discussed in the published literature by Axford and Hines (1961). However, the sharpness of the observed boundary between the 'polar cap' and 'auroral zone' currents in the S_D system convinced him that it corresponded to a topological boundary, between open and closed field lines.

Given this initial picture, much further development was clearly possible, in two particular directions. The first concerns the motion of particles in the system and the nature of the plasma populations and auroral precipitation that would follow, it already being noted by Dungey (1961) that there would be a 'mixing of the interplanetary plasma into the outer atmosphere' (the latter term meaning the magnetosphere as noted above). The second concerns the dynamics of the system resulting from its interaction with the variable solar wind. Numerous aspects of these two topics were discussed during the 1960s, as will now be overviewed in turn.

1.3.2 Particle Motion and Plasma Populations in the 'Open' Model

Initial discussion of particle motions centred on the acceleration of particles in the current sheets downstream from the X-type neutral lines in the system, with particular attention being focused on the tail current sheet due to its relative simplicity. The magnetic geometry of the outer magnetosphere had initially been examined to distances of ~40 R_E in the post-dusk sector by the Explorer-10 spacecraft in March 1961 (Heppner et al. 1963), and to distances of ~30 R_E in the dawn and midnight sectors by IMP-1 between November 1963 and May 1964 (Ness 1965), the latter clearly delineating the lobes of the tail and the central current sheet. At this time, Jim Dungey had a tie-up with the Ionospheric Research Laboratory at Pennsylvania State University, in which he supervised graduate students mainly by mail (see e.g. Dungey (1994)). The second of these students, Ted Speiser, was set the task of computing and analysing the motion of particles in such current sheet systems, with initial results being reported at the Plasma Space Science Symposium held in Washington in June 1963 (Dungey 1965b). (The principal activities of the first of these students, Don Fairfield, who also reportedly undertook related computations using an analog computer, will be mentioned below in the discussion of dynamics.) As illustrated in panel (a) of Fig. 1.4 for the case of strictly anti-parallel fields on either side of the sheet, particles of both signs drift into the current sheet under the action of the cross-tail electric field, begin to oscillate about the field reversal region, and are accelerated indefinitely along it by the electric field, ions

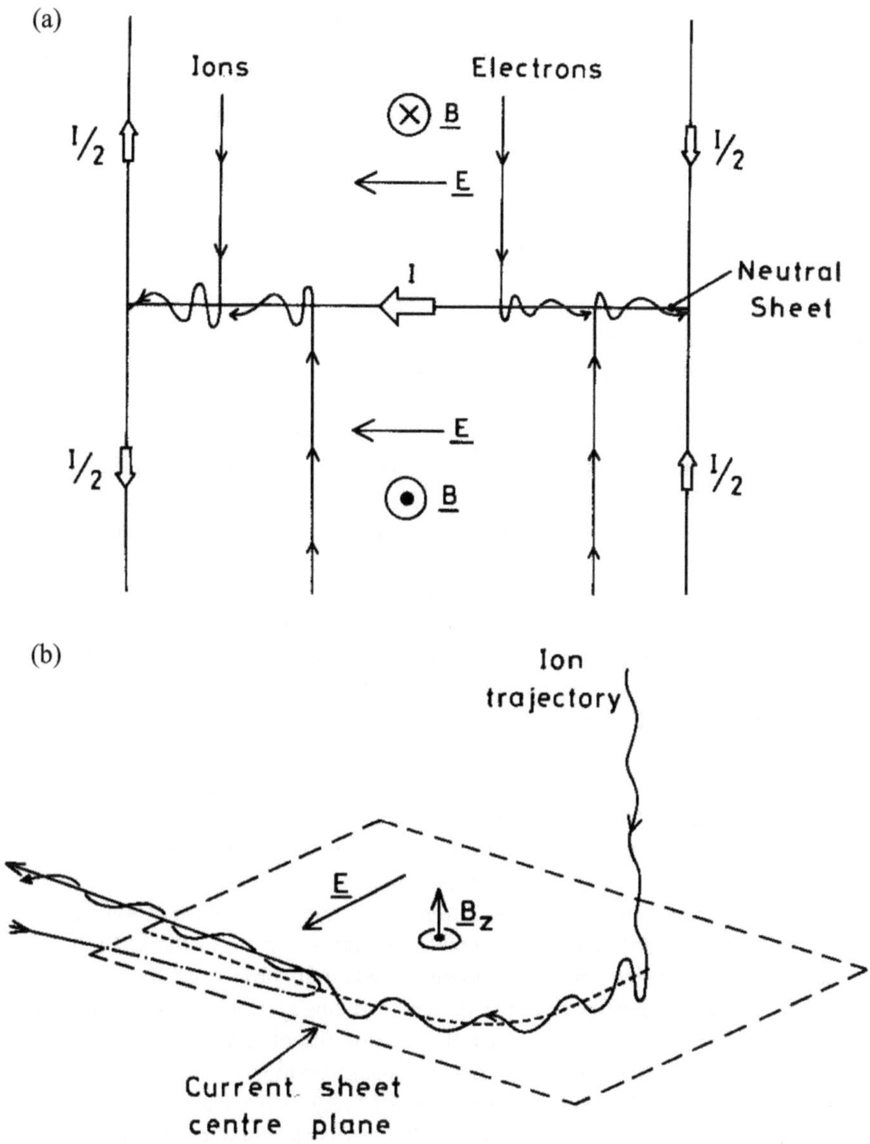

Fig. 1.4 (**a**) Sketch of particle trajectories in a neutral sheet system in which anti-parallel fields ***B*** reverse direction across the sheet, which carries current per unit length $I = 2B/\mu_o$ as required by Ampère's law. A transverse electric field ***E*** is also present in the direction of the current. Charged particles $\boldsymbol{E} \times \boldsymbol{B}$ drift into the current sheet from both sides, then oscillate about it while being accelerated along it by the electric field, ions and electrons in opposite directions, both contributing to the sheet current. The particles remain within the sheet until they reach the boundaries of the system, illustrated schematically by the current-carrying vertical planes. (**b**) Sketch of an ion trajectory in a current sheet lying in the *x-y* plane, with the *x* magnetic field component reversing from negative to positive with increasing *z*, and threaded by a uniform field in the *z* direction (B_z), so that the field lines form sharply bent hairpins as shown by the arrowed curve. A uniform electric field ***E*** is also present in the *y* direction (which may be transformed away throughout by moving in the +*x* direction (to the *left*) with the $\boldsymbol{E} \times \boldsymbol{B}$ speed of the field lines in the centre plane). Particles in

and electrons in opposite directions, both contributing to the current *I*, until they reach the boundaries of the system. When a small field component is added normal to the sheet as shown in panel (b) of Fig. 1.4 (the B_z field), such that the field lines form sharply-bent hairpins as in the region downstream from an X-type neutral line, the initial motion is the same as before, but the particles are then turned round in the plane of the sheet by the B_z field and are ejected from it along the field lines, in the direction away from the X-line (Speiser 1965a, b, 1967). For sufficiently uniform conditions along the sheet, the particles are ejected with a speed of essentially twice the speed of contraction of the field lines away from the X-line on either side, the latter speed being E_y/B_z using usual magnetospheric coordinates (*z* northward and *y* from dawn to dusk). The electrons thus gain much smaller energies than the ions, except close to the neutral line where B_z is very weak, the total energization being limited, of course, by the total potential drop across the system. This was estimated by Dungey (1966b) as being 'tens of kilovolts'. We note that brief qualitative descriptions of these orbits had previously been given by Dungey [1953, 1958a (§6.4)] in discussion of reconnection-related particle acceleration, the current sheet acceleration representing a detailed physical realisation of the auroral acceleration mechanism suggested in outline by Hoyle [1949 (§27)].

With regard to the nature of the plasma populations that follow, Dungey (1968a, b) noted that when dayside reconnection is in progress, the magnetopause boundary should become more 'diffuse' at high latitudes as the magnetosheath and magnetospheric plasmas mix across the boundary along the reconnected open field lines. The rapidly out-flowing magnetospheric plasma forms 'spikes' of energetic particles on new open field lines within the magnetosheath, possibly also crossing the shock into the upstream solar wind (Dungey 1968a, b), while the magnetosheath plasma will continue to flow away from the Earth along outer tail field lines at some fraction of the solar wind speed. The latter plasma also drifts slowly towards the centre plane, forming a wedge-shaped shadow zone behind the Earth whose angle is determined by the ratio of the convection speed and the down-tail flow [Dungey 1967b (§IV.6), 1968b]. The region inside the shadow-zone in the tail lobes will contain only low-energy plasma flowing slowly outwards from the ionosphere, estimated to have a density of only ~0.1 cm^{-3} [Dungey 1967b (§IV.6)]. It may be noted that these discussions significantly precede the observational discoveries in the polar magnetosphere of dayside magnetopause boundary layers and the plasma mantle made by the HEOS-2 spacecraft in 1972 (e.g., Rosenbauer et al. 1975).

The combined plasma in the tail lobes then drifts slowly into the central current sheet under the action of the cross-tail electric field, where it is accelerated downstream from the tail neutral line as discussed above, and further compressed

---

Fig. 1.4 (continued) the untransformed frame (as shown) drift into the current sheet and start to oscillate about it and accelerate along it in the electric field, as for the neutral sheet system in this figure, but are then turned round within the sheet by the B_z field and are ejected from the sheet along the field lines away from the direction of the X-line (off the diagram to the *right*). From Cowley (1980)

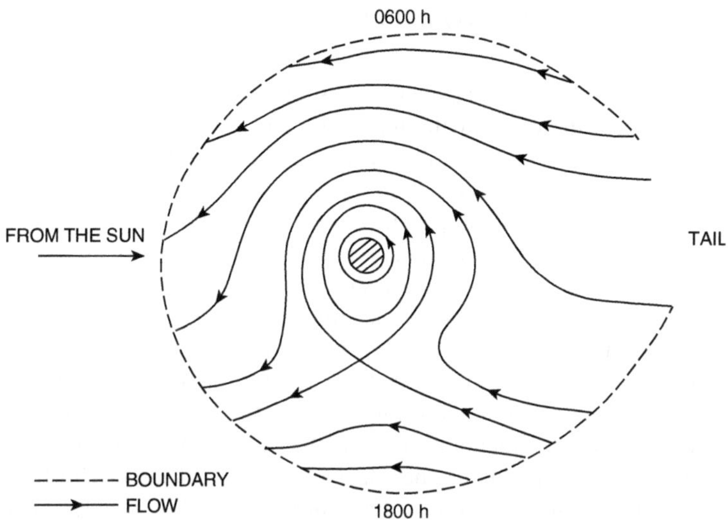

Fig. 1.5 Closed field line flow in the equatorial plane of the magnetosphere formed by the combined action of planetary rotation and sunward flow generated by the interaction with the solar wind. The two regimes of flow are separated by the streamline passing through a stagnation point in the dusk sector. From Dungey (1967b)

on newly-closed field lines contracting toward the Earth. A region of hot earthward-moving plasma is thus formed between the tail lobes, bounded by field lines that go to the tail neutral line, which Dungey (1968a, b) took to correspond to the tail 'plasma sheet' newly discovered in Vela satellite data by Bame et al. (1966). The neutral line itself was thought to lie 'some hundreds of earth radii downwind of the Earth' (Dungey 1968a, b), the entire length of the 'connected' tail having previously been estimated as ~1,000 R_E on the basis of an anti-sunward travel time of open field lines across the polar ionosphere of several hours, combined with a solar wind speed of several hundred km s^{-1} (Dungey 1965c). The hot plasma sheet, together with further cold plasma 'evaporating' from the ionosphere into the magnetosphere, then flows round the Earth towards the dayside boundary, though avoiding a central region of flux tubes that is dominated instead by Earth's rotation [Dungey 1967b (§IV.6)]. As shown in the sketch of equatorial flows in Fig. 1.5 taken from the latter paper, the regions of rotating and sunward flow are separated by a flow stagnation point at dusk, with the plasma in the steady-state system being in equilibrium with the ionosphere in the region of rotating flow, while being of lesser density in the hot plus cold plasma region outside. It was noted that this flow pattern thus accounts for the plasma density 'knee' derived from whistler observations by Carpenter (1966) (i.e. the plasmapause boundary of the plasmasphere), paralleling related discussions by Nishida (1966) and Brice (1967). It is thus seen that the reconnection magnetosphere model accounts in a natural way for all of the major steady-state plasma populations observed within the magnetosphere.

1.3.3 Magnetospheric Dynamics

We now turn to magnetospheric dynamics, associated particularly with the response to variations in the direction of the magnetic field in the upstream medium, the potential significance of which for the location of the neutral points and consequent auroras had already been noted by Hoyle [1949 (§27)] and Dungey (1950, Chapter VI). While the initial papers by Dungey (1961, 1962b) specified an approximately southward interplanetary magnetic field (IMF), the observations from a growing number of spacecraft showed that the IMF can be variable in direction on time scales down to several minutes, thus suggesting the need for a more general approach. Dungey (1963c) thus examined the 'dipole plus uniform field' model in more detail, showing that the general 'doughnut plus two cylinder' topology is the same as for a southward field, with the sheets of field lines that emanate from the two neutral points defining the surfaces that separate the 'closed', 'open', and 'interplanetary' field lines. Isometric views of these field lines for an IMF directed from dusk to dawn ($-y$ direction) computed by Cowley (1973b) are shown in Fig. 1.6, in a paper which discusses the detailed changes in field line connection that occur during the reconnection process in this field geometry. Panel (a) shows the field lines from the southern dusk-side neutral point ('N.P.') that form the northern segment of the surface separating open and closed field lines, while panel (b) shows the field lines from the northern dawn-side neutral point that form the boundary between northern open field lines and interplanetary field lines. The only exception to this structure occurs for a strictly northward IMF, when the cylinders of open flux shrink to two singular field lines connecting the northern and southern neutral points on the magnetopause surface to the poles of the planet, at which point the magnetosphere becomes magnetically closed. Alfvén's model corresponds to the latter magnetic condition, as does the Chapman-Ferraro picture inside the magnetosphere.

The above considerations suggest that the structure of the magnetospheric field, and the plasma flow and currents within it should depend strongly on the direction of the IMF. The first of Jim Dungey's students at Pennsylvania State University, Don Fairfield as mentioned above, had been studying the S_D current system using high-latitude ground magnetic data from the IGY (Fairfield 1963), and as a follow-on project, Jim suggested that he should examine magnetograms at the times when the Explorer 12 spacecraft lay outside the magnetosphere. This spacecraft operated between August and December 1961, with initial apogee near noon at a radial distance of ~13 R_E, thus generally extending beyond the magnetopause into the dayside 'transition region' (i.e. the magnetosheath), but not into the solar wind and interplanetary field proper. Despite the variability of the field in this region, Fairfield and Cahill (1966) found an excellent correspondence between southward fields outside the magnetopause and geomagnetic disturbance at high-latitude, and northward fields and magnetic quiet, thus constituting the first clear observational support for the dynamical reconnection model. An example from their paper is shown in Fig. 1.7, with format details being given in the caption.

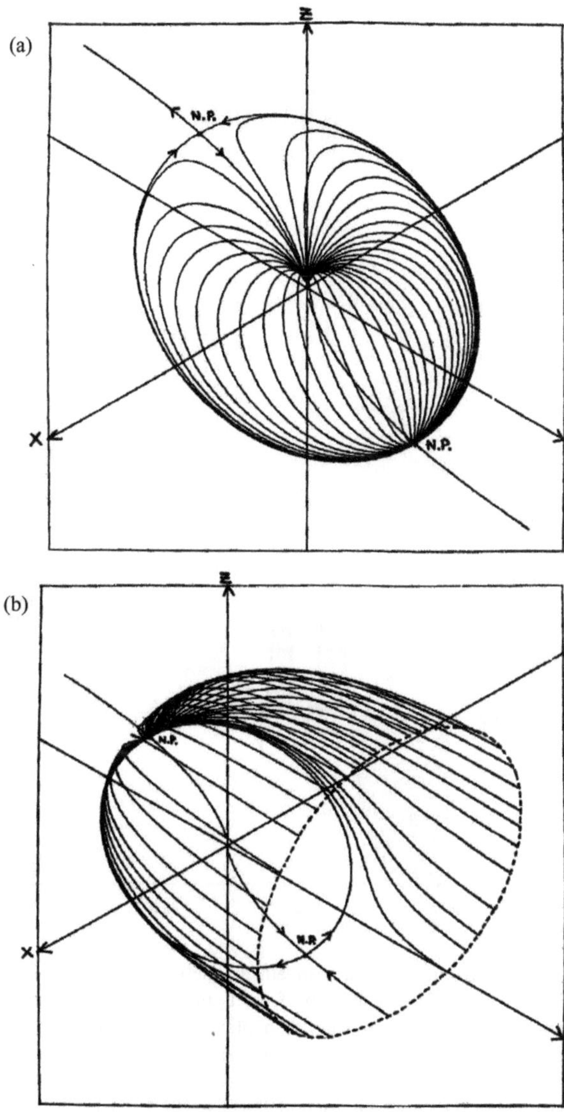

Fig. 1.6 Illustration of field line surfaces in the 'dipole plus uniform field' paradigm separating lines of differing connectivity, shown in isometric (30°) projection. The uniform field is taken to be directed from dusk to dawn ($-y$ direction in usual magnetospheric co-ordinates). (**a**) Field lines from the southern dusk-side neutral point ('N.P.') that form the northern segment of the boundary between closed and open field lines. (**b**) Field lines from the northern dawn-side neutral point that form the boundary between northern open field lines and interplanetary field lines. From Cowley (1973b)

More specifically, Fairfield and Cahill (1966) noted that 'the typical situation seems to be that either a north-to-south change is followed almost immediately by a bay event [i.e., substorm], or else there is a small, gradual increase in the polar-cap disturbance with a bay event following after a time delay of as much as an hour or more'. They argued that 'the dramatic suddenness with which bay events often begin suggests some kind of instability', and quote a personal communication from Dungey in 1965 envisaging 'a build-up of the flux connected to the polar cap until an instability in the tail produces rapid disconnection of the interplanetary lines

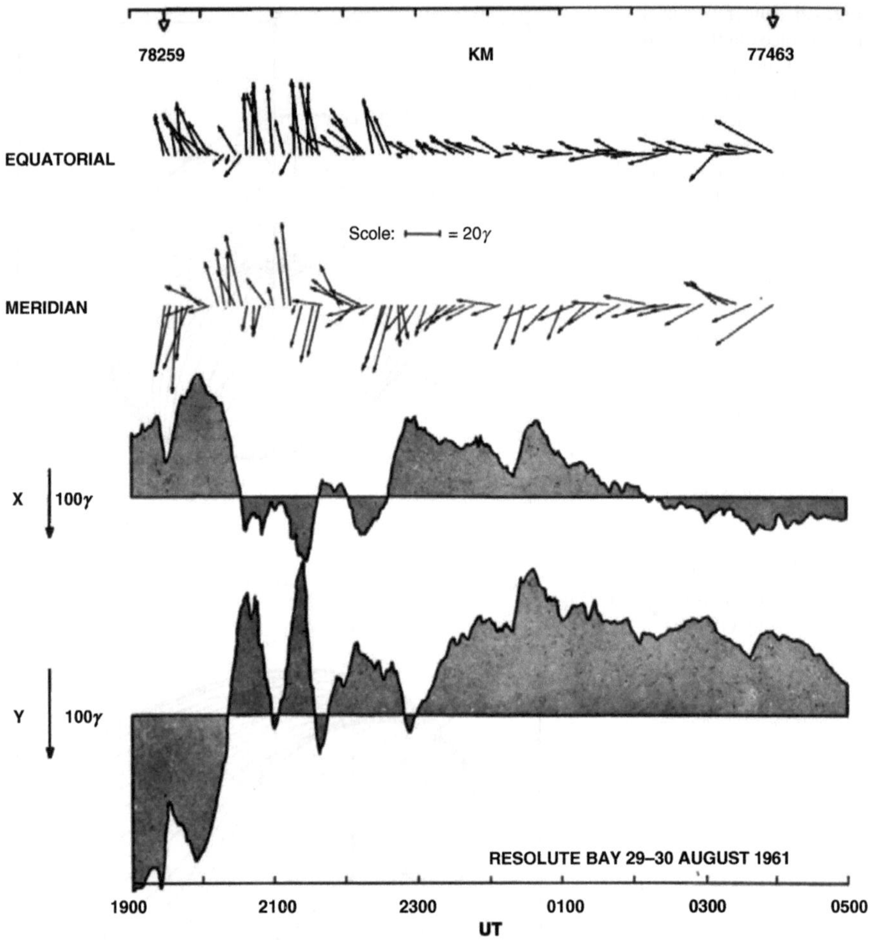

Fig. 1.7 Averaged magnetosheath magnetic field vectors from the Explorer 12 spacecraft are plotted versus time at the top of the figure (γ equals nT), projected into both the geomagnetic equatorial and magnetic meridian planes as indicated. The Sun is towards the right in both projections, while in the equatorial projection east is to the top and west to the bottom, and in the meridian projection north is up and south is down. Time is indicated in hours at the bottom, between 19 UT on 29 August and 05 UT on 30 August 1961, and spacecraft distance from Earth is indicated at the *top*. At the *bottom* of the figure the variations of the north, *X*, and east, *Y*, components of the Resolute Bay horizontal magnetic disturbance field are shown, plotted relative to a quiet-day baseline. From Fairfield and Cahill (1966)

with a resulting substorm', this representing the earliest discussion of dynamics within the framework of the reconnection model. However, other 'tentative' possibilities for substorm onset were also discussed around this time, involving the 'tangling' of tail field lines connected sequentially to interplanetary fields of differing orientation, resulting in increased field tension in the tail, which is then released (Dungey 1966a).

In subsequent publications the basic dynamics of the open magnetosphere became more clearly stated. When the interplanetary field turns from north to south dayside reconnection is enhanced, leading to increasing open flux in the tail and equatorward expansion of the polar cap and adjacent auroral oval (Dungey 1968a). The latter expansion should be greatest on the dayside, but the anti-sunward motion of open field lines over the polar cap will cause the boundary to move equatorward on the nightside too (Dungey 1968b, c). At some point rapid reconnection then begins in the tail corresponding to substorm (expansion phase) onset (Dungey 1966a, b), which can be delayed by an hour or more relative to the southward turning according to Fairfield and Cahill (1966). This correspondingly causes a reduction in the open flux in the tail and poleward contraction of the polar cap and adjacent auroral oval on the nightside (Dungey 1966b, 1968c), as had been found in IGY auroral data by Akasofu (1964). It may be noted that the dynamics of the reconnection model thus requires there to be a 'growth phase' of open flux accumulation in the tail prior to the onset of 'expansion', these considerations considerably preceding the phenomenological discussion by McPherron (1970), who does not refer to Dungey's ideas. The enhancement of sunward convection on closed field lines due to rapid tail reconnection also results in contraction of the plasmasphere in the inner magnetosphere, with high density cold plasma being swept towards the dayside, followed by plasmasphere expansion as the sunward flow abates, then incorporating flux from the tail into the rotating region (Dungey 1967b). The 'outstanding problem' for the open model, however, concerned the mechanism for the sharp onset of tail reconnection, possibly involving an 'explosive instability', or else some sharp trigger in the solar wind such as a shock or sudden change in the direction of the IMF (Dungey 1968c). The sharpness of the onset is indicated by the transient 'pt' pulsation (i.e., Pi2) that is observed to accompany the onset of every 'bay' disturbance (Dungey 1966a, b, 1968c).

1.3.4 International Reception

Following the initial study of the connection between geomagnetic disturbance and southward IMF by Fairfield and Cahill (1966), numerous further studies confirmed this relationship, both on statistical and case study bases (e.g., Scatten and Wilcox 1967; Wilcox et al. 1967; Rostoker 1968; Hirshberg and Colburn 1969; Arnoldy 1971). One might then have supposed rapid acceptance of the basic ideas of the reconnection model by the space plasma research community. However, this was by no means the case, as can be gathered in some measure from the published 'discussions' recorded at the end of a number of symposium publications. Thus at the conclusion of the talk given at the Plasma Space Science Symposium held in Washington in 1963 we find Alfvén doubting 'very much' that the aurora has any connection with magnetic null points on the basis of terrella experiments (Dungey 1965b), while at the Advanced Study Institute held in Bergen in 1965 we find that 'Dessler emphasized that in the theory that does not allow magnetic field merging, the neutral sheath [tail neutral sheet] was predicted, whereas the Dungey model is modified to fit the experiments after the fact' (Dungey 1966a).[2] It was perhaps natural for those who had proposed their own models to challenge Dungey's, but on the positive side Axford and colleagues were early to recognise the theoretical attractions of the model (e.g., Axford et al. 1965).

It should nevertheless be realised that during the 1960s and 1970s, at the least, magnetosphere research was strongly coloured by controversy surrounding 'reconnection' and its application to the Earth's magnetosphere. An appropriate flavour of such opinion may be obtained from the research career 'reflection' by Heikkila (2011, p. xiii), who records that 'At the AGU fall meeting in 1972, Bill Olson, Juan Roederer, and I shook hands to fight reconnection as presently conceived'. In his personal historical review Akasofu [2007 (§1.9)] also writes '... the resulting neutral line or the X-line has become a *magic* line. Many phenomena are blindly ascribed to unknown and unproved physical processes associated with the X-line. ..... I avoided this particular paradigm and decided to go my own way. I must confess that this decision was not based on any rational thinking.' Growing numbers within the international research community thought otherwise, however, influenced ultimately by successful comparison between theoretical expectation and observations as outlined above. Thus despite even common usage of synonyms such as 'merging', 'annihilation', or 'erosion' to soften opposition through distancing from the original term 'reconnection',[3] the 1961 paper eventually became, and continues to be, a 'citation classic' (see Dungey 1983).

1.4 Expansion

1.4.1 Current Sheet Equilibria and Wave Noise

During the 1970s and 1980s research relevant to our topic by Jim Dungey and his students and post-doc fellows at Imperial College focussed on further development of a number of the areas outlined above, which we now go on to discuss. The first such area was the physics of magnetic current sheets, directed principally towards the geomagnetic tail, based on the earlier particle trajectory analyses of Speiser (1965a, b, 1967). The overall goals of the programme were set out by Dungey and Speiser (1969), involving first the construction of models of current sheets in which the particle distribution functions are consistent with the fields via Liouville's theorem and Maxwell's equations, and second the determination of the nature of

[2] Dessler (1964) had proposed a magnetosphere model that contained an essentially fixed tail extending to the outer boundary of the heliosphere. We mention this only to give some idea of the nature of contemporary commentary.

[3] The story in the last section of David Southwood's contribution concerning publication in 1970 of the paper by Aubry and co-workers is telling in this regard.

the wave 'noise' that would result from potentially unstable current sheet velocity distributions, and its effect on the behaviour of the particles. Initial discussion of these topics had been given by Dungey [1963c (§3.4)], and had been considered further through a minor feud with Axford concerning the necessity or otherwise for 'dissipation' within the current sheet. Dungey (1968a, b) pointed out that although $\boldsymbol{j}.\boldsymbol{E}$ will indeed be positive within the current sheet, implying energy transfer from electromagnetic field to particles via Poynting's theorem, this does not necessarily imply that entropy is increased in the process, the latter requiring randomisation of the particle motion via 'collisions'. Rather, the $\boldsymbol{j}.\boldsymbol{E} > 0$ condition could apply equally to direct acceleration of particles within the current sheet as in Speiser's trajectory calculations illustrated in Fig. 1.4 [see also Alfvén (1968) discussed below], though 'noise' might then be a natural consequence as indicated above. The individuals mentioned below who were involved in this and directly related work at Imperial, in addition to Jim Dungey and the present author, included Jim Eastwood, John Tendys, Einar Gjøen, Eamon Bowers, Colette Robertson, Denis Hayward, and Peter Smith.

The nature of the current sheet particle trajectories computed by Speiser indeed implies that a specific finite current will flow within the sheet even in the noise-free case when no collisions are present (Speiser 1970). In the case of a strict magnetic neutral sheet illustrated in panel (a) of Fig. 1.4, Alfvén (1968) showed that a specific voltage

$$\Phi = \left(\frac{B^2}{\mu_0 en}\right) \tag{1.2}$$

must exist across the system to provide the correct total current, where B and n are the field strength and ion and electron number densities within the tail lobes on either side. This result is independent of the details of the structure of the current layer or of the particle motion within it, though it does depend on the assumption that none of the current is provided by particles entering from the boundaries (i.e. from the magnetosheath in the tail), which would act to reduce the voltage. The self-consistent structure of such a neutral sheet system was subsequently investigated by Cowley (1973a), who showed that the condition for charge-neutrality required the plasma inflow to be concentrated near the boundary where the ions leave the current sheet, associated with an electric field directed away from the current sheet on either side. Dungey considered that the flow perturbation produced by such a field would launch an Alfvén wave along the field lines as the flux tubes reconnect in the tail, associated with a strong field-aligned current, which in the steady state would stand in the flow on the outer surface of the plasma sheet as observations had indicated (Dungey 1975a, b; Hayward and Dungey 1983). Analysis then showed that the wave and field-aligned current would thin as it approaches Earth, down to the kilometer-scales of individual auroral arcs (Dungey 1982). It had become evident from observational studies in the 1970s and 1980s that discrete auroral arcs are directly associated with such field-aligned currents and the field-aligned voltages that result. The issue of the origins of the aurora was thus found to involve significantly more complication than, though still being connected with, the physical picture first proposed by Hoyle.

In the case of a current sheet downstream from the reconnection site which is threaded by a weak near-uniform magnetic field B_z, with trajectories illustrated in panel (b) of Fig. 1.4, the electric field E_y in the direction of the current can be transformed away by moving in the x-direction with speed E_y/B_z (the de Hoffman-Teller speed), equal to the $\boldsymbol{E} \times \boldsymbol{B}$ field line contraction speed at the centre of the current sheet as indicated above. Equilibrium then requires a given particle flux along the field lines into and out of the current sheet in the transformed frame, equivalent to the marginal fire-hose stability condition, together with a matching of the magnetic and electric field structure of the sheet to the current and charge produced by the particles oscillating within it (Eastwood 1972, 1974). In the simple case of a cold plasma drifting normally into the current sheet with near-zero field-aligned velocity in the untransformed (Earth's) frame, the equilibrium condition requires the field line contraction speed to be

$$E_y/B_z \simeq V_A/\sqrt{2}, \tag{1.3}$$

where V_A is the Alfvén speed in the inflow region. The accelerated ions then jet out along the field lines away from the current sheet on either side at twice this speed, $\sim \sqrt{2}V_A$, the acceleration being equivalent to an elastic collision with a mirror moving at the de Hoffmann-Teller speed (e.g., Cowley and Southwood 1980).

The subsequent issue of the nature of the waves generated by instability of the particle distributions set up within the current sheet is then significantly complicated by the non-standard nature of the zeroth-order particle trajectories illustrated in Fig. 1.4, and by the thinness of the current layer in one of its dimensions. The latter configuration means in particular that instability within the current sheet can be damped by coupling to waves that radiate energy into the surrounding medium, this aspect being investigated by Dungey (1969) and Gjøen (1971). Wave growth and instability conditions were then considered within a variety of contexts by Tendys (1970), Gjøen (1972), Bowers (1973a, b), Robertson et al. (1981), Smith et al. (1984), and Smith (1984). The latter studies culminated in an approximate quasi-linear treatment by Dungey and Smith (1984) which suggested that such 'noise' would not produce a major effect on the current-carrying electron distributions, this result being completed shortly before Jim Dungey's retirement from Imperial in 1984.

1.4.2 Large-Scale Structure and Dynamics: Further Developments

During the 1980s and early 1990s, development of large-scale aspects of the reconnection model at Imperial, based on Dungey's initial concepts outlined in Sect. 1.3, was largely undertaken by the present author, brief consideration of which concludes the present survey. Large-scale properties of plasma populations within the magnetosphere were discussed by Cowley (1980), comparing expectations with the rapidly increasing body of data on the Earth's space environment that had been obtained from space missions such as HEOS-2, IMP-6 and -7, ATS-6, Pioneer-7 and -8, and Prognoz-1 and -2. This discussion was subsequently up-dated by Cowley (1993) to encompass results from later space missions, particularly with regard to energetic ion outflow from the auroral ionosphere. Theoretical discussion centred on the acceleration of ions in the current sheets downstream from the reconnection sites at the magnetopause and in the tail, as outlined above, and the consequent plasma populations that are formed depending on the connectivity of the field lines. On the dayside, the downstream field lines are connected to the Earth at one end only (i.e. open field lines), with the accelerated plasmas forming boundary layers on either side of the dayside magnetopause, the dayside cusp as the particles 'bounce' in their mirror motion near the Earth, and the tail plasma mantle beyond. In the tail, the downstream field lines are connected to the Earth at both ends earthward of the neutral line (i.e. closed field lines), forming the hot trapped plasma sheet population with an Earthward-flowing boundary layer on its outer surface, or at no ends tailward of the neutral line (i.e. 'disconnected' interplanetary field lines), forming a thin plasma sheet population at the tail centre that flows continuously tailward at speeds comparable to the lobe Alfvén speed as indicated in the above discussion, eventually re-joining the solar wind. Figure 1.8 illustrates these populations in a cut through the noon-midnight meridian plane with the Sun to the left, taken from Cowley et al. (2003). The solar wind and populations derived directly from it are shown by green dots, the out-flowing ionospheric polar wind plasma and equilibrium plasmasphere by blue dots, and the hot ring current and plasma sheet plasma by red dots (including a tailward-flowing plasmoid structure tailward of the tail neutral line pinched off during a substorm). The tell-tale nature of the accelerated ion velocity distributions resulting from current sheet acceleration was elucidated with particular reference to the dayside magnetopause by Cowley (1982), having the form of 'mushroom caps' aligned (nearly) along the field direction, truncated at the de Hoffman-Teller speed. Such distributions were subsequently observed in detailed ion data returned by the AMPTE-UKS (Smith and Rodgers 1991) and AMPTE-CCE (Fuselier et al. 1991) spacecraft. Accelerated flows at the dayside magnetopause consistent with reconnection between magnetospheric and magnetosheath fields (outlined above) had first been reported in ion data from the ISEE-1 and -2 spacecraft by Paschmann et al. (1979) and Sonnerup et al. (1981), while few-minute field perturbations interpreted as being due to

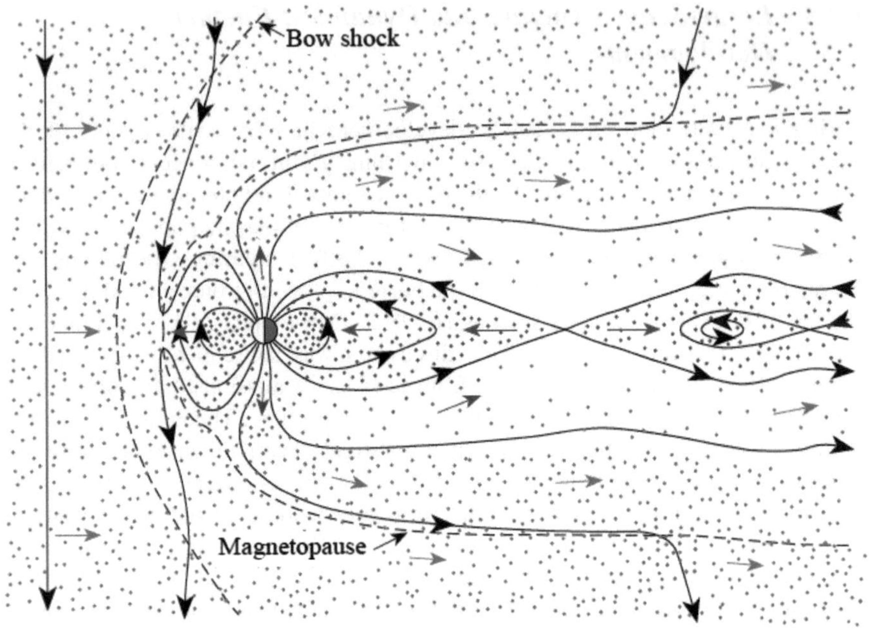

Fig. 1.8 Noon-midnight meridian cross-section through the Earth's magnetosphere, where the *arrowed solid lines* show magnetic field lines, while the *dashed lines* show the bow shock and magnetopause as marked. The *coloured dots* show the principal plasma populations originating in the solar wind (*green*) and the Earth's ionosphere (*blue*). Both sources contribute to the hot plasma sheet population located at the centre plane of the tail (*red*). From Cowley et al. (2003) (Color figure online)

transient reconnection producing 'flux transfer events' had been reported by Russell and Elphic (1979).

A second topic involved further development of ideas concerning the dependence of magnetospheric structure and flow on the direction and strength of the IMF. The first detailed observations of electric fields over the polar regions obtained by the OGO-6 spacecraft in 1969 had shown that the $\boldsymbol{E} \times \boldsymbol{B}$ flow is indeed usually of twin-vortex form compatible with the S_D current system, but that there are also strong dawn-dusk asymmetries depending on the east-west component of the interplanetary field, IMF B_y, that have opposite senses in the northern and southern hemispheres (Heppner 1972). Cowley (1981a), updated by Cowley et al. (1991), showed that the ensemble of such effects observed in the dayside cusp, polar flows, auroral zone location, and plasma mantle that had been observed up to those times results from the east-west tension forces on open flux tubes associated with IMF B_y, opposite in opposite hemispheres, as sketched (specifically for IMF B_y positive) in panel (a) of Fig. 1.9. This results in asymmetric addition of open flux tubes to the tail lobes as sketched in panel (b) of the figure, where for IMF B_y positive open flux is added preferentially to the dawn side of the northern lobe, and to the dusk side of the southern lobe (and vice-versa for IMF B_y negative). Since

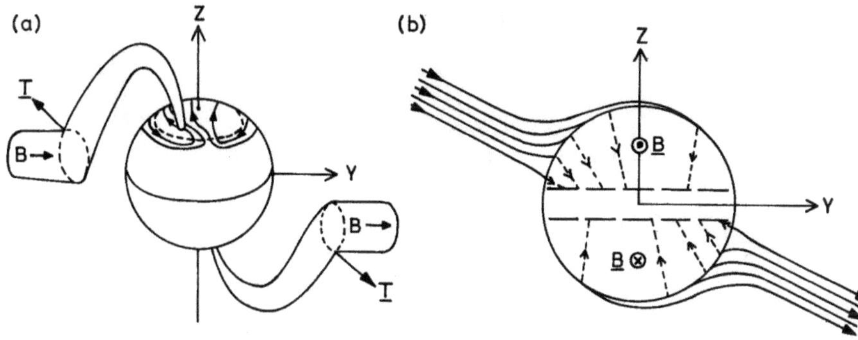

Fig. 1.9 Sketches illustrating the effects associated with the y component of the IMF, specifically for B_y positive, in views looking from the direction of the Sun. Panel (**a**) sketches newly-opened flux tubes on the dayside, showing the east-west tension forces $\boldsymbol{T}$ that are oppositely-directed in the northern and southern hemispheres, resulting in asymmetrical addition of open flux tubes to the tail lobes, illustrated in panel (**b**). Panel (**a**) also illustrates the resulting asymmetrical flows that are driven on open flux tubes in the polar ionosphere, also in opposite senses in the two hemispheres. From Cowley (1981b)

open flux thus 'enters' the northern lobe preferentially on one side, and 'leaves' the southern lobe preferentially on the other, it follows that a cross-tail field will exist having the same sense as, but weaker than, the IMF B_y field, as had been found previously in IMP-6 tail field data by Fairfield (1979). A related effect was subsequently found in geostationary orbit magnetic field data by Cowley and Hughes (1983). In effect, as a result of the magnetic 'opening' of the magnetosphere through reconnection, the IMF partially penetrates into the magnetosphere, as had first been imagined by Hoyle in the summer of 1947.

A further complication arises for strongly northward IMF, when reconnection ceases at the low-latitude magnetopause, but can then occur at high latitudes beyond the cusp, resulting in 'lobe reconnection'. This process results in regions of sunward convection appearing within the polar cap (e.g., Burke et al. 1979), associated with either one or two additional cells of flow depending upon concurrent IMF B_y. Theoretically, however, a number of possibilities arise as illustrated in Fig. 1.10, taken from Cowley (1981b), depending on whether the tail is taken to be open or closed, and whether the reconnection occurs at both lobes, simultaneously or in sequence, or at only one. Panel (a) in the figure corresponds to the closed two-lobe system briefly discussed by Dungey (1963c), panel (d) to the closed single-lobe system discussed by McDiarmid et al. (1980), and panel (e) to the open single-lobe system discussed by Russell (1972) in the context of significant simultaneous IMF B_y.

Observations such as those outlined above showed that the flow in the Earth's magnetosphere responds dynamically to the direction of the IMF in the manner expected for the reconnection model. In addition, studies using low-altitude polar-orbiting spacecraft data showed that the voltage drop across the polar cap is also well-ordered by the B_z component of the IMF, becoming larger for increasingly

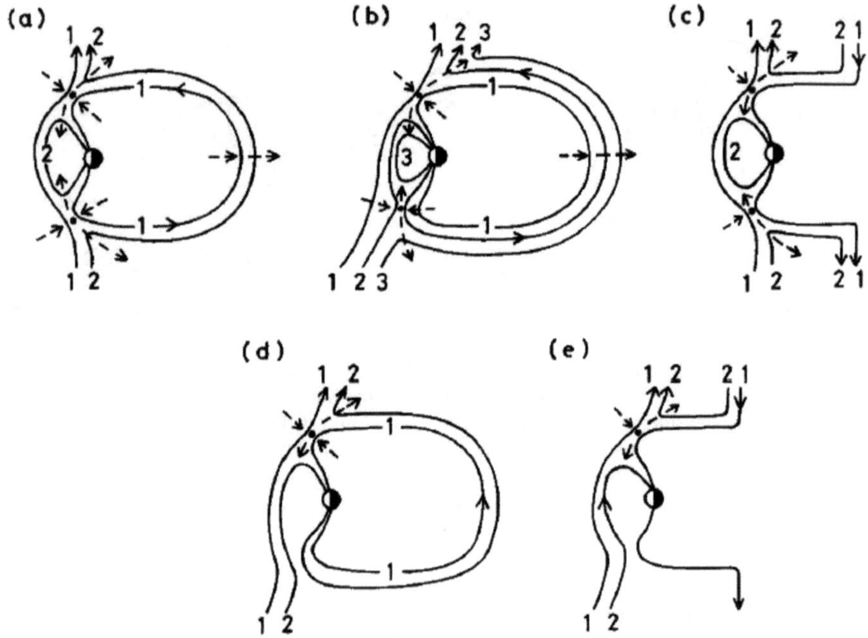

Fig. 1.10 Sketches of reconnection between northward-directed IMF lines and tail field lines poleward of the cusp. The *numbers* indicate the field line sequence and the *heavy dot* the location of the neutral line. The *dashed arrows* indicate the plasma flow in the vicinity of the latter. From Cowley (1981b)

negative (southward) fields (Reiff et al. 1981). However, with a ~90 min cadence for observations in a given hemisphere, low-altitude spacecraft cannot be employed to investigate the temporal responses of the flow to variations in the IMF, which occur on shorter time scales. More detailed investigation thus requires high-cadence measurements of the ionospheric flow using ground-based radar systems, combined with spacecraft observations of the IMF upstream of the magnetosphere.

An ideal opportunity to undertake such joint observations occurred in 1984 when the AMPTE-IRM and -UKS spacecraft were launched into orbits with apoapsides just upstream of the bow shock near noon, while the newly-operational EISCAT-UHF radar at Tromsø made measurements of the flow at high latitudes in the dayside sector. Due to the too-lengthy observation sequences planned for the EISCAT 'common programme' measurements, a fast beam-swinging experiment ('UK-Polar') had been developed at the Rutherford Appleton Laboratory at the instigation of the present author, in which the radar beam was pointed at low elevation and swung between two positions either side of north using 2 min dwells. Flow data were then obtained at 15 s cadence in a sequence of range gates that spanned the cusp ionosphere between ~71° and ~75° magnetic latitude, with the data from the two pointing directions being combined to produce flow vectors every 2.5 min. Figure 1.11 shows one such interval, taken from Willis et al. (1986), where

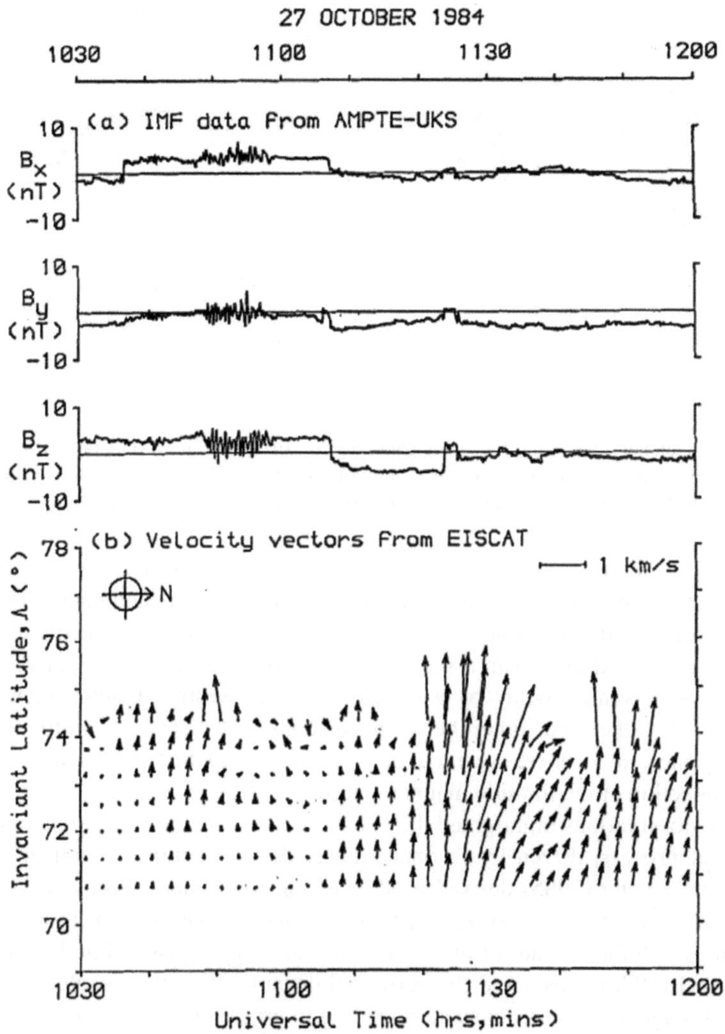

Fig. 1.11 AMPTE-UKS magnetic field data and EISCAT UK-Polar ionospheric flow vectors obtained during a 1.5 h interval on 27 October 1984. The *top three panels* show the three components of the IMF in GSM coordinates, while the *lower panel* shows latitude profiles of the ionospheric flow, where for purposes of visibility the vectors have been rotated clockwise by 90° into the direction of the electric field associated with the flow. After Willis et al. (1986)

we show 1.5 h of joint data in which the three components of the IMF measured by AMPTE-UKS are shown at the top of the plot, and the beam-swung EISCAT flow vectors at the bottom. For reasons of visibility the latter vectors have been rotated 90° clockwise into the direction of the electric field associated with the flow, such that the largely northward-pointing vectors represent a westward-directed flow. The local time of the radar spans ~13–14.5 h in the early post-noon sector during the interval. A sharp reversal of IMF B_z from positive to negative is observed to occur at

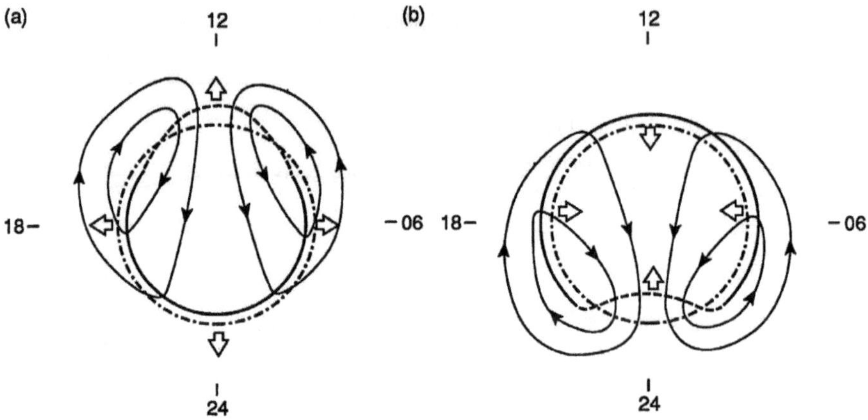

Fig. 1.12 Sketches illustrating the flows generated by (**a**) open flux production at the dayside magnetopause, and (**b**) open flux closure in the tail, in views looking down on the northern pole with noon at the top and dusk to the left. From Cowley and Lockwood (1992)

~11:08 UT, which is followed by a sudden onset of ionospheric flow ~5 min after the change was estimated to have arrived at the subsolar magnetopause. The latter 'delay' corresponds to a couple of Alfvén wave transit times between the equator and ionosphere. After ~15 min the field then returns to smaller negative values, followed with similar delay by a reduction in the westward flow. Such results illustrated for the first time the rapid response of the flow in the coupled magnetosphere-ionosphere system to changes in the direction of the IMF, mediated by reconnection at the dayside magnetopause.

These observations highlighted the need to discuss the flow as a dynamic time-dependent process, driven by reconnection at the magnetopause and in the tail. Siscoe and Huang (1985) derived models of the ionospheric flow that would be driven during intervals of dayside reconnection with no tail reconnection, and vice versa, in which the amount of open flux thus varies with time, while Freeman and Southwood (1988) considered the perturbed dayside flows that would result from localised magnetopause reconnection leading to erosion of the boundary. Building on these discussions, Cowley and Lockwood (1992) proposed a simple conceptual framework which allows one to discuss the general time-dependent case. The central point is that the flow within the system is not associated with the existence of open flux directly, but with changes in the open flux caused by reconnection events occurring at the magnetopause and in the tail. After a long interval with no such events the flow dies away to zero as the system approaches equilibrium in the near-Earth region, independent of how much open flux is actually present. In the polar ionosphere the equilibrium open flux is taken for simplicity to occupy a circular region whose radius depends on the amount of open flux present. If a reconnection event then perturbs the system, represented by a displacement of the boundary in the local time sector where the reconnection takes place, equatorward for open flux production on the dayside, poleward for open flux closure on the nightside, the ensuing flow then carries the system towards its new equilibrium configuration with the changed amount of open flux. This is illustrated in Fig. 1.12

for the cases of (a) open flux production near noon, and (b) open flux closure near midnight, in views looking down on the northern pole with noon at the top and dusk to the left in each case. The original open-closed boundary is shown by the solid circular segment, with the perturbed boundary resulting from the reconnection event indicated by the dashed line, displaced equatorward on the dayside in case (a), and poleward on the nightside in case (b). The equilibrium boundary containing the modified amount of open flux is shown by the dot-dashed lines, expanded from the original in case (a), and contracted in case (b) as indicated by the large arrows. The flow streamlines, shown by the arrowed solid lines then take the perturbed boundary into the new equilibrium boundary, and when that has happened, the flow stops until another event occurs. The time scale for an individual flow event was estimated to be ~15 min. Steady flow only occurs when both processes are present at equal rates, in which case the field configuration is also steady, and the electric field given by a potential, constant (to a first approximation) along field lines. Evidently a number of additional complexities arise more generally, such as the east-west effects of IMF B_y and lobe reconnection for IMF B_z positive as outlined above, but we will not revisit these aspects here.

1.5 Recovery

It is evident from the above discussion that a central observational goal of recent magnetospheric research has been to monitor the rates of open flux production and closure within the system on a long-term basis, and to relate such observations to upstream interplanetary conditions, and to the occurrence of magnetospheric substorms and main phase storms. In practical terms this is only possible using ionospheric data, and requires global knowledge of the position of the open-closed field line boundary, and of the plasma flow across it. Such observations only became possible in the late 1990s and 2000s, however, using global auroral images from spacecraft to locate the position of the boundary, and data from the international SuperDARN radar network to measure the large-scale ionospheric flow. At this point we therefore end the present survey, the thread of which is picked up in the contribution to these proceedings by Steve Milan, who carries the story through to the present time.

In conclusion, it is evident that Jim Dungey's paper published in *Phys. Rev. Lett.* in 1961, based on his PhD researches stimulated by Fred Hoyle during 1947–1950, represents the key foundation of the mature research areas of 'solar-terrestrial physics' and 'space weather' that exist today. As mentioned above, however, this was initially by no means widely recognised by the research community at large, but eventually became so in response to the success of the model in accounting for observations. The ramifications of the 1961 paper have subsequently kept the international community productively engaged for more than the past 50 years. This was indeed a massive achievement.

Acknowledgements Thanks are due to David Southwood for useful commentary on an earlier draft of this contribution.

References

Akasofu, S.-I.: The development of the auroral substorm. Planet. Space Sci. **12**, 273–282 (1964)

Akasofu, S.-I.: Exploring the Secrets of the Aurora (Second Edition). Springer Science + Business Media, New York (2007)

Alfvén, H.: Cosmical Electrodynamics. Oxford University Press, Oxford (1950)

Alfvén, H.: Some properties of magnetospheric neutral surfaces. J. Geophys. Res. **73**, 4379–4381 (1968)

Arnoldy, R.L.: Signature in the interplanetary medium for substorms. J. Geophys. Res. **76**, 5189–5201 (1971)

Axford, W.I., Hines, C.O.: A unifying theory of high-latitude geophysical phenomena and geomagnetic storms. Can. J. Phys. **39**, 1433–1464 (1961)

Axford, W.I., Petschek, H.E., Siscoe, G.L.: Tail of the magnetosphere. J. Geophys. Res. **70**, 1231–1236 (1965)

Bame, S.J., Asbridge, J.R., Felthauser, H.E., Olson, R.A., Strong, I.B.: Electrons in the plasma sheet of the earth's magnetic tail. Phys. Rev. Lett. **16**, 138–142 (1966)

Birkeland, K.: The Norwegian Aurora Polaris Expedition 1902-1903, Vol. 1, On the Cause of Magnetic Storms and the Origin of Terrestrial Magnetism, Sect. 1. Aschehoug, Kristiana (1908)

Bowers, E.C.: A theoretical prediction of ion plasma oscillations in a neutral sheet. Astrophys. Space Sci. **21**, 399–423 (1973a)

Bowers, E.C.: A short wavelength instability in the neutral sheet of the earth's magnetic tail. Astrophys. Space Sci. **24**, 349–369 (1973b)

Brice, N.M.: Bulk motion of the magnetosphere. J. Geophys. Res. **72**, 5193–5211 (1967)

Burke, W.J., Kelley, M.C., Sagalyn, R.C., Smiddy, M., Lai, S.T.: Polar cap electric field structures with a northward interplanetary magnetic field. Geophys. Res. Lett. **6**, 21–24 (1979)

Carpenter, D.L.: Whistler studies of the plasmasphere in the magnetosphere 1. Temporal variations in the the position of the knee and some evidence on plasma motions near the knee. J. Geophys. Res. **71**, 693–709 (1966)

Chapman, S., Bartels, J.: Geomagnetism, Vol 1 Geomagnetic and Related Phenomena. Clarendon Press, London (1940)

Chapman, S., Ferraro, V.C.A.: A new theory of magnetic storms: Part 1. The initial phase. Terr. Magn. Atmos. Electr. **36**, 77–97 (1931). 171–186

Chapman, S., Ferraro, V.C.A.: A new theory of magnetic storms: Part 1. The initial phase (continued). Terr. Magn. Atmos. Electr. **37**, 147–156 (1932). 421–429

Coleman Jr., P.J., Davis, L., Sonett, C.P.: Steady component of the interplanetary magnetic field: Pioneer V. Phys. Rev. Lett. **5**, 43–46 (1960)

Cowley, S.W.H.: A self-consistent model of a simple magnetic neutral sheet system surrounded by a cold collisionless plasma. Cosmic Electrodycn. **3**, 448–501 (1973a)

Cowley, S.W.H.: A qualitative study of the reconnection between the earth's magnetic field and an interplanetary field of arbitrary orientation. Radio Sci. **8**, 903–913 (1973b)

Cowley, S.W.H.: Plasma populations in a simple open model magnetosphere. Space Sci. Rev. **26**, 217–275 (1980)

Cowley, S.W.H.: The causes of convection in the earth's magnetosphere—a review of developments during the IMS. Rev. Geophys. Space Phys. **20**, 531–565 (1982)
Cowley, S.W.H.: The magnetosphere and its interaction with the solar wind and with the ionosphere. In: DeWitt, R.N., Duston, D.P., Hyder, A.K. (eds.) The Behaviour of Systems in the Space Environment, pp. 147–181. Kluwer Academic, Dordrecht (1993)
Cowley, S.W.H., Hughes, W.J.: Observation of an IMF sector effect in the Y magnetic field component at geostationary orbit. Planet. Space Sci. **31**, 73–90 (1983)
Cowley, S.W.H., Lockwood, M.: Excitation and decay of solar wind-driven flows in the magnetosphere-ionosphere system. Ann. Geophys. **10**, 103–115 (1992)
Cowley, S.W.H., Southwood, D.J.: Some properties of a steady-state geomagnetic tail. Geophys. Res. Lett. **7**, 833–836 (1980)
Cowley, S.W.H.: Magnetospheric asymmetries associated with the Y-component of the IMF. Planet. Space Sci. **29**, 79–96 (1981)
Cowley, S.W.H.: Magnetospheric and ionospheric flow and the interplanetary magnetic field. In: *The Physical Basis of the Ionosphere in the Solar-Terrestrial System*, AGARD-CP-295, pp. (4-1)–(4-14) (1981b)
Cowley, S.W.H., Morelli, J.P., Lockwood, M.: Dependence of convective flows and particle precipitation in the high-latitude dayside ionosphere on the X and Y components of the interplanetary magnetic field. J. Geophys. Res. **96**, 5557–5564 (1991)
Cowley, S.W.H., Davies, J.A., Grocott, A., Khan, H., Lester, M., McWilliams, K.A., Milan, S.E., Provan, G., Sandholt, P.E., Wild, J.A., Yeoman, T.K.: Solar wind-magnetosphere-ionosphere interactions in the earth's plasma environment. Phil. Trans. R. Soc. A **361**, 113–126 (2003)
Dessler, A.J.: Length of magnetospheric tail. J. Geophys. Res. **69**, 3913–3918 (1964)
Dungey, J.W.: Some researches in cosmic magnetism. PhD Thesis, Cambridge University (1950)
Dungey, J.W.: Conditions for the occurrence of electrical discharges in astrophysical systems. Philos. Mag. **44**, 725–738 (1953)
Dungey, J.W.: Cosmic Electrodynamics. Cambridge University Press, Cambridge (1958a)
Dungey, J.W.: The neutral point discharge theory of solar flares. A reply to Cowling's criticism. In: Lehnert, B. (ed.) Electromagnetic Phenomena in Cosmical Physics, pp. 135–139. Cambridge University Press, Cambridge (1958b)
Dungey, J.W.: Interplanetary magnetic field and the auroral zones. Phys. Rev. Lett. **6**, 47–48 (1961)
Dungey, J.W.: The origin of irregularities in the F region. J. Phys. Soc. Jpn. **17**(Suppl. A-I), 300–301 (1962a)
Dungey, J.W.: The interplanetary field and auroral theory. J. Phys. Soc. Jpn. **17**(Suppl. A-II), 15–19 (1962b)
Dungey, J.W.: Resonant effect of plasma waves on charged particles in a magnetic field. J. Fluid Mech. **15**, 74–82 (1963a)
Dungey, J.W.: Loss of Van Allen electrons due to whistlers. Planet. Space Sci. **11**, 591–595 (1963b)
Dungey, J.W.: The structure of the exosphere or adventures in velocity space. In: DeWitt, C., Hieblot, J., Lebeau, A. (eds.) Geophysics the Earth's Environment, pp. 504–550. Gordon and Breach, New York (1963c)
Dungey, J.W.: Effects of electromagnetic perturbations on particles trapped in the radiation belts. Space Sci. Rev. **4**, 199–222 (1965a)
Dungey, J.W.: Null points in space plasma. In: Chang, C.C., Huang, S.S. (eds.) Proceedings of Plasma Space Science Symposium, pp. 160–169. D. Reidel, Dordrecht (1965b)
Dungey, J.W.: The length of the magnetospheric tail. J. Geophys. Res. **70**, 1753 (1965c)
Dungey, J.W.: Hydromagnetic waves. In: Matsushita, S., Campbell, W.J. (eds.) Physics of Geomagnetic Phenomena, pp. 913–933. Academic Press, New York (1967)
Dungey, J.W.: The reconnection model of the magnetosphere. In: McCormac, B.M. (ed.) Earth's Particles and Fields, pp. 385–392. Reinhold Book, New York (1968a)

Dungey, J.W.: Waves and particles in the magnetosphere. In: Carovillano, R.L., McClay, J.F., Radoski, H.R. (eds.) Physics of the Magnetosphere, pp. 218–259. D. Reidel, Dordrecht (1968b)

Dungey, J.W.: Polar substorms—theoretical review. In: Mitra, A.P., Jacchia, L.G., Newman, W.S. (eds.) Space Research VIII, pp. 243–252. North Holland, Amsterdam (1968c)

Dungey, J.W.: Damping of waves in the current sheet in the geomagnetic tail by radiation. Planet. Space Sci. **17**, 1291–1296 (1969)

Dungey, J.W.: Some remaining mysteries in the aurora. Q. J. R. Astron. Soc. **16**, 117–127 (1975a)

Dungey, J.W.: Neutral sheets. Space Sci. Rev. **17**, 173–180 (1975b)

Dungey, J.W.: Thinning of field-aligned currents. Geophys. Res. Lett. **9**, 1243–1245 (1982)

Dungey, J.W.: Citation classic. Curr. Contents Phys. Chem. Earth Sci. **23**(49), 20 (1983)

Dungey, J.W.: Memories, maxims, and motives. J. Geophys. Res. **99**, 19189–19197 (1994)

Dungey, J.W., Smith, P.R.: Quasilinear effect of noise in a magnetic neutral sheet. Planet. Space Sci. **32**, 1201–1204 (1984)

Dungey, J.W., Speiser, T.W.: Electromagnetic noise in the current sheet in the geomagnetic tail. Planet. Space Sci. **17**, 1285–1290 (1969)

Dungey, J.W.: Electrodynamics of the outer atmosphere. Scientific Report 69, Ionospheric Research Laboratory, Pennsylvania State University (1954)

Dungey, J.W.: Electrodynamics of the outer atmosphere. In: Physics of the Ionosphere, pp. 229–236. The Physical Society, London (1955)

Dungey, J.W.: Survey of acceleration and diffusion. In: McCormac, B.M. (ed.) Radiation Trapped in the Earth's Magnetic Field, pp. 389–397. D. Reidel, Dordrecht (1966)

Dungey, J.W.: The magnetosphere. In: *Professorial Inaugural Lectures,* pp. 273–288. Imperial College, London (1966b)

Dungey, J.W.: The theory of the quiet magnetosphere. In: Solar-Terrestrial Physics, pp. 91–106. Academic Press, London (1967b)

Eastwood, J.W.: Consistency of fields and particle motion in the 'Speiser' model of the current sheet. Planet. Space Sci. **20**, 1555–1568 (1972)

Eastwood, J.W.: The warm current sheet model, and its implications on the temporal behaviour of the geomagnetic tail. Planet. Space Sci. **22**, 1641–1668 (1974)

Egeland, A., Burke, W.J.: Kristian Birkeland—The First Space Scientist. Springer, Dordrecht (2005)

Fairfield, D.H.: Ionosphere current patterns in high latitudes. J. Geophys. Res. **68**, 3589–3602 (1963)

Fairfield, D.H.: On the average configuration of the geomagnetic tail. J. Geophys. Res. **84**, 1950–1958 (1979)

Fairfield, D.H., Cahill Jr., L.J.: Transition region magnetic field and polar magnetic disturbances. J. Geophys. Res. **71**, 155–169 (1966)

Freeman, M.P., Southwood, D.J.: The effect of magnetospheric erosion on mid- and high-latitude ionospheric flows. Planet. Space Sci. **36**, 509–522 (1988)

Fuselier, S.A., Klumpar, D.M., Shelley, E.G.: Ion reflection and transmission during reconnection at the earth's subsolar magnetopause. Geophys. Res. Lett. **18**, 139–142 (1991)

Giovanelli, R.G.: A theory of chromospheric flares. Nature **158**, 81–82 (1946)

Giovanelli, R.G.: Magnetic and electric phenomena in the sun's atmosphere associated with sunspots. Mon. Not. R. Astron. Soc. **107**, 338–355 (1947)

Giovanelli, R.G.: Magnetic and electric phenomena in the sun's atmosphere associated with sunspots. Mon. Not. R. Astron. Soc. **108**, 163–176 (1948)

Gjøen, E.: Radiation damping of waves in the current sheet in the geomagnetic tail. Planet. Space Sci. **19**, 635–642 (1971)

Gjøen, E.: Resonant interaction of an electrostatic wave with electrons in a current sheet. Planet. Space Sci. **20**, 1205–1212 (1972)

Hayward, D., Dungey, J.W.: An Alfvén wave approach to auroral field-aligned currents. Planet. Space Sci. **31**, 579–585 (1983)

Heikkila, W.J.: Earth's Magnetosphere: Formed by the Low-Latitude Boundary Layer. Elsevier Science, Amsterdam (2011)

Heppner, J.P.: Polar-cap electric field distributions related to the interplanetary magnetic field direction. J. Geophys. Res. **77**, 4877–4887 (1972)

Heppner, J.P., Ness, N.F., Skillman, T.L., Scearce, C.S.: Explorer X magnetic field measurements. J. Geophys. Res. **68**, 1–46 (1963)

Hirshberg, J., Colburn, D.S.: The interplanetary field and geomagnetic variations—a unified view. Planet. Space Sci. **17**, 1183–1206 (1969)

Hoyle, F.: Some Recent Researches in Solar Physics. Cambridge University Press, Cambridge (1949)

McDiarmid, I.B., Burrows, J.R., Wilson, M.D.: Comparison of magnetic field perturbations and solar electron profiles in the polar cap. J. Geophys. Res. **85**, 1163–1170 (1980)

McPherron, R.L.: Growth phase of magnetospheric substorms. J. Geophys. Res. **75**, 5592–5599 (1970)

Ness, N.F.: The earth's magnetic tail. J. Geophys. Res. **70**, 2989–3005 (1965)

Nishida, A.: Formation of plasmapause, or magnetospheric plasma knee, by the combined action of magnetospheric convection and plasma escape from the tail. J. Geophys. Res. **71**, 5669–5679 (1966)

Paschmann, G., Sonnerup, B.U.Ö., Papamastorakis, I., Sckopke, N., Haerendel, G., Bame, S.J., Asbridge, J.R., Gosling, J.T., Russell, C.T., Elphic, R.C.: Plasma acceleration at the earth's magnetopause: evidence for reconnection. Nature **282**, 243–246 (1979)

Reiff, P.H., Spiro, R.W., Hill, T.W.: Dependence of polar cap potential drop on interplanetary parameters. J. Geophys. Res. **86**, 6739–6748 (1981)

Robertson, C., Cowley, S.W.H., Dungey, J.W.: Wave-particle interactions on a magnetic neutral sheet. Planet. Space Sci. **29**, 399–403 (1981)

Rosenbauer, H., Grünwaldt, H., Montgomery, M.D., Paschmann, G., Sckopke, N.: Heos 2 plasma observations in the distant polar magnetosphere: the plasma mantle. J. Geophys. Res. **80**, 2723–2737 (1975)

Rostoker, G.: Relationship between the onset of a geomagnetic bay and the configuration of the interplanetary magnetic field. J. Geophys. Res. **73**, 4382–4387 (1968)

Russell, C.T.: The configuration of the magnetosphere. In: Dyer, E.R. (ed.) Critical Problems of Magnetospheric Physics, pp. 1–16. National Academy of Sciences, Washington, DC (1972)

Russell, C.T., Elphic, R.C.: ISEE observations of flux transfer events at the dayside magnetopause. Geophys. Res. Lett. **6**, 33–36 (1979)

Scatten, K.H., Wilcox, J.M.: Response of the geomagnetic activity index K_p to the interplanetary magnetic field. J. Geophys. Res. **72**, 5185–5192 (1967)

Siscoe, G.S., Huang, T.S.: Polar cap inflation and deflation. J. Geophys. Res. **90**, 543–547 (1985)

Smith, P.R.: Electrostatic wave noise in the neutral sheet in the earth's geomagnetic tail. Planet. Space Sci. **32**, 1147–1167 (1984)

Smith, M.F., Rodgers, D.J.: Ion distributions at the dayside magnetopause. J. Geophys. Res. **96**, 11617–11624 (1991)

Smith, P.R., Cowley, S.W.H., Dungey, J.W.: Dispersion relations in the electrostatic approximation for waves in a magnetic neutral sheet. Planet. Space Sci. **32**, 1135–1145 (1984)

Sonnerup, B.U.Ö., Paschmann, G., Papamastorakis, I., Sckopke, N., Haerendel, G., Bame, S.J., Asbridge, J.R., Gosling, J.T., Russell, C.T.: Plasma acceleration at the earth's magnetopause: evidence for magnetic field line reconnection at the Earth's magnetopause. J. Geophys. Res. **86**, 10049–10067 (1981)

Speiser, T.W.: Particle trajectories in a model current sheet, based on the open model of the magnetosphere, with applications to auroral particles. J. Geophys. Res. **70**, 1717–1728 (1965a)

Speiser, T.W.: Particle trajectories in model current sheets 1. Analytical solutions. J. Geophys. Res. **70**, 4219–4226 (1965b)

Speiser, T.W.: Particle trajectories in model current sheets 2. Application to auroras using a geomagnetic tail model. J. Geophys. Res. **72**, 3919–3932 (1967)

Speiser, T.W.: Conductivity without collisions or noise. Planet. Space Sci. **18**, 613–622 (1970)
Storey, L.R.O.: An investigation of whistling atmospherics. Phil. Trans. R. Soc. A **246**, 113–141 (1953)
Tendys, J.: Amplification of electrostatic waves in the neutral sheet. Cosmic Electrodyn. **1**, 328–347 (1970)
Wilcox, J.M., Scatten, K.H., Ness, N.F.: Influence of interplanetary magnetic field and plasma on geomagnetic activity during quiet-sun conditions. J. Geophys. Res. **72**, 19–26 (1967)
Willis, D.M., Lockwood, M., Cowley, S.W.H., Van Eyken, A.P., Bromage, B.J.I., Rishbeth, H., Smith, P.R., Crothers, S.R.: A survey of simultaneous observations of the high latitude ionosphere and interplanetary magnetic field with EISCAT and AMPTE-UKS. J. Atmos. Terr. Phys. **48**, 987–1008 (1986)

Chapter 2
Sun et Lumière: Solar Wind-Magnetosphere Coupling as Deduced from Ionospheric Flows and Polar Auroras

S.E. Milan

Abstract The Dungey (Phys. Rev. Lett. 6:47–48, 1961) open model of the magnetosphere, and especially its time-dependent form, the expanding/contracting polar cap (ECPC) paradigm, has provided an important theoretical framework within which to understand solar wind-magnetosphere-ionosphere coupling. This paper reviews the evidence supporting the open and ECPC models, and discusses developments that have arisen in the last 20 years, concentrating on the contributions made by measurements of the ionospheric convection pattern and global auroral imagery. Various magnetospheric phenomena are discussed within the context of the open model, including substorms, geomagnetic storms, steady magnetospheric convection, sawtooth events, cusp auroral spots, and transpolar arcs. The review concludes with a discussion of avenues for future research in the field of solar wind-magnetosphere-ionosphere coupling.

2.1 Introduction

As outlined in Stan Cowley's preceding paper (chapter "Dungey's Reconnection Model of the Earth's Magnetosphere: The First 40 Years"), the pioneering work of Jim Dungey has provided a coherent theoretical framework in which many aspects of the large-scale structure and dynamics of the magnetosphere-ionosphere system can be understood. Development of the open model of the magnetosphere from the original "Dungey cycle" picture (Dungey 1961), to the fully time-dependent "expanding/contracting polar cap" paradigm (ECPC) of solar wind-magnetosphere-ionosphere coupling (Lockwood et al. 1990; Cowley and Lockwood 1992; Lockwood and Cowley 1992) took 30 years. The subsequent 20 years have seen a

S.E. Milan (✉)
Department of Physics and Astronomy, University of Leicester, Leicester LE1 7RH, UK

Birkeland Centre for Space Science, Box 7803, 5020 Bergen, Norway
e-mail: ets@leicester.ac.uk

D. Southwood et al. (eds.), *Magnetospheric Plasma Physics: The Impact of Jim Dungey's Research*, Astrophysics and Space Science Proceedings 41,
DOI 10.1007/978-3-319-18359-6_2

steady accumulation of observational evidence for both Dungey's open model and the ECPC. Alongside this, there has been a growing appreciation that magnetospheric phenomena beyond those originally considered by Dungey and his co-workers fit within the framework provided by the open model. Furthermore, Dungey's ideas have gained increasing application in the study of the magnetospheres of solar system bodies other than the Earth.

The body of literature that provides support for Dungey's picture is too large to be summarized in a brief review. To be focussed, this review confines itself to a discussion of how the availability of global auroral imagery and measurements of the ionospheric convection pattern have helped cement the ideas Dungey put forward at a time when there was a general dearth of observational evidence to support them. What evidence there was came largely from ground-based observations, as described by Cowley (chapter "Dungey's Reconnection Model of the Earth's Magnetosphere: The First 40 Years"), and these helped develop Dungey's ideas. This review is divided into several sections: Sect. 2.2 describes the current understanding of the time-dependent Dungey cycle, the ECPC; Sect. 2.3 describes the observations of the ionospheric convection pattern and auroras which provide evidence for the ECPC; Sect. 2.4 describes how the ECPC and substorm cycle are related; Sect. 2.5 explains how magnetic reconnection rates can be quantified; Sect. 2.6 discusses the current understanding of how magnetotail reconnection is controlled; Sect. 2.7 looks at the role of reconnection in magnetospheric dynamics when the interplanetary magnetic field is directed northwards; and Sect. 2.8 concludes with a brief discussion of future research directions.

2.2 The Modern View of the Dungey Cycle: The Expanding/Contracting Polar Cap Paradigm

Dungey proposed that magnetospheric dynamics were driven largely by magnetic reconnection occurring at the magnetopause between the interplanetary magnetic field (IMF) and the terrestrial field. IMF orientation will be discussed in terms of the usual Geocentric Solar Magnetic (GSM) coordinate system, in which the X-axis points towards the Sun, the X-Z plane contains the Earth's magnetic axis, and Y is perpendicular to this, pointing in a generally duskwards direction. In this system, "northwards" and "southwards" directed field relate to IMF $B_Z > 0$ and $B_Z < 0$, respectively. Reconnection was expected to occur most efficiently where the magnetic shear across the magnetopause was high, that is near the subsolar point for southwards-directed IMF (Dungey 1961), and at high latitudes for northward IMF (Dungey 1963). Magnetic reconnection was also proposed to occur between the oppositely-directed magnetic fields either side of the neutral sheet in the magnetotail, especially when the IMF is directed southwards (Dungey 1961). The combined action of subsolar and magnetotail reconnection leads to a circulation of magnetic field and plasma in the magnetosphere in what is now known as the

Dungey cycle (see Fig. 1.2 of chapter "Dungey's Reconnection Model of the Earth's Magnetosphere: The First 40 Years"). As remarked by Stan Cowley at the end of his paper, by the mid-1990s it was clear that time dependence of magnetic reconnection at the magnetopause, due to variations in interplanetary conditions, and in the magnetotail lead to a highly dynamic system, in which the proportion of the terrestrial magnetic flux that is open can vary considerably on timescales of minutes and hours. For instance, studies of the location of the dayside magnetopause showed that inward motion occurs as a consequence of erosion by reconnection during southwards IMF, without immediate return of newly-closed flux to the dayside by reconnection in the magnetotail (e.g., Aubry et al. 1970; Haerendel et al. 1978). This decoupling of the dayside and nightside processes leads to a new view of how the Dungey cycle is powered: many workers contributed to the development of this new view, notably Siscoe and Huang (1985) and Freeman and Southwood (1988), culminating in the "expanding/contracting polar cap" paradigm (ECPC), first fully elucidated by Cowley and Lockwood (1992). We describe the ECPC in the rest of this section.

The top row of Fig. 2.1 presents schematics of the magnetic field orientation in the magnetopause of the Earth (white lines), looking from the Sun, and the locations where the magnetic shear is high for different orientations of the IMF (purple lines): $B_Y > 0$, $B_Z < 0$; $B_Y > 0$, $B_Z > 0$; $B_Y = 0$, $B_Z > 0$. The magnetopause is roughly paraboloid in shape (with indentations near the cusps) due to stress-balance between the ram pressure of the solar wind and magnetic pressure inside the magnetosphere. In this section, we concentrate on the left column of Fig. 2.1 in which the region of high shear (red) is located across the magnetopause at low latitudes, where terrestrial magnetic flux is closed (green region) rather than open lobe flux (blue region) as found at higher latitudes, tailward of the cusp openings. Once reconnection occurs between the IMF and the terrestrial field, highly-kinked new open field lines evolve across the magnetopause under the influence of the magnetosheath flow (directed radially away from the subsolar point) and the magnetic tension force (Cowley 1981a, b) (see also Fig. 1.9 of chapter "Dungey's Reconnection Model of the Earth's Magnetosphere: The First 40 Years"). Figure 2.2 shows the expected motion of the intersection points of newly reconnected field lines with the magnetopause as they join the existing open flux of the lobes (Cowley and Owen 1989; Cooling et al. 2001). For the $B_Y > 0$ case, tension forces pull northern (southern) hemisphere flux towards the dawn (dusk) sector. The left-middle panel of Fig. 2.1 shows a cross-section of the magnetotail, with the open flux (blue) of the northern and southern lobes, and the closed flux associated with the plasma sheet (green). Across the top of the lobes is indicated the region of newly-opened flux created by reconnection, pulled towards dawn in the northern hemisphere by tension forces, indicating that the tail magnetopause has been deformed from a cylinder; equally, the subsolar magnetopause is eroded by the action of reconnection. Cowley and Lockwood (1992) realized that such deformation of the magnetopause from a paraboloid would result in stress-imbalances which lead to motions of the plasma within the magnetosphere to return the system to equilibrium with the solar wind flow (see Fig. 2.3).

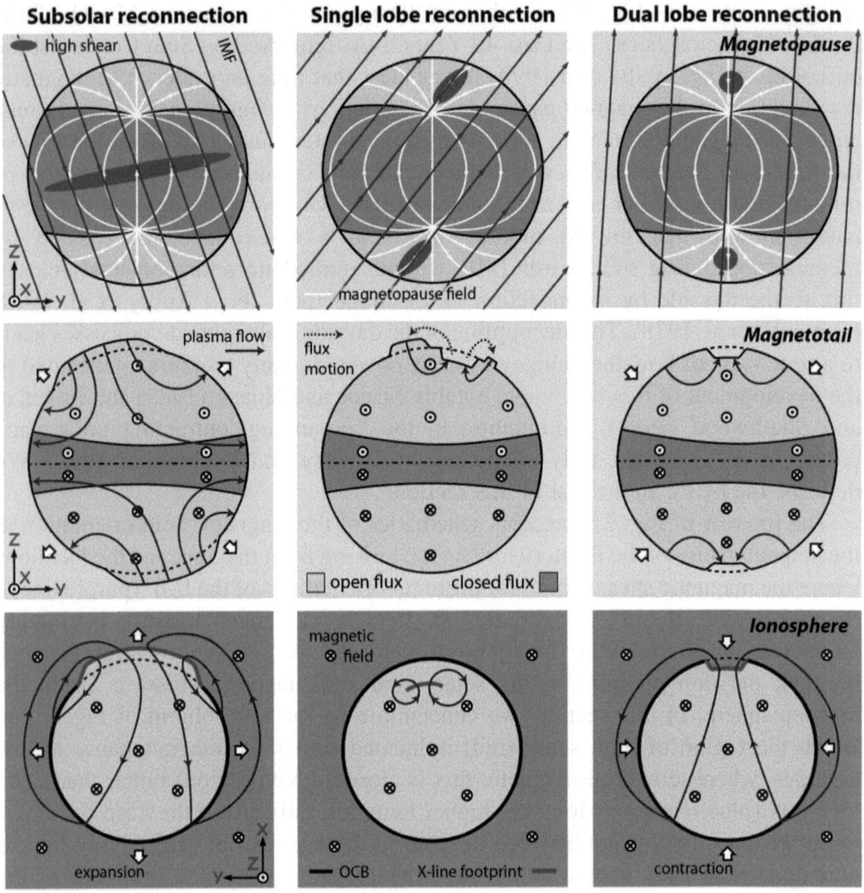

Fig. 2.1 Schematics of solar wind-magnetosphere-ionosphere coupling. The *upper row* shows the magnetopause looking from the Sun, the *middle row* a cross-section of the magnetotail, the *bottom row* a view of the northern hemisphere ionosphere from above. *Green* (*blue*) *regions* indicate magnetic field lines that are closed (open). *Purple arrows* indicate the direction of the interplanetary magnetic field, *white arrows* the direction of the magnetic field in the magnetopause. *Dotted* and *crossed circles* indicate the magnetic field direction into or out of the plane of the diagram. *Red areas* in the *upper panels* show where the IMF/magnetopause magnetic shear is high and reconnection is likely to occur. In the *lower panels*, the *thick black line* indicates the location of the open/closed field line boundary (OCB); the *thick red line* indicates the ionospheric footprint of the magnetopause reconnection X-line. *Blue arrows* indicate the direction of plasma motions (and magnetic flux transport) in the magnetotail and in the ionosphere. In the ionosphere, the OCB moves with these plasma flows, but there is relative motion of the flows with respect to the X-line (Color figure online)

The left-bottom panel of Fig. 2.1 shows the magnetic flux of the northern lobe (blue) and plasma sheet (green) mapped into the northern hemisphere. The ionospheric projection of the lobe is known as the polar cap, usually of roughly circular cross-section, centred somewhat antisunwards of the geomagnetic pole; the polar

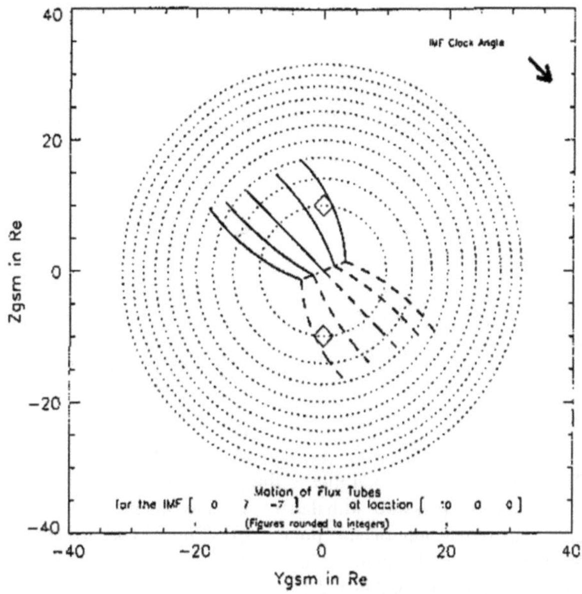

Fig. 2.2 The motion of the intersection points of five pairs of newly-reconnected field lines away from a subsolar reconnection X-line for IMF $B_Z < 0$, $B_Y > 0$, looking from the Sun. *Concentric circles* represent the surface of the magnetopause, and *diamonds* indicate the openings of the cusps [from Cooling et al. (2001)]

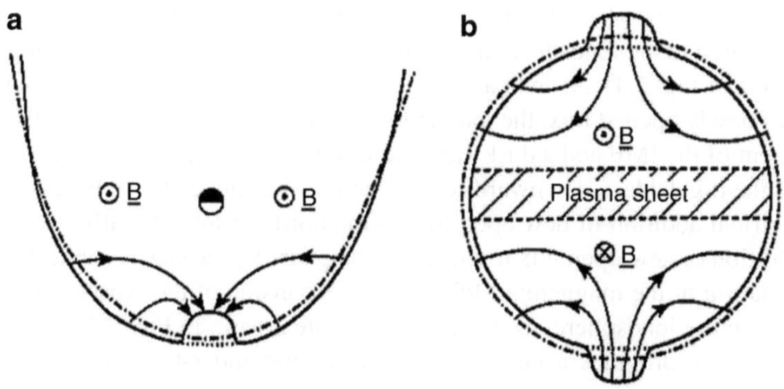

Fig. 2.3 (**a**) The magnetopause location in the equatorial plane, showing erosion by a burst of subsolar reconnection and the subsequent plasma flows which return the magnetopause to stress balance with the solar wind flow. (**b**) Deformation of the magnetotail cross-section by the addition of new open flux to the lobes, and the plasma flows which return it to equilibrium [from Cowley and Lockwood (1992)]

cap boundary (thick black line) is also known as the open/closed field line boundary (OCB). The closed field lines adjacent to the dayside OCB map out through the cusps to the dayside magnetopause whereas on the nightside the OCB maps to the boundary between open and closed field lines in the magnetotail. In this figure, the

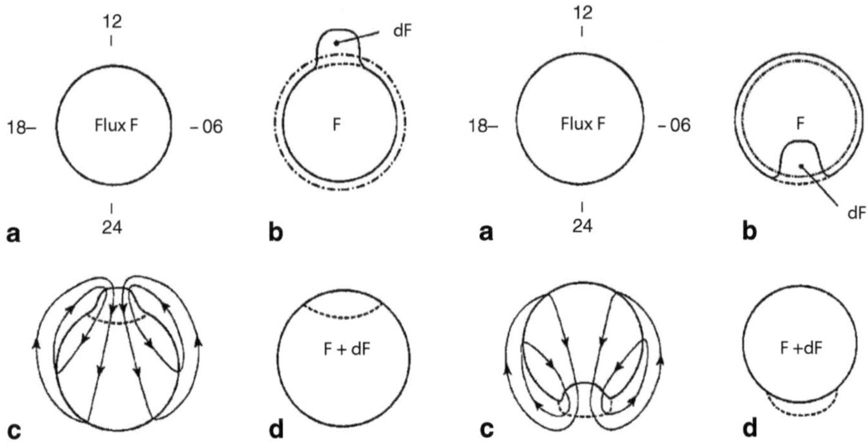

Fig. 2.4 (*Left*) The addition of new open flux to the polar cap by dayside reconnection, and the flows which are excited by pressure imbalances on the magnetopause. Bursts of dayside reconnection lead to an expansion of the polar cap. (*Right*) The corresponding picture for magnetotail reconnection in which open flux is closed at the nightside OCB and the polar cap contracts [from Cowley and Lockwood (1992)]

region of previously closed flux adjacent to the dayside OCB has been opened by the action of reconnection (the footprint of the active X-line is indicated as a thick red portion of the OCB) and the polar cap is deformed in a similar manner to the tail magnetopause. Cowley and Lockwood (1992) again realized that the flows excited within the magnetosphere by the deformation of the magnetopause (Fig. 2.3) are communicated to the ionosphere and return the polar cap to a circular configuration (left panels of Fig. 2.4). As indicated in the left-bottom panel of Fig. 2.1, within the region of newly opened flux, the magnetic tension forces associated with the $B_Y > 0$ component of the IMF add a duskwards component to the ionospheric flow; further towards the nightside, a dawnwards component is introduced to the flows due to the asymmetrical addition of new open flux to the northern lobe. Finally, we note that each addition of new open flux to the magnetosphere by dayside reconnection leads to an inflation of the magnetotail lobe and an expansion of the cross-section of the polar cap in the ionosphere, as indicated by white arrows in Fig. 2.1.

This picture provides a unifying framework for understanding the motion of plasma in the magnetosphere and ionosphere as a consequence of low latitude magnetopause reconnection. In this picture, reconnection acts to change the topology of magnetic field lines—an increasing proportion of the flux associated with the Earth's dipole becoming open—but plasma motions are driven by pressure exerted on the magnetopause by the flow of the solar wind, with a contribution from tension forces on newly opened field lines.

Magnetic reconnection in the magnetotail acts to decrease the proportion of open flux in the magnetosphere: closed field lines are able to return towards the dayside magnetosphere, initially through magnetic tension forces and subsequently through pressure imbalances associated with changes in magnetospheric shape as erosion of magnetospheric flux on the dayside continues [right panels of Fig. 2.4, from Cowley and Lockwood (1992)]. In this way, each burst of dayside or nightside reconnection leads to a slow shuffling of open flux antisunwards across the polar cap, or equivalently from the high-latitude lobe magnetopause towards the magnetotail neutral sheet.

This picture describes magnetospheric dynamics in terms of changes in the proportion of open flux and the stresses exerted by deviations of the magnetopause from hydrodynamic equilibrium with the solar wind. Stresses are transmitted within the magnetosphere by electric currents: Chapman-Ferraro currents generated at the magnetopause are diverted into the magnetosphere as "region 1" Birkeland currents that flow along magnetic field lines into the ionosphere (field-aligned currents or FACs), Pedersen currents flow across the auroral zone ionosphere, and "region 2" Birkeland currents then connect into the inner magnetosphere where they are subsequently closed through a partial ring current (e.g., Iijima and Potemra 1978; Cowley 2000). These current systems are then a fundamental part of the magnetosphere-ionosphere coupling that delivers stress to the ionosphere to cause it to move in response to changes in magnetospheric structure caused by magnetic reconnection.

The rate of magnetic reconnection at the dayside is dependent on conditions in the interplanetary medium, most importantly the orientation and strength of the IMF, and this will vary on timescales as short as minutes. The rate of magnetic reconnection in the magnetotail cannot instantaneously adjust itself to match the magnetopause rate, resulting in a significant variation in the amount of open flux in the magnetosphere. Changes in the proportion of open flux, and hence the size of the polar cap, are a measure of the global reconnection rates. Quantitatively, the rate of magnetic reconnection—the magnetic flux transferred from a closed topology to an open topology at the magnetopause (or open to closed in the magnetotail) in unit time—has dimensions of Wb s^{-1} or equivalently volts (V). The continuity equation for open flux can be expressed as

$$\frac{dF_{PC}}{dt} = \Phi_D - \Phi_N \tag{2.1}$$

where Φ_D is the dayside (subsolar magnetopause) reconnection rate and Φ_N is the nightside (magnetotail) reconnection rate (e.g., Siscoe and Huang 1985; Lockwood and Cowley 1992; Milan et al. 2007; Lockwood et al. 2009). As discussed in detail by Chisham et al. (2008), the dayside reconnection voltage is equal to the line integral of the motional electric field of the ionospheric plasma convection across the dayside portion of the OCB (indicated in red in the left-bottom panel of Fig. 2.1); similarly, the nightside reconnection rate is equal to the magnetic flux transported across the nightside OCB in unit time. Lockwood (1991) deduced that if it was assumed that the polar cap remains circular as it expands and contracts then the rate of antisunwards transport of magnetic flux across the dawn-dusk meridian—known as the transpolar voltage (TPV) or cross polar cap potential (CPCP), expressed as Φ_{PC}—will be given by

$$\Phi_{PC} = \tfrac{1}{2}(\Phi_D + \Phi_N). \tag{2.2}$$

This transpolar voltage can be determined from observations of the ionospheric convection pattern and is used as a measure of the overall strength of the Dungey cycle, a combination of contributions from both magnetopause and magnetotail reconnection. With suitable assumptions, the global electrostatic potential pattern associated with the ionospheric convection can be solved analytically. Siscoe and Huang (1985) were the first to deduce the relationship between ionospheric convection and an expanding polar cap, driven by dayside reconnection. Increasingly sophisticated models have subsequently been developed, by Freeman and Southwood (1988) to include the effect of non-uniform motion of the OCB, by Freeman (2003) to include the influence of both region 1 and 2 Birkeland FAC systems, and by Milan et al. (2012) and Milan (2013) to include both dayside and nightside reconnection contributions and to model the FAC magnitudes.

This association between dayside and nightside reconnection, magnetic flux transport within the magnetosphere, changes in size of the polar cap, and accompanying ionospheric flows has come to be known as the expanding/contracting polar cap paradigm (ECPC) (e.g., Freeman 2003; Milan et al. 2007). Testing the predictions of the ECPC has been a major endeavour since the 1990s when the ideas were first expressed as a coherent picture by Cowley and Lockwood (1992). In the following sections we discuss the observations that have validated the major predictions of the ECPC.

2.3 Observations of Ionospheric Convection and Polar Cap Area

Primary observables for testing the ECPC are the polar ionospheric convection pattern and the location of the polar cap boundary (or OCB) at all local times. There are significant challenges inherent in making observations over the majority of the polar regions at sufficient temporal resolution to capture the time-dependent aspects of the behaviour, one of the major reasons for the relatively slow universal adoption of the ECPC paradigm.

The presence of an ionospheric convection pattern had been inferred by Dungey (1961) from magnetic perturbations associated with the S_D ionospheric current system, as discussed by Cowley (chapter "Dungey's Reconnection Model of the Earth's Magnetosphere: The First 40 Years"). Direct measurement of the electric field driving ionospheric convection, however, was made first in the 1970s by satellites (e.g., Heppner 1977; Heppner and Maynard 1987; Rich and Hairston 1994), and later by radars (e.g., Greenwald et al. 1978; Evans et al. 1980; Foster 1983; Alcayde et al. 1986; Ruohoniemi and Greenwald 1996). Necessarily statistical in nature, these studies were not able to easily investigate time-dependencies in the convection, though they were able to determine the influence of the orientation

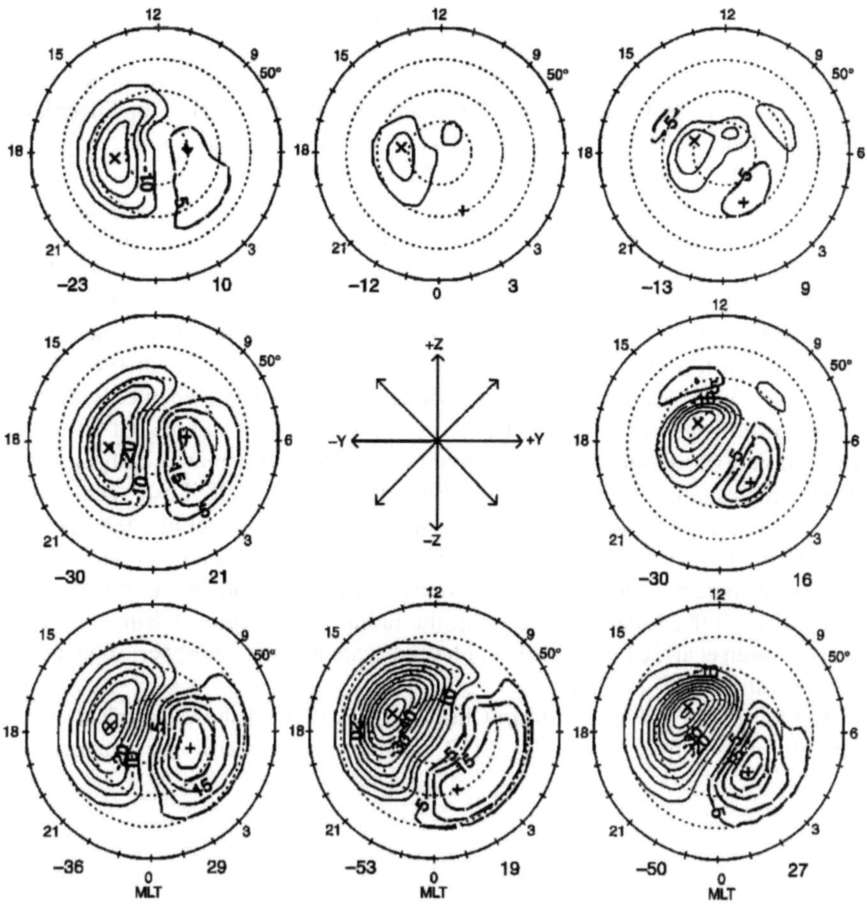

Fig. 2.5 Average ionospheric electrostatic potential patterns for differing orientations of the interplanetary magnetic field, when the total field magnitude is between 6 and 12 nT, deduced from SuperDARN radar observations. The potential contours are also streamlines of the flow [from Ruohoniemi and Greenwald (1996)]

of the IMF on the average morphology of the convection pattern, as demonstrated in Fig. 2.5 from Ruohoniemi and Greenwald (1996). For southward IMF, the convection pattern is twin-celled with antisunwards flow across the polar cap and return flow in the auroral zone; for northward IMF the pattern is more complicated but often features sunward flow in the dayside polar cap, as will be discussed in Sect. 2.7. For southward IMF conditions, observations showed that antisunwards convections speeds in excess of 1 km s^{-1} can be achieved, but are more typically a few 100 m s^{-1}, $\Phi_{PC} = 30$–50 kV, with a transport time across the polar cap of a few hours. Return flow takes a commensurate period of time, leading to a full Dungey cycle time of 8 or more hours during typical solar wind conditions. During extreme solar wind conditions, observations suggest that the ionospheric convection potential saturates and cannot exceed 250 kV (e.g., Siscoe et al. 2002; Hairston et al. 2003, 2005).

Initially, only spatially-limited observations of the time-dependence of convection were available (see for example Fig. 1.11 of chapter "Dungey's Reconnection Model of the Earth's Magnetosphere: The First 40 Years") (e.g., Willis et al. 1986; Etemadi et al. 1988; Lockwood and Freeman 1989). More recently, the Super Dual Auroral Radar Network (SuperDARN) (Greenwald et al. 1995; Chisham et al. 2007) has grown in extent and, under favourable conditions, synoptic maps of the global convection pattern are available. This allows the time-dependence of ionospheric convection flows to be determined, as will be discussed in Sect. 2.4.

Several techniques have been used to determine the location of the open/closed field line boundary, including the poleward edge of the electrojets (e.g., Akasofu and Kamide 1975), the poleward edge of auroral luminosity, determined either from space or from the ground (Meng et al. 1977; Craven and Frank 1987; Frank and Craven 1988; Kamide et al. 1999; Brittnacher et al. 1999; Hubert et al. 2006a), the poleward edge of the region 1 FAC region (e.g. Clausen et al. 2013a, b), the convection reversal boundary (e.g., Taylor et al. 1996), coherent radar backscatter characteristics (e.g., Baker et al. 1995, 1997; Lewis et al. 1998; Lester et al. 2001; Chisham et al. 2008), or a combination of these (e.g., Milan et al. 2003; Boakes et al. 2008). For work when only an approximate size of the polar cap is necessary, this can be characterized from a knowledge of the approximate radius of the auroral oval (Milan 2009; Milan et al. 2009b), the radius of the region 1 Birkeland FAC oval (Clausen et al. 2012), or the lower latitude extent of the ionospheric convection pattern (Imber et al. 2013a).

The left panel of Fig. 2.6 [from Chisham et al. (2008)] shows a typical proton auroral image from the FUV/SI12 instrument onboard the Imager for

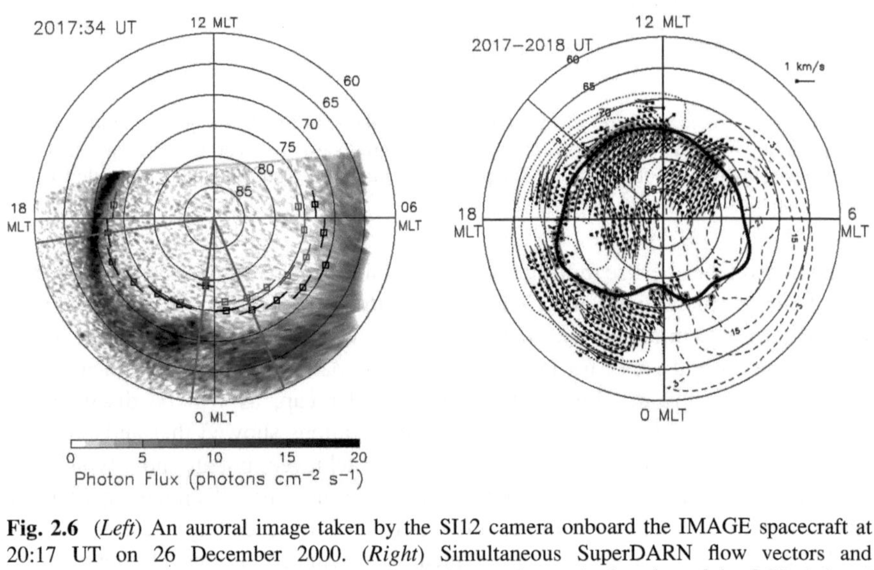

Fig. 2.6 (*Left*) An auroral image taken by the SI12 camera onboard the IMAGE spacecraft at 20:17 UT on 26 December 2000. (*Right*) Simultaneous SuperDARN flow vectors and reconstructed electrostatic potential pattern superimposed over the location of the OCB deduced from the auroral image [from Chisham et al. (2008)]

Magnetopause-to-Aurora Global Exploration (IMAGE) spacecraft (Mende et al. 2000a, b). Red/black lines and squares indicate different estimates of the location of the poleward boundary of emission, using two different techniques, a proxy for the OCB as discussed by Milan et al. (2003). In this particular example, a complete view of the auroral oval is not afforded, and so the full extent of the polar cap cannot be determined. However, when the OCB can be determined (or assumptions made about its location) at all local times, it is possible to use a suitable model of the terrestrial magnetic field, **B**, to determine the magnetic flux threading the polar cap, F_{PC}, as the surface integral of **B** over the polar cap area:

$$F_{PC} = \int_{PC} \mathbf{B} \cdot d\mathbf{s}. \tag{2.3}$$

Measurements show that F_{PC} is typically of the order of 0.4 GWb, but can vary between 0.2 and 1 GWb, that is between 2.5 and 12 % of the 8 GWb associated with the terrestrial magnetic dipole (Milan et al. 2007; Coumans et al. 2007; DeJong et al. 2007; Boakes et al. 2009; Huang et al. 2009). The polar caps occupy two roughly circular regions surrounding the magnetic poles with radii close to 1,500 km, but this can vary markedly, especially during geomagnetic storms when expansion of the polar caps drive the auroral zone down to mid-latitudes (Milan 2009), as demonstrated in Fig. 2.7. Long time-scale observations of the size of the polar cap show that there is a considerable solar cycle dependence of the average open flux content of the magnetosphere, with the ionospheric convection pattern, and by inference the polar cap boundary, being expanded to lower latitudes during solar maximum (Imber et al. 2013b).

The full potential of the observations is realized when both the ionospheric flow and the OCB can be imaged. The right panel of Fig. 2.6 shows an interval when both can be characterized. As discussed by Chisham et al. (2008), the dayside and nightside reconnection voltages are the integrals of the motional electric field associated with the ionospheric convection across dayside and nightside portions of the OCB. This is discussed more fully in Sect. 2.5, after first discussing the behaviour of the ECPC during substorm cycles.

2.4 The Substorm Cycle and the ECPC

The auroral substorm (Akasofu 1964) is the fundamental mode through which the magnetosphere responds to its interaction with the solar wind. Substorms display two main phases (McPherron 1970; Rostoker et al. 1980; Akasofu et al. 1992): the "growth phase" when the IMF is directed southwards, the polar cap expands, and the auroras move to lower latitudes; the "expansion phase" when nightside auroras

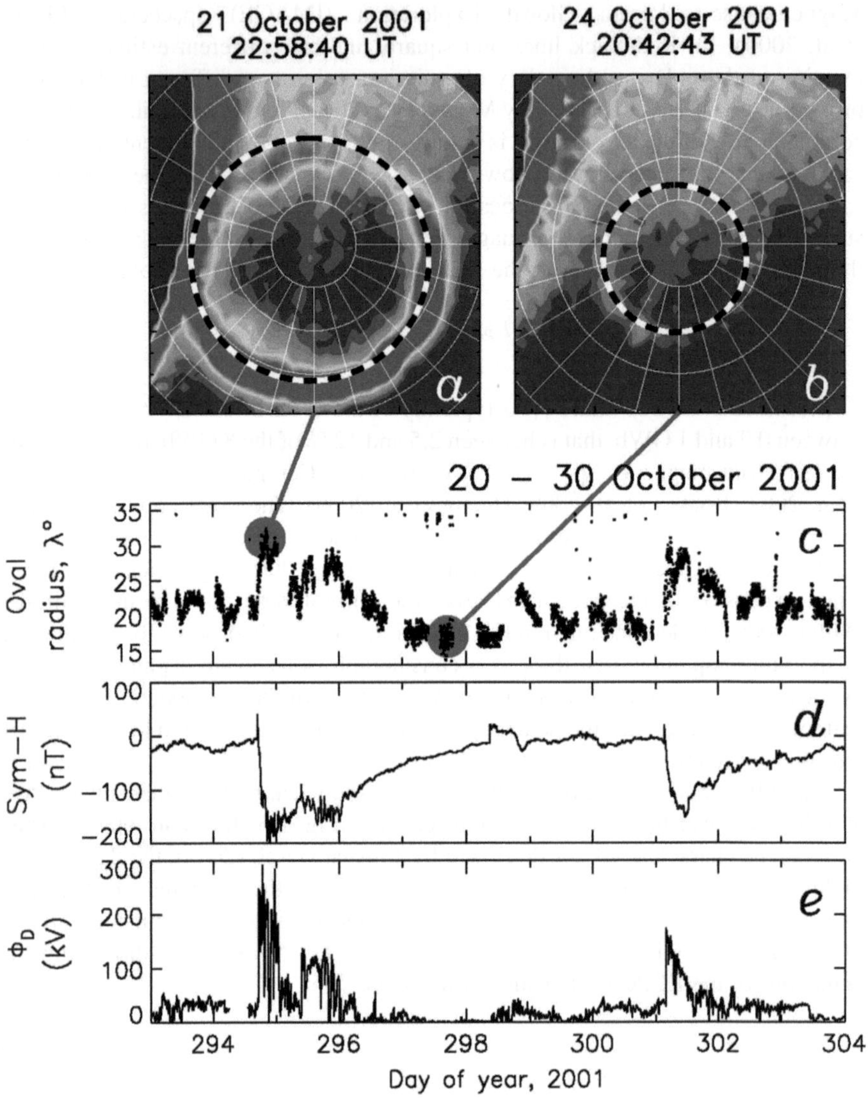

Fig. 2.7 (**a**, **b**) Images of the proton auroral oval from the FUV/SI12 instrument onboard the IMAGE spacecraft. *Dashed circles* indicate circles that have been fitted to the main intensity of the oval. (**c**) Variations in the radius of the fitted circles, used as a proxy for variations in polar cap size, for the 11-day period 20–30 October 2001. (**d**) The Sym-H index indicating the occurrence of two geomagnetic storms during this period. (**e**) A proxy for the dayside reconnection rate, parameterized by upstream interplanetary conditions [from Milan (2009)]

are intensified, the polar cap contracts, and the auroras retreat to higher latitudes. Understanding the substorm in the context of the ECPC is a central plank of validating the paradigm.

Figure 2.8 [from Milan et al. (2007)] shows two examples of the variation of F_{PC} over several hours (thick grey lines, panel a: 09–16:30 UT on 5 June 1998; panel g: 00–12 UT on 26 August 1998), along with supporting observations. During both intervals, F_{PC} is seen to vary between approximately 0.2 and 1.0 GWb. When the IMF is directed southwards, that is $B_Z < 0$ nT (panel d: 09–10, 12–13, 14–16 UT; panel j: 02–05, 10–11 UT) and subsolar reconnection is expected to occur, F_{PC}

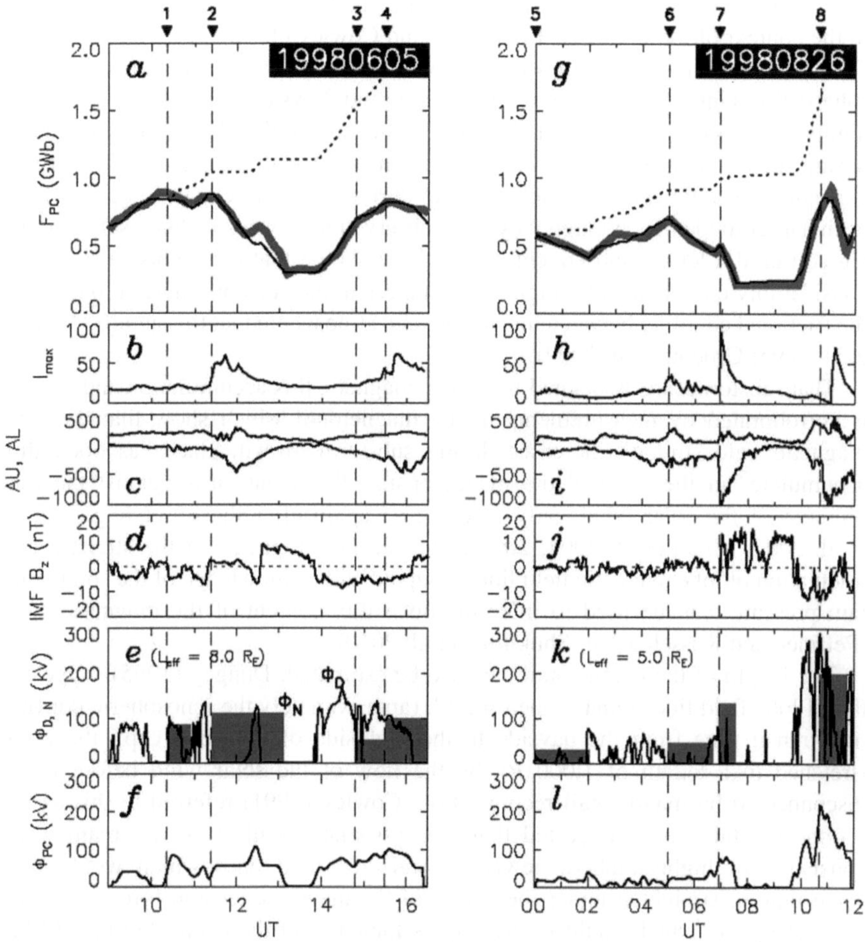

Fig. 2.8 (**a**, **g**) Estimates of the changing polar cap flux, F_{PC}, for the intervals 09:00–16:30 UT, 5 June 1998, and 00:00–12:00 UT, 26 August 1998 (*thick grey lines*); superimposed is the time-integral of the dayside and nightside reconnection rates [see Eq. (2.1) and panels **e** and **k**]. (**b**, **h**) Changes in the maximum auroral brightness in the images (arb. units). (**c**, **i**) The AU and AL indices representing the strengths of the ionospheric auroral electrojets. (**d**, **j**) The B_Z component of the IMF as measured by the ACE spacecraft. (**e**, **k**) Estimated dayside (*black line*) and nightside reconnection rates (*grey rectangles*). The dayside rate, Φ_D, is estimated from interplanetary parameters, whereas the nightside rate, Φ_N, is fitted to match changes in F_{PC}. (**f**, **l**) The cross-polar cap potential, Φ_{PC}, calculated using Eq. (2.2) [from Milan et al. (2007)]

tends to increase, corresponding to an expansion of the polar cap. Significant and rapid contractions of the polar cap, that is decreases in F_{PC} (panel a: 11:30–13,13:30–16:30 UT; panel g: 05–07, 07–07:30, 11–12 UT), are accompanied by enhancements of the nightside auroral emission intensity seen in the auroral images (panels b and h) and negative excursions of the AL index (panels b and h), both indicative of substorm activity. This indicates that most episodes of rapid reconnection in the magnetotail are associated with the occurrence of substorms, that is substorms play the major role in closing the Dungey cycle, as first suggested in the context of the ECPC by Lockwood and Cowley (1992).

Expansions and contractions of the polar cap should be accompanied by ionospheric flows into the polar cap on the dayside and flows out of the polar cap on the nightside, respectively. Milan et al. (2003) demonstrated that flow into the polar cap during substorm growth phase was indeed consistent with the rate at which it was expanding. The flows out of the polar cap during substorm expansion phase have been harder to identify, though they were observed for a weak isolated substorm by Grocott et al. (2004). Subsequent studies have shown that the nightside flows are considerably complicated by high conductances in the substorm auroral bulge and the frictional coupling between ionosphere and atmosphere that this entails (Morelli et al. 1995; Grocott et al. 2009).

That substorms are associated with open magnetic flux accumulation and closure is corroborated by measurements in the magnetotail which show that the lobe magnetic field strength increases during substorm growth phase, as open flux accumulates in the tail and the magnetopause flares outwards against the ram pressure of the solar wind, only to decrease again after the onset (e.g., Slavin et al. 2002; Milan et al. 2004, 2008). Indeed, measurements of the strength and orientation of lobe magnetic field lines, coupled with a knowledge of the solar wind ram pressure, can be used to infer the open flux content of the magnetosphere (Petrinec and Russell 1996; Shukhtina et al. 2010).

The length of the magnetotail can also be estimated. Dungey (1965) suggested that if lobe field lines remain open for 4 h (approximately the ionospheric convection transit time from the dayside to the nightside of the polar cap), these are stretched to a length of 1,000 R_E by the flow of the solar wind before being disconnected by magnetotail reconnection. Cowley (1991) referred to this as the "connected tail", and suggested that it is associated with a down-stream wake consisting of highly-kinked, newly-disconnected field lines which would take some time to straighten under the action of the magnetic tension force, the "disconnected tail" which could be five times longer than the connected tail. Milan (2004a) showed that knowledge of the recent history of the size of the polar cap and the dayside and nightside reconnection rates allows the length and flux content of the magnetotail lobes to be quantified, and showed that, somewhat counterintuitively, the magnetotail is longer during quiet magnetospheric periods than disturbed periods. During northwards IMF conditions dayside and nightside reconnection rates are low and any pre-existing open flux lengthens at the solar wind flow speed. When dayside reconnection recommences, following a southward turning of the IMF, nightside reconnection will eventually be triggered, and the oldest, longest open field lines will be removed from the system.

2.5 Quantifying Reconnection Rates

As dayside and nightside reconnection are key to understanding the dynamics of the magnetosphere, we wish to quantify the rates at which these processes occur, either through (in)direct measurement or through the use of empirically-determined relations ("proxies") between, say, interplanetary conditions and the magnetopause reconnection rate. As the variation in F_{PC} is a competition between the creation of open flux at the magnetopause and the subsequent closure of flux in the magnetotail [Eq. (2.1)], observations of F_{PC} can be used to assess reconnection rates. However, observations of dF_{PC}/dt alone are not sufficient to determine both Φ_D and Φ_N independently, only the difference between them. Two different approaches are available to quantify either Φ_D or Φ_N, or both. In the first, Φ_D or Φ_N can be determined if assumptions are made regarding the value of the other, for instance

$$\Phi_N \approx \Phi_D^* - \frac{dF_{PC}}{dt} \tag{2.4}$$

where Φ_D^* is a proxy for Φ_D. It has long been known that the dayside reconnection rate is controlled by the interplanetary conditions upstream of the Earth. One of the first and simplest proxies used is the "half wave rectified solar wind electric field",

$$\Phi_D^* \propto V_X B_S, \tag{2.5}$$

where V_X is the solar wind flow speed and B_S is the southward component of the IMF, that is $B_S = |B_Z|$ if $B_Z < 0$ nT and $B_S = 0$ otherwise (Burton et al. 1975; Holzer and Slavin 1978, 1979). This relates the reconnection rate to the interplanetary magnetic flux transported towards the Earth per unit length along the GSM Y-axis. The constant of proportionality in Eq. (2.5) should be related to the width of the channel in the solar wind that impinges on the magnetopause and reconnects, which was estimated to be between 10 and 20 % of the width of the magnetosphere (Reiff et al. 1981).

Milan (2004b) and Milan et al. (2007) used Eq. (2.5) with an "effective length" L_{eff} between 5 and 8 R_E (Earth radii) to estimate Φ_D^* (Fig. 2.8e, k, black line labelled Φ_D) and hence the expected accumulation of open flux, the increase in F_{PC}, in the absence of nightside reconnection (Fig. 2.8a, g, dashed lines). Discrepancies between the observed F_{PC} and the modelled allows periods of nightside reconnection to be identified, and a rough estimate of the rate and duration of reconnection (Fig. 2.8e, k, grey rectangles). The episodes of nightside reconnection so identified match periods of substorm activity as described above, as well as smaller events when auroral brightness or geomagnetic indices indicate magnetotail activity. In a survey of 25 flux closure events, Milan et al. (2007) estimated that on average 0.25 GWb of flux were closed over a duration of 70 min at a reconnection rate near 90 kV; similarly, DeJong et al. (2007) found a 30 % decrease in polar cap area during substorms.

A similar argument was used by Milan et al. (2012) to determine the optimum functional form for Φ_D^*. Substorm growth phases were studied, during which it was assumed that $\Phi_N = 0$. Various interplanetary parameters including solar wind speed, density, IMF magnitude and orientation were fitted to the observed expansion of the polar cap, yielding the proxy

$$\Phi_D^* = 3.3 \times 10^5 V_X^{4/3} (B_Y^2 + B_Z^2)^{1/2} \sin^{9/2} \tfrac{1}{2}\theta \qquad (2.6)$$

where θ is the IMF clock angle. Interestingly, it was found that solar wind density did not play a role in the parameterization.

Both Φ_D and Φ_N can be measured with observations of the ionospheric plasma flow across the OCB, as described in detail by Chisham et al. (2008). Local estimates of the reconnection electric field can be found along limited portions of the OCB, or the overall dayside and nightside reconnection voltages can be determined if observations are available at all local times around the OCB. Unfortunately, the requirement of global auroral images (or another method of OCB location) and excellent determination of the global ionospheric convection pattern means that this technique is only applicable in a limited number of cases. Local estimates of the dayside reconnection electric field were provided by Baker et al. (1997) and Blanchard et al. (2001); an estimate of the voltage along the whole dayside portion of the OCB by Milan et al. (2003) demonstrated that the dayside reconnection voltage was consistent with the observed expansion rate of the polar cap during a substorm growth phase (assuming $\Phi_N = 0$). Local nightside reconnection rates were investigated by de la Beaujardiere et al. (1991) confirming that flow out of the polar cap across nightside OCB was elevated during substorm expansion phase. Full determination of both dayside and nightside reconnection rates using combined global auroral imaging and SuperDARN radar flows, similar to the right panel of Fig. 2.6, by Hubert et al. (2006a), allowed them to show that substorm expansion phase could be associated with nightside reconnection rates as high as 120 kV and that pseudo-breakups occurring during substorm growth phase were associated with modest tail reconnection as well.

An alternative means of determining the contributions of dayside and nightside reconnection to changes in the size of the polar cap and ionospheric convection is to measure the spatial- and time-dependence of the Birkeland current system using the Active Magnetosphere and Planetary Electrodynamics Response Experiment or AMPERE (Anderson et al. 2000, 2002; Clausen et al. 2012). Clausen et al. (2012, 2013a, b) have demonstrated that the region 1 and 2 current "ovals" expand and contract in response to substorms and storms in a manner consistent with the ECPC. Moreover, the strengths of the currents, which are related to the strength of the ionospheric convection, are seen to respond to both dayside and nightside

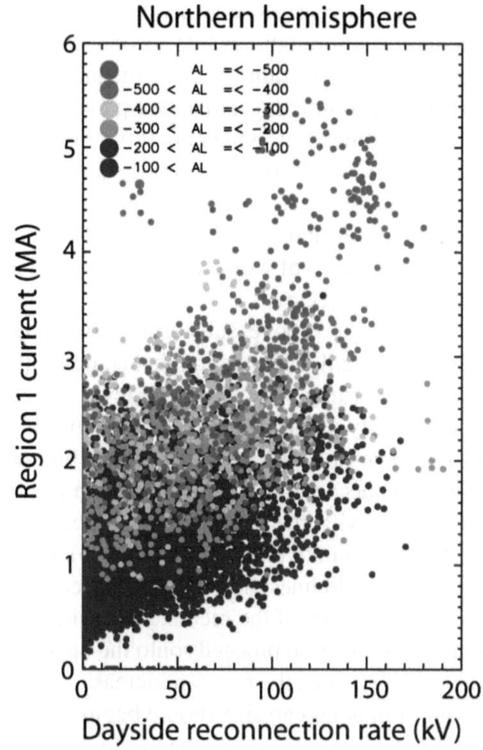

Fig. 2.9 The magnitude of the current flowing in the region 1 Birkeland current system in the northern hemisphere during the month of February 2010, as determined from the AMPERE experiment. The position of each *dot* represents the magnitude of the current as a function of a proxy for the dayside reconnection rate; the *colour* represents magnitude of the AL index, which is used as a proxy for the nightside reconnection rate (figure courtesy of J. C. Coxon) (Color figure online)

contributions (J. C. Coxon, private communication). Figure 2.9 presents measurements of the region 1 current strength in the northern hemisphere from each 10 min interval during February 2010, plotted as a function of a dayside reconnection proxy [Eq. (2.6)], and colour-coded with the concurrent AL index, more negative values of which indicate substorm activity and hence nightside reconnection. The currents increase with Φ_D^*, but there is a large spread in values of current for a particular reconnection rate, with higher currents at increasingly negative AL. These data suggest, then, that convection is greater when both dayside and nightside reconnection contribute, consistent with Eq. (2.2) and models of the relation between currents and reconnection (e.g., Milan 2013).

2.6 What Triggers and Controls Nightside Reconnection?

Studies to date, including those described above, suggest that the dayside subsolar magnetic reconnection rate is closely determined by conditions in the interplanetary medium, though there is some debate as to why this should be (e.g., Borovsky et al. 2008) and there is still not a good characterization of transpolar voltage saturation during extreme solar wind conditions and a variety of possible explanations (e.g., Siscoe et al. 2002, 2004; Shepherd 2006). Once open flux accumulates through dayside reconnection the magnetosphere must close it again through reconnection in the magnetotail. Although the Dungey cycle and expanding/contracting polar cap paradigms make this clear, and allow a quantitative treatment of magnetic flux transport within the magnetosphere and the attendant ionospheric convection, they do not provide a means of determining when and at what rate the magnetosphere will do this. Much work has been undertaken to understand magnetic flux release in the magnetotail, including the related questions of why it occurs largely in an episodic manner and what triggers it when it does occur.

As described above, substorms play a major role in flux closure. It is thought that the magnetotail reconnection that achieves this occurs in two phases. There is some debate regarding whether a low level of reconnection between lobe field lines is continuously on-going at a distant neutral line or X-line (DXL), perhaps many 10 s of R_E down-tail. The apparent lack of flux closure during non-substorm times (see Fig. 2.8) puts a rather stringent limit on the rate at which this occurs. Substorms are associated with the formation of a near-Earth X-line (NEXL) in the vicinity of 20 R_E down-tail (e.g., Baker et al. 1996). Initially, reconnection must occur between the closed field lines of the plasma sheet, at what is expected to be a low rate due to the mass-loading of the reconnection site and the corresponding low Alfvén speed. Once reconnection proceeds onto the open field lines of the lobes, which are largely devoid of plasma, the rate can increase. It is at this stage that open flux is closed and changes in polar cap size should begin to be evident. As the substorm proceeds, the NEXL may migrate down the tail to occupy the posited location of the DXL.

The flux closure during a substorm is accompanied by charged particle precipitation producing auroral brightenings, enhancements of the ionospheric conductivity, ionospheric convection enhancements, and auroral electrojet activations. The magnitude of these signatures can be used as an indicator of the “energy” of the substorm. It is well known that substorms come in many sizes and that the location of the initial auroral brightening associated with the substorm can occur over a wide range of latitudes (e.g., Frey et al. 2004). The onset latitude can be considered a proxy for the expansion of the auroral oval, the size of the polar cap, and hence the open flux content of the magnetosphere at the time of onset. It has been demonstrated that substorm energy is closely correlated to the open flux content of the magnetosphere prior to onset (e.g., Akasofu and Kamide 1975; Kamide et al. 1999; Milan et al. 2009a). The amount of flux closed during substorms is also related to the open flux content prior to onset (Shukhtina et al. 2005; Milan et al. 2009a).

Energetic substorms which close a lot of flux might be expected to occur when the dayside accumulation of open flux is rapid, though it is not clear if enhanced flux closure should be effected by a few large substorms or many smaller substorms. A study of the occurrence rate of substorms and the flux closed in each substorm suggested that both increased in approximate proportion to ${\Phi_D}^{1/2}$ such that the nightside flux closure rate matched the open flux production rate at the dayside (Milan et al. 2008). It can also be shown that the probability that a substorm will initiate in the near future increases as the open flux content increases (Milan et al. 2007; Boakes et al. 2009). However, these observations do not explain why "weak" substorms occur on a contracted oval while on other occasions the magnetosphere allows itself to accumulate a large quantity of open flux before initiating an energetic substorm.

It is natural to think that as the magnetosphere accumulates open flux, and the magnetic and plasma pressure in the magnetotail increase, conditions in the vicinity of the neutral sheet become more favourable for the onset of reconnection and that at some point reconnection acts as a "pressure release valve". However, the studies described above clearly indicate that there is no fixed open flux "threshold" at which substorms are triggered (Boakes et al. 2009); equally, studies of magnetic pressure build-up in the magnetotail prior to onset show that a fixed pressure threshold does not exist either (e.g., Milan et al. 2008).

The level of open flux at which substorms occur has been shown to increase during geomagnetic storms (Milan et al. 2008, 2009b), the signature of which is an enhanced ring current and the magnetic perturbation produced by this as measured by negative excursions of the D_{st} and Sym-H indices (see Fig. 2.7). It has been speculated that geomagnetic storms produce conditions in the magnetotail which disfavour the onset of reconnection and hence lead to greater accumulation of open flux before it is released. For instance, Kistler et al. (2006) have shown that the concentration of heavy ions in the plasma sheet increases during storms, and is has been suggested that the associated decrease in Alfvén velocity impedes fast reconnection (e.g., Ouellette et al. 2013). Alternatively, Milan et al. (2008, 2009b) have suggested that the magnetic perturbation associated with the enhanced ring current dipolarizes the magnetotail, halting the onset of reconnection until the lobe pressure builds up to produce a sufficiently "tail-like" field once again.

The substorm cycle is not the only "mode" by which the magnetosphere releases open magnetic flux accumulated at the dayside. Other modes that have been described include sawtooth events and steady magnetospheric convection (SMC) events. Sawtooth events display a large and very regular ~3 h expansion and contraction cycle of the polar cap (DeJong et al. 2007; Hubert et al. 2008; Huang et al. 2009), and are named after the characteristic appearance of geosynchronous orbit trapped particle fluxes—a gradual drop out of fluxes during the growth phase and a sudden increase at onset (Belian et al. 1995). At present, it is not clear if these are a fundamentally different mode of coupling, or whether they are an extreme example of the substorm cycle occurring during strong solar wind driving conditions in the main phase of geomagnetic storms (Cai et al. 2011). SMCs, on the other hand, do not show an expansion/contraction cycle, the polar cap remaining of approximately uniform size (DeJong and Clauer 2005; DeJong et al. 2008), even though dayside driving and magnetospheric and ionospheric convection are on-going (e.g., McWilliams et al. 2008). In this case, Eq. (2.1) indicates that the nightside reconnection must closely match the dayside rate—hence lending them the alternative name "balanced reconnection intervals" (BRIs) (DeJong et al. 2008)—and by Eq. (2.2) the transpolar voltage must be equal to both. Many SMCs appear to start with a substorm (Sergeev 1977; Kissinger et al. 2012a). Milan et al. (2006) investigated a substorm that displayed repeated dipolarizations during a ~2 h expansion phase which only subsided once the IMF turned northwards and dayside reconnection abated: this suggests that SMCs may be prolonged substorms that maintain reconnection in the tail because dayside reconnection continues after onset, a "driven recovery phase" as described by DeJong et al. (2008). How the magnetosphere achieves this, and if the nightside rate adjusts itself if the dayside rate changes are unclear. Kissinger et al. (2012b) have demonstrated that magnetospheric convection avoids the inner magnetosphere during SMCs and Juusola et al. (2013) suggest that the magnetic perturbation associated with an enhanced ring current leads to reconnection occurring further down-tail than during substorms.

Some workers have suggested that substorms, SMCs, and sawtooth events represent a spectrum of responses of the magnetosphere to different conditions in the interplanetary medium and differing levels of solar wind-magnetosphere coupling (Cai et al. 2006; Partamies et al. 2009). On the other hand, it is possible that preconditioning of the magnetosphere is necessary to drive it into one particular mode of response (e.g., Kissinger et al. 2012b; Juusola et al. 2013). Alternatively, the work of Grocott et al. (2009) suggests that enhanced ionospheric conductivity in the substorm auroral bulge during very disturbed conditions leads to frictional coupling between the ionosphere and atmosphere ("line-tying") that can inhibit steady ionospheric flows and, as a consequence, steady magnetospheric convection.

As discussed earlier, the accumulation of open magnetic flux in the magnetotail lobes increases the internal pressure of the tail as it inflates against the flow of the solar wind on the outside (Coroniti and Kennel 1972; Petrinec and Russell 1996), which is thought to play a role in the initiation of nightside reconnection. A sudden increase in solar wind ram pressure also acts to increase the internal tail pressure, and there are many documented cases of this triggering reconnection and substorm onset (Boudouridis et al. 2003; Milan et al. 2004; Hubert et al. 2006b, 2009). Northward turnings of the IMF have also been implicated in the triggering of substorms (e.g., Caan et al. 1978; Lyons et al. 1997; Hsu and McPherron 2002), but several refutations also exist (Morley and Freeman 2007; Wild et al. 2009) and many cases where substorms occur without apparent external triggers (e.g., Huang 2002).

2.7 Reconnection During Northward IMF

The most active magnetospheric conditions tend to occur during periods of prolonged southward IMF, but many interesting phenomena occur when the IMF is directed northwards as well. Dungey (1963) proposed that reconnection would occur between northward IMF and terrestrial field lines tailwards of the cusps. Cowley (1981b) proposed several scenarios (see Fig. 1.10 of chapter "Dungey's Reconnection Model of the Earth's Magnetosphere: The First 40 Years") in which reconnection took place with closed or open magnetospheric field lines and independently or with the same interplanetary magnetic field line in the two hemispheres. These different scenarios lead to different predictions regarding the dynamics driven in the magnetosphere.

In an open magnetosphere, it is likely that northwards IMF reconnection will take place with open lobe field lines, termed "lobe reconnection". If the IMF has a significant B_Y component, different interplanetary field lines will reconnect at northern and southern reconnection sites (see Fig. 2.1, middle column). This is known as "single lobe reconnection" or SLR, even though it may be happening in both hemispheres simultaneously, though possibly at different rates. In the event that there is only a small B_Y component the same interplanetary field line might reconnect in both hemispheres, "dual lobe reconnection" or DLR (Fig. 2.1, right column). In the case of SLR, flux is neither opened nor closed and the polar cap size remains of constant size. However, the combination of tension forces on newly-reconnected field lines and deformations of the magnetopause by the redistributed open flux result in "lobe stirring" in the ionosphere, sunward flow at the footprint of the reconnection line and the formation of "reverse lobe convection cells" (Cowley and Lockwood 1992; Huang et al. 2000; Milan et al. 2005b); such reverse cells are just discernible in the average convection pattern for northward IMF in Fig. 2.5. Magnetosheath plasma injected on the newly reconnected, sunward moving field lines displays a "reverse ion dispersion" (Woch and Lundin 1992), and auroras associated with this precipitation can form a "cusp auroral spot" (Fig. 2.10a) (Milan et al. 2000; Frey et al. 2002). Changes in the B_Y component of the IMF change the location on the magnetopause at which the antiparallel condition is met, and the local time of the cusp spot, indicating the mapping from the magnetopause, moves accordingly (Fig. 2.10b). In the central panel of Fig. 2.1, the effect of reconnection is shown only in the northern lobe, to emphasize that the rate of reconnection in each hemisphere is independent of the other, and may even be absent in one.

As IMF B_Y becomes small, DLR becomes possible, closing open flux (Fig. 2.1, right column). Imber et al. (2006) suggested that a signature of DLR should be ionospheric convection out of the polar cap across the dayside OCB and a corresponding contraction of the polar cap as open flux is closed. Examples of this signature have been reported (Imber et al. 2006, 2007; Marcucci et al. 2008). The observed length of the footprint of the X-line in the ionosphere allowed Imber et al. (2006) to estimate that DLR should only occur for IMF clock angles less than 10°; they were also able to demonstrate that DLR had the potential to be an extremely efficient method of solar wind capture by the magnetosphere, and could easily supply the plasma seen to accumulate in the "cold dense plasma sheet" during prolonged periods of northwards IMF (e.g., Øieroset et al. 2005).

Episodic nightside closure of flux occurs during northward IMF, but at a much reduced rate. These events were first identified as bursts of rapid azimuthal ionospheric convection in the midnight sector close to the boundary of the polar cap (Senior et al. 2002; Grocott et al. 2003); although modest auroral brightenings are associated with the flows (Milan et al. 2005b), the magnetic perturbation produced is small and so the significance of this phenomenon was not realized until radar measurements of the flows were available. These events have been termed "tail reconnection during IMF northwards, non-substorm intervals" or TRINNIs.

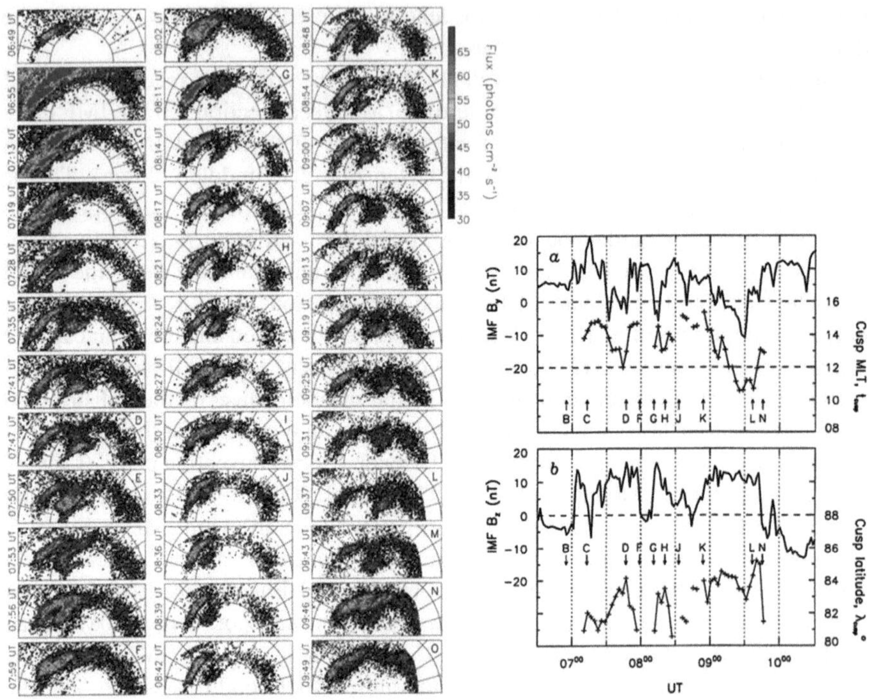

Fig. 2.10 (*Left*) Observations of the dayside auroras from the Polar UVI instrument on 26 August 1998, showing the appearance of a cusp auroral spot poleward of the main auroral zone. (*Right*) The magnetic local time and latitude of the centroid of the cusp spot during this period, along with the corresponding IMF B_Y and B_Z components measured by the Wind spacecraft [from Milan et al. (2000)]

Grocott et al. (2004) demonstrated that the eastwards/westwards sense of the flows were associated with the B_Y component of the IMF, and postulated that they were driven by reconnection occurring in a magnetotail that is twisted by the tension forces imposed on lobe field lines interlinked with northwards IMF (e.g., Cowley 1981a). In this scenario, the flows should have opposite senses in the two hemispheres, subsequently verified by Grocott et al. (2005). Each TRINNI burst is thought to close ~0.1 GWb of flux at a rate of 30 kV over a period of a few 10s min, compared to ~0.25 GWb at ~100 kV over an hour or more during substorms (Milan et al. 2007).

Another phenomenon associated with northwards IMF is the formation of theta auroras or transpolar arcs (TPAs), in which the polar cap becomes bisected by a tongue of auroras (Frank et al. 1982, 1986). These auroral features can form at the dawn or dusk sides of the polar cap, or appear to grow into the polar cap from the nightside auroral oval, and can subsequently move dawnwards or duskwards as IMF B_Y changes (Craven and Frank 1991; Craven et al. 1991; Cumnock et al. 1997; Cumnock and Blomberg 2004; Kullen 2000; Kullen et al. 2002; Fear and Milan 2012a). Several mechanisms for creating TPAs have been discussed in the literature, but two main competing ideas gained ground, as discussed by Zhu et al. (1997). In the first, precipitation is associated with a field-aligned current sheet, which is itself associated with a large-scale shear in the ionospheric convection flow within the open polar cap (e.g., Burke et al. 1982). In the second, precipitation is associated with a tongue of closed field lines which protrudes into the polar cap from the nightside, effectively plasma sheet extending to much higher latitudes than usual (e.g., Frank et al. 1982). Milan et al. (2005b), Goudarzi et al. (2008) and Fear and Milan (2012b) demonstrated that TPA formation was sometimes associated with the occurrence of TRINNI flows, and proposed that the arc was associated with closed field lines that were created by reconnection in a twisted magnetotail and which could not easily convect to the dayside and so accumulated in the midnight sector, eventually protruding upwards into the lobes. This mechanism explains the observed local time dependence of TPA formation, dusk-side for IMF $B_Y > 0$ and dawn-side for $B_Y < 0$ in the northern hemisphere, due to the sense in the twist of the magnetotail. Subsequent motion of the arcs dawnwards or duskwards was postulated to be participation of these closed field lines embedded within the lobe in lobe stirring driven by single lobe reconnection (Milan et al. 2005b) or due to the asymmetrical addition of new open flux to the lobes by B_Y-dominated southwards IMF (Goudarzi et al. 2008).

2.8 Concluding Remarks

Dungey's open magnetosphere paradigm of magnetospheric dynamics provides a powerful theoretical framework within which to understand most aspects of the structure and large-scale dynamics of the magnetosphere. Observations of the expanding/contracting polar cap and the associated ionospheric convection pattern provide a means of quantitatively exploring magnetic reconnection rates, and the magnetospheric response to reconnection. Although the basic mechanisms are well-understood and placed on a firm observational footing, there are still several fundamental outstanding questions. It is not understood what leads to the onset of magnetotail reconnection during substorms. The role(s) of feedback loops within the coupled solar wind-magnetosphere-ionosphere system is (are) poorly understood: for instance, do heavy ion mass-loading of the plasma sheet or the magnetic perturbation produced by an enhanced ring current play a role in controlling magnetotail onset thresholds and rates?; do storm-time plasmaspheric plumes play a role in modulating the dayside reconnection rate through heavy ion mass-loading of the magnetopause? There is a significant body of work investigating mechanisms by which the transpolar voltage of the magnetosphere may saturate at values near ~250 kV when solar wind driving is extreme, though there is perhaps a dearth of observations that allow this to be investigated in detail.

Two complementary pictures of magnetospheric dynamics exist. The ECPC describes dynamics in terms of changes in open flux and the stresses exerted by deviations of the magnetopause from hydrodynamic equilibrium with the solar wind. Alternatively, the dynamics can be described in terms of the current systems that transmit stress throughout the magnetosphere. These two paradigms have yet to be fully reconciled, but new observational techniques to measure the spatial- and time-dependence of the currents allows their response in the context of the ECPC to be explored, and will allow the relationship between reconnection rates and FAC dynamics to be better understood.

This review has concentrated on the ramifications of the open model for the terrestrial magnetosphere, but magnetic reconnection clearly plays an important role in the magnetospheres of other planets. Although observational evidence is more difficult to acquire at Mercury or the outer planets, our understanding of the dynamics of those magnetospheres is indebted to the work of Dungey. We conclude by remarking that the Dungey cycle and ECPC (suitably modified for local conditions) have been invoked to explain behaviour at Mercury (e.g., Milan and Slavin 2011; Slavin et al. 2010), Jupiter (e.g., Cowley et al. 2003), Saturn (e.g., Badman et al. 2005; Cowley et al. 2005; Milan et al. 2005a), and Uranus (Cowley 2013). Inevitably, many exciting developments of Dungey's work will arise as observations of those distant systems improve.

Acknowledgements SEM was supported by the Science and Technology Facilities Council (STFC), UK, grant no. ST/K001000/1 and the Research Council of Norway/CoE under contract 223252/F50. SEM would like to thank the PIs of the instruments from which data are presented in this review article, including S. B. Mende of the University of California Berkeley (IMAGE FUV), G. K. Parks of the University of Washington (Polar UVI), B. J. Anderson of JHU/APL (AMPERE), and the PIs of the Super Dual Auroral Radar Network (SuperDARN). SuperDARN is a collection of radars funded by national scientific funding agencies of Australia, Canada, China, France, Japan, South Africa, United Kingdom and United States of America. SEM also thanks J. C. Coxon, of the University of Leicester, for producing Fig. 2.9 of this paper.

References

Akasofu, S.-I.: The development of the auroral substorm. Planet. Space Sci. **12**, 273–282 (1964)

Akasofu, S.-I., Kamide, Y.: Substorm energy. Planet. Space Sci. **23**, 223 (1975)

Akasofu, S.-I., Meng, C.-I., Marita, K.: Changes in the size of the open field line region during substorms. Planet. Space Sci. **40**, 1513–1524 (1992)

Alcayde, D., Caudal, G., Fontanari, J.: Convection electric fields and electrostatic potential over $61° < \Lambda < 72°$ invariant latitude observed with the European incoherent scatter facility. J. Geophys. Res. **91**, 233 (1986)

Anderson, B.J., Takahashi, K., Toth, B.A.: Sensing global Birkeland currents with Iridium® engineering magnetometer data. Geophys. Res. Lett. **27**, 4045–4048 (2000). doi:10.1029/2000GL000094

Anderson, B.J., Takahashi, K., Kamei, T., Waters, C.L., Toth, B.A.: Birkeland current system key parameters derived from Iridium observations: method and initial validation results. J. Geophys. Res. **107**, 1079 (2002). doi:10.1029/2001JA000080

Aubry, M.P., Russell, C.T., Kivelson, M.G.: Inward motion of the magnetopause before a substorm. J. Geophys. Res. **75**, 7018–7031 (1970). doi:10.1029/JA075i034p07018

Badman, S.V., Bunce, E.J., Clarke, J.T., Cowley, S.W.H., Gérard, J.-C., Grodent, D., Milan, S.E.: Open flux estimates in Saturn's magnetosphere during the January 2004 Cassini-HST campaign, and implications for reconnection rates. J. Geophys. Res. **110**, A11216 (2005). doi:10.1029/2005JA011240

Baker, K.B., Dudeney, J.R., Greenwald, R.A., Pinnock, M., Newell, P.T., Rodger, A.S., Mattin, N., Meng, C.-I.: HF radar signatures of the cusp and low-latitude boundary layer. J. Geophys. Res. **100**, 7671–7695 (1995)

Baker, D.N., Pulkkinen, T.I., Angelopoulos, V., Baumjohann, W., McPherron, R.L.: J. Geophys. Res. **101**, 12975–13010 (1996)

Baker, K.B., Rodger, A.S., Lu, G.: HF-radar observations of the dayside magnetic merging rate: a geospace environment modeling boundary layer campaign study. J. Geophys. Res. **102**, 9603–9617 (1997)

Belian, R.D., Cayton, T.E., Reeves, G.D.: Quasi-periodic global substorm-generated variations observed at geosynchronous orbit. In: Ashour-Abdalla, M., Chang, T., Dusenburg, P. (eds.) Space Plasmas Coupling Between Small and Medium Scale Processes. Geophysical Monograph Series, vol. 86, p. 143. AGU, Washington, DC (1995)

Blanchard, G.T., Ellington, C.L., Lyons, L.R., Rich, F.J.: Incoherent scatter radar identification of the dayside magnetic separatrix and measurement of magnetic reconnection. J. Geophys. Res. **106**, 8185–8195 (2001)

Boakes, P.D., Milan, S.E., Abel, G.A., Freeman, M.P., Chisham, G., Hubert, B., Sotirelis, T.: On the use of IMAGE FUV for estimating the latitude of the open/closed field line boundary in the ionosphere. Ann. Geophys. **26**, 2579–2769 (2008)

Boakes, P.D., Milan, S.E., Abel, G.A., Freeman, M.P., Chisham, G., Hubert, B.: A statistical study of the open magnetic flux content of the magnetosphere at substorm onset. Geophys. Res. Lett. **36**, L04105 (2009). doi:10.1029/2008GL037059

Borovsky, J.E., Hesse, M., Birn, J., Kuznetsova, M.M.: What determines the reconnection rate at the dayside magnetosphere? J. Geophys. Res. **113**, A07210 (2008)

Boudouridis, A., Zesta, E., Lyons, R., Anderson, P.C., Lummerzheim, D.: Effect of solar wind pressure pulses on the size and strength of the auroral oval. J. Geophys. Res. **108** (2003). doi: 10.1029/2002JA009373

Brittnacher, M., Fillingim, M., Parks, G., Germany, G., Spann, J.: Polar cap area and boundary motion during substorms. J. Geophys. Res. **104**, 12251–12262 (1999)

Burke, W.J., Gussenhoven, M.S., Kelley, M.C., Hardy, D.A., Rich, F.J.: Electric and magnetic field characteristics of discrete arcs in the polar cap. J. Geophys. Res. **87**, 2431–2443 (1982). doi:10.1029/JA087iA04p02431

Burton, R.K., McPherron, R.L., Russell, C.T.: An empirical relationship between interplanetary conditions and Dst. J. Geophys. Res. **80**, 4204–4212 (1975)

Caan, M.N., McPherron, R.L., Russell, C.T.: The statistical magnetic signature of magnetospheric substorms. Planet. Space Sci. **26**, 269–279 (1978)

Cai, X., Clauer, C.R., Ridley, A.J.: Statistical analysis of ionospheric potential patterns for isolated substorms and sawtooth events. Ann. Geophys. **24**, 1977–1991 (2006)

Cai, X., Zhang, J.-C., Clauer, C.R., Liemohn, M.W.: Relationship between sawtooth events and magnetic storms. J. Geophys. Res. **116**, A07208 (2011). doi:10.1029/2010JA016310

Chisham, G., Lester, M., Milan, S.E., Freeman, M.P., Bristow, W.A., Grocott, A., McWilliams, K. A., Ruohoniemi, J.M., Yeoman, T.K., Dyson, P.L., Greenwald, R.A., Kikuchi, T., Pinnock, M., Rash, J.P.S., Sato, N., Sofko, G.J., Villain, J.-P., Walker, A.D.M.: A decade of the Super Dual Auroral Radar Network (SuperDARN): scientific achievements, new techniques and future directions. Surv. Geophys. **28**, 33–109 (2007). doi:10.1007/s10712-007-9017-8

Chisham, G., Freeman, M.P., Abel, G.A., Lam, M.M., Pinnock, M., Coleman, I.J., Milan, S.E., Lester, M., Bristow, W.A., Greenwald, R.A., Sofko, G.J., Villain, J.-P.: Remote sensing of the spatial and temporal structure of magnetopause and magnetotail reconnection from the ionosphere. Rev. Geophys. **46**, RG1004 (2008). doi:10.1029/2007RG000223

Clausen, L.B.N., Baker, J.B.H., Ruohoniemi, J.M., Milan, S.E., Anderson, B.J.: Dynamics of the region 1 Birkeland current oval derived from the Active Magnetosphere and Planetary

Electrodynamics Response Experiment (AMPERE). J. Geophys. Res. **117**, A06233 (2012). doi:10.1029/2012JA017666
Clausen, L.B.N., Baker, J.B.H., Ruohoniemi, J.M., Milan, S.E., Coxon, J.C., Wing, S., Ohtani, S., Anderson, B.J.: Temporal and spatial dynamics of the region 1 and 2 Birkeland currents during substorms. J. Geophys. Res. Space Phys. **118**, 3007–3016 (2013a). doi:10.1002/jgra.50288
Clausen, L.B.N., Milan, S.E., Baker, J.B.H., Ruohoniemi, J., Glassmeier, K.-H., Coxon, J.C., Anderson, B.J.: On the influence of open magnetic flux on substorm intensity: ground- and space-based observations. J. Geophys. Res. Space Phys. **118**, 2958–2969 (2013b). doi:10.1002/jgra.50308
Cooling, B.M.A., Owen, C.J., Schwartz, S.J.: Role of magnetosheath flow in determining the motion of open flux tubes. J. Geophys. Res. **106**, 18763–18775 (2001)
Coroniti, F.V., Kennel, C.F.: Changes in magnetospheric configuration during the substorm growth phase. J. Geophys. Res. **77**, 3361–3370 (1972). doi:10.1029/JA077i019p03361
Coumans, V., Blockx, C., Gérard, J.-C., Hubert, B., Connors, M.: Global morphology of substorm growth phases observed by the IMAGESI12 imager. J. Geophys. Res. **112**, A11211 (2007). doi:10.1029/2007JA012329
Cowley, S.W.H.: Magnetospheric asymmetries associated with the Y-component of the IMF. Planet. Space Sci. **29**, 79–96 (1981)
Cowley, S.W.H.: Magnetospheric and ionospheric flow and the interplanetary magnetic field. In: The Physical Basis of the Ionosphere in the Solar-Terrestrial System, AGARD-CP-295, pp. (4–1)–(4–14) (1981b)
Cowley, S.W.H.: The structure and length of tail-associated phenomena in the solar wind downstream from the Earth. Planet. Space Sci. **7**, 1039 (1991)
Cowley, S.W.H.: Magnetosphere-ionosphere interactions: a tutorial review. In: Ohtani, S., et al. (eds.) Magnetospheric Current Systems. Geophysical Monograph Series, vol. 118, pp. 91–106. AGU, Washington, DC (2000). doi:10.1029/GM118p0091
Cowley, S.W.H.: Response of Uranus' auroras to solar wind compressions at equinox. J. Geophys. Res. Space Phys. **118**, 2897–2902 (2013). doi:10.1002/jgra.50323
Cowley, S.W.H., Lockwood, M.: Excitation and decay of solar wind-driven flows in the magnetosphere-ionosphere system. Ann. Geophys. **10**, 103–115 (1992)
Cowley, S.W.H., Owen, C.J.: A simple illustrative model of open flux tube motion over the dayside magnetopause. Planet. Space Sci. **27**, 1461–1475 (1989)
Cowley, S.W.H., Bunce, E.J., Stallard, T.S., Miller, S.: Jupiter's polar ionospheric flows: theoretical interpretation. Geophys. Res. Lett. **30**, 1220 (2003). doi:10.1029/2002GL016030
Cowley, S.W.H., Badman, S.V., Bunce, E.J., Clarke, J.T., Gérard, J.-C., Grodent, D., Jackman, C. M., Milan, S.E., Yeoman, T.K.: Reconnection in a rotation-dominated magnetosphere and its relation to Saturn's auroral dynamics. J. Geophys. Res. **110**, A02201 (2005). doi:10.1029/2004JA010796
Craven, J.D., Frank, L.A.: Latitudinal motions of the aurora during substorms. J. Geophys. Res. **92**, 4565 (1987)
Craven, J.D., Frank, L.A.: Diagnosis of auroral dynamics using global auroral imaging with emphasis on large-scale evolution. In: Meng, C.-I., Rycroft, M.J., Frank, L.A. (eds.) Auroral Physics, pp. 273–288. Cambridge University Press, Cambridge, UK (1991)
Craven, J.D., Murphree, J.S., Frank, L.A., Cogger, L.L.: Simultaneous optical observations of transpolar arcs in the two polar caps. Geophys. Res. Lett. **18**, 2297–2300 (1991). doi:10.1029/91GL02308
Cumnock, J.A., Blomberg, L.G.: Transpolar arc evolution and associated potential patterns. Ann. Geophys. **22**, 1213–1231 (2004). doi:10.5194/angeo-22-1213-2004
Cumnock, J.A., Sharber, J.R., Heelis, R.A., Hairston, M.R., Craven, J.D.: Evolution of the global aurora during positive IMF B_Z and varying IMF B_Y conditions. J. Geophys. Res. **102**, 17489–17497 (1997). doi:10.1029/97JA01182
de la Beaujardiere, O., Lyons, L.R., Friis-Christensen, E.: Sondrestrom radar measurements of the reconnection electric field. J. Geophys. Res. **96**, 13907–13912 (1991)

DeJong, A.D., Clauer, C.R.: Polar UVI images to study steady magnetospheric convection events: initial results. Geophys. Res. Lett. **32**, L24101 (2005). doi:10.1029/2005GL024498

DeJong, A.D., Cai, X., Clauer, R.C., Spann, J.F.: Aurora and open magnetic flux during isolated substorms, sawteeth, and SMC events. Ann. Geophys. **25**, 1865–1876 (2007)

DeJong, A.D., Ridley, A.J., Clauer, C.R.: Balanced reconnection intervals: four case studies. Ann. Geophys. **26**, 3897–3912 (2008)

Dungey, J.W.: Interplanetary magnetic fields and the auroral zones. Phys. Rev. Lett. **6**, 47–48 (1961)

Dungey, J.W.: The structure of the ionosphere, or adventures in velocity space. In: DeWitt, C., Hiebolt, J., Lebeau, A. (eds.) Geophysics: The Earth's Environment, pp. 526–536. Gordon and Breach, New York (1963)

Dungey, J.W.: The length of the magnetospheric tail. J. Geophys. Res. **70**, 1753 (1965)

Etemadi, A., Cowley, S.W.H., Lockwood, M., Bromage, B.J.I., Willis, D.M., Luhr, H.: The dependence of high-latitude dayside ionospheric flows on the north-south component of the IMF—a high time resolution correlation-analysis using EISCAT POLAR and AMPTE UKS and IRM data. Planet. Space Sci. **36**, 471–498 (1988)

Evans, J.V., Holt, J.M., Oliver, W.L., Wand, R.H.: Millstone Hill incoherent scatter observations of auroral convection over $60° < \Lambda < 75°$, 2, Initial results. J. Geophys. Res. **85**, 41 (1980)

Fear, R.C., Milan, S.E.: The IMF dependence of the local time of transpolar arcs: implications for formation mechanism. J. Geophys. Res. **117**, A03213 (2012a). doi:10.1029/2011JA017209

Fear, R.C., Milan, S.E.: Ionospheric flows relating to transpolar arc formation. J. Geophys. Res. **117**, A09230 (2012b). doi:10.1029/2012JA017830

Foster, J.C.: An empirical electric field model derived from Chatanika radar data. J. Geophys. Res. **88**, 981 (1983)

Frank, L.A., Craven, J.D.: Imaging results from Dynamics Explorer 1. Rev. Geophys. **26**, 249 (1988)

Frank, L.A., Craven, J.D., Burch, J.L., Winningham, J.D.: Polar views of the Earth's aurora with Dynamics Explorer. Geophys. Res. Lett. **9**, 1001–1004 (1982)

Frank, L.A., Craven, J.D., Gurnett, D.A., Shawhan, S.D., Weimer, D.R., Burch, J.L., Winningham, J.D., Chappell, C.R., Waite, J.H., Heelis, R.A., Maynard, N.C., Sugiura, M., Peterson, W.K., Shelley, E.G.: The theta aurora. J. Geophys. Res. **91**, 3177–3224 (1986)

Freeman, M.P.: A unified model of the response of ionospheric convection to changes in the interplanetary field. J. Geophys. Res. **108**, 1024 (2003). doi:10.1029/2002JA009385

Freeman, M.P., Southwood, D.J.: The effect of magnetospheric erosion on mid- and high-latitude ionospheric flows. Planet. Space Sci. **36**, 509–522 (1988)

Frey, H.U., Mende, S.B., Immel, T.J., Fuselier, S.A., Claflin, E.S., Gerard, J.C., Hubert, B.: Proton aurora in the cusp. J. Geophys. Res. **107**, 1091 (2002). doi:10.1029/2001JA900161

Frey, H.U., Mende, S.B., Angelopoulos, V., Donovan, E.F.: Substorm onset observations by IMAGE-FUV. J. Geophys. Res. **109**, 2004 (2004). doi:10.1029/2004JA010607

Goudarzi, A., Lester, M., Milan, S.E., Frey, H.U.: Multi-instrument observations of a transpolar arc in the northern hemisphere. Ann. Geophys. **26**, 201–210 (2008)

Greenwald, R.A., Weiss, W., Nielsen, E., Thomson, N.R.: STARE: a new radar auroral backscatter experiment in northern Scandinavia. Radio Sci. **13**, 1021–1039 (1978). doi:10.1029/RS013i006p01021

Greenwald, R.A., Baker, K.B., Dudeney, J.R., Pinnock, M., Jones, T.B., Thomas, E.C., Villain, J.-P., Cerisier, J.-C., Senior, C., Hanuise, C., Hunsucker, R.D., Sofko, G., Koehler, J., Nielsen, E., Pellinen, R., Walker, A.D.M., Sato, N., Yamagishi, H.: DARN/SuperDARN: a global view of the dynamics of high-latitude convection. Space Sci. Rev. **71**, 761–796 (1995)

Grocott, A., Cowley, S.W.H., Sigwarth, J.B.: Ionospheric flow during extended intervals of northward but By-dominated IMF. Ann. Geophys. **21**, 509–538 (2003)

Grocott, A., Badman, S.V., Cowley, S.W.H., Yeoman, T.K., Cripps, P.J.: The influence of IMF B_y on the nature of the nightside high latitude ionospheric flow during intervals of positive IMF B_z. Ann. Geophys. **22**, 1755–1764 (2004)

Grocott, A., Yeoman, T.K., Milan, S.E., Cowley, S.W.H.: Interhemispheric observations of the ionospheric signature of tail reconnection during IMF-northward non-substorm intervals. Ann. Geophys. **23**, 1763–1770 (2005)

Grocott, A., Wild, J.A., Milan, S.E., Yeoman, T.K.: Superposed epoch analysis of the ionospheric convection evolution during substorms: onset latitude dependence. Ann. Geophys. **27**, 591–600 (2009)

Haerendel, G., Paschmann, G., Sckopke, N., Rosenbauer, H., Hedgecock, P.C.: The frontside boundary layer of the magnetosphere and the problem of reconnection. J. Geophys. Res. **83**, 3195 (1978)

Hairston, M.R., Hill, T.W., Heelis, R.A.: Observed saturation of the ionospheric polar cap potential during the 31 March 2001 storm. Geophys. Res. Lett. **30**, 1325 (2003). doi:10.1029/2002GL015894

Hairston, M.R., Drake, K.A., Skoug, R.: Saturation of the ionospheric polar cap potential during the October–November 2003 superstorms. J. Geophys. Res. **110**, A09S26 (2005). doi:10.1029/2004JA010864

Heppner, J.P.: Empirical models of high-latitude electric fields. J. Geophys. Res. **82**, 1115 (1977)

Heppner, J.P., Maynard, N.C.: Empirical high-latitude electric-field models. J. Geophys. Res. **92**, 4467–4489 (1987)

Holzer, R.E., Slavin, J.A.: Magnetic-flux transfer associated with expansions and contractions of dayside magnetosphere. J. Geophys. Res. **83**, 3831–3839 (1978)

Holzer, R.E., Slavin, J.A.: A correlative study of magnetic flux transfer in the magnetosphere. J. Geophys. Res. **84**, 2573–2578 (1979)

Hsu, T.-S., McPherron, R.L.: An evaluation of the statistical significance of the association between northward turnings of the interplanetary magnetic field and substorm expansion onsets. J. Geophys. Res. **107**, 1398 (2002). doi:10.1029/2000JA000125

Huang, C.-S.: Evidence of periodic (2–3 hour) near-tail magnetic reconnection and plasmoid formation: geotail observations. Geophys. Res. Lett. **29**, 2189 (2002). doi:10.1029/2002GL016162

Huang, C.-S., Sofko, G.J., Koustov, A.V., Andre, D.A., Ruohoniemi, J.M., Greenwald, R.A., Hairston, M.R.: Evolution of ionospheric multicell convection during northward interplanetary magnetic field with $|B_z/B_y| > 1$. J. Geophys. Res. **105**, 27095–27107 (2000)

Huang, C.-S., DeJong, A.D., Cai, X.: Magnetic flux in the magnetotail and polar cap during sawteeth, isolated substorms, and steady magnetospheric convection events. J. Geophys. Res. **114**, A07202 (2009). doi:10.1029/2009JA014232

Hubert, B., Milan, S.E., Grocott, A., Cowley, S.W.H., Gérard, J.-C.: Dayside and nightside reconnection rates inferred from IMAGE-FUV and SuperDARN data. J. Geophys. Res. **111**, A03217 (2006a). doi:10.1029/2005JA011140

Hubert, B., Palmroth, M., Laitinen, T.V., Janhunen, P., Milan, S.E., Grocott, A., Cowley, S.W.H., Pulkkinen, T., Gérard, J.-C.: Compression of the Earth's magnetotail by interplanetary shocks drives magnetic flux closure. Geophys. Res. Lett. **33**, L10105 (2006b). doi:10.1029/2006GL026008

Hubert, B., Milan, S.E., Grocott, A., Cowley, S.W.H., Gérard, J.C.: Open magnetic flux and magnetic flux closure during sawtooth events. Geophys. Res. Lett. **35**, L23301 (2008). doi:10.1029/2008GL036374

Hubert, B., Blockx, C., Milan, S.E., Cowley, S.W.H.: Statistical properties of flux closure induced by solar wind dynamic pressure fronts. J. Geophys. Res. **114**, A07211 (2009). doi:10.1029/2008JA013813

Iijima, T., Potemra, T.A.: Large-scale characteristics of field-aligned currents associated with substorms. J. Geophys. Res. **83**, 599 (1978)

Imber, S.M., Milan, S.E., Hubert, B.: Ionospheric flow and auroral signatures of dual lobe reconnection. Ann. Geophys. **24**, 3115–3129 (2006)

Imber, S.M., Milan, S.E., Hubert, B.: Observations of significant flux closure by dual lobe reconnection. Ann. Geophys. **25**, 1617–1627 (2007)

Imber, S.M., Milan, S.E., Lester, M.: The SuperDARN Heppner-Maynard Boundary as a proxy for the latitude of the auroral oval. J. Geophys. Res. Space Phys. **118**, 685–697 (2013a). doi:10.1029/2012JA018222

Imber, S.M., Milan, S.E., Lester, M.: Solar cycle variations in polar cap area measured by the SuperDARN radars. J. Geophys. Res. Space Phys. **118**, 6188–6196 (2013b). doi:10.1002/jgra.50509

Juusola, L., Partamies, N., Tanskanen, E.: Effect of the ring current on preconditioning the magnetosphere for steady magnetospheric convection. Geophys. Res. Lett. **40**, 1917–1921 (2013). doi:10.1002/grl.50405

Kamide, Y., Kokubun, S., Bargatze, L.F., Frank, L.A.: The size of the polar cap as an indicator of substorm energy. Phys. Chem. Earth C **24**, 119 (1999)

Kissinger, J., McPherron, R.L., Hsu, T.-S., Angelopoulos, V., Chu, X.: Necessity of substorm expansions in the initiation of steady magnetospheric convection. Geophys. Res. Lett. **39**, L15105 (2012a). doi:10.1029/2012GL052599

Kissinger, J., McPherron, R.L., Hsu, T.-S., Angelopoulos, V.: Diversion of plasma due to high pressure in the inner magnetosphere during steady magnetospheric convection. J. Geophys. Res. **117**, A05206 (2012b). doi:10.1029/2012JA017579

Kistler, L.M., Mouikis, C.G., Cao, X., Frey, H., Klecker, B., Dandouras, I., Korth, A., Marcucci, M.F., Lundin, R., McCarthy, M., Friedel, R., Lucek, E.: Ion composition and pressure changes in storm time and nonstorm substorms in the vicinity of the near-Earth neutral line. J. Geophys. Res. **111**, A11222 (2006). doi:10.1029/2006JA011939

Kullen, A.: The connection between transpolar arcs and magnetotail rotation. Geophys. Res. Lett. **27**, 73–76 (2000). doi:10.1029/1999GL010675

Kullen, A., Brittnacher, M., Cumnock, J.A., Blomberg, L.G.: Solar wind dependence of the occurrence and motion of polar auroral arcs: a statistical study. J. Geophys. Res. **107**, 1362 (2002). doi:10.1029/2002JA009245

Lester, M., Milan, S.E., Besser, V., Smith, R.: A case study of HF radar spectra and 630.0 nm auroral emission. Ann. Geophys. **19**, 327–340 (2001)

Lewis, R.V., Freeman, M.P., Reeves, G.D.: The relationship of HF backscatter to the accumulation of open magnetic flux prior to substorm onset. J. Geophys. Res. **103**, 26613–26619 (1998)

Lockwood, M.: On flow reversal boundaries and transpolar voltage in average models of high-latitude convection. Planet. Space Sci. **39**, 397–409 (1991)

Lockwood, M., Freeman, M.P.: Recent ionospheric observations relating to solar wind-magnetosphere coupling. Philos. Trans. R. Soc. Lond. A. **328**, 93 (1989)

Lockwood, M., Cowley, S.W.H., Freeman, M.P.: The excitation of plasma convection in the high-latitude ionosphere. J. Geophys. Res. **95**, 7961–7972 (1990)

Lockwood, M., Cowley, S.W.H.: Ionospheric convection and the substorm cycle. In: Proceedings of the International Conference on Substorms (ICS-1), ESA SP-335, pp. 99–109 (1992)

Lockwood, M., Hairston, M., Finch, I., Rouillard, A.: Transpolar voltage and polar cap flux during the substorm cycle and steady convection events. J. Geophys. Res. **114**, A01210 (2009). doi:10.1029/2008JA013697

Lyons, L.R., Blanchard, G.T., Samson, J.C., Lepping, R.P., Yamamoto, T., Moretto, T.: Coordinated observations demonstrating external substorm triggering. J. Geophys. Res. **102**, 27039–27051 (1997)

Marcucci, M.F., Coco, I., Ambrosino, D., Amata, E., Milan, S.E., Bavassano Cattaneo, M.B., Retinò, A.: Extended SuperDARN and IMAGE observations for northward IMF: evidence for dual lobe reconnection. J. Geophys. Res. **113**, A02204 (2008). doi:10.1029/2007JA012466

McPherron, R.L.: Growth phase of magnetospheric substorms. J. Geophys. Res. **75**, 5592–5599 (1970)

McWilliams, K.A., Pfeifer, J.B., McPherron, R.L.: Steady magnetospheric convection selection criteria: implications of global SuperDARN convection measurements. Geophys. Res. Lett. **35**, L09102 (2008)

Mende, S.B., Heetderks, H., Frey, H.U., Lampton, M., Geller, S.P., Habraken, S., Renotte, E., Jamar, C., Rochus, P., Spann, J., Fuselier, S.A., Gerard, J.-C., Gladstone, R., Murphree, S., Cogger, L.: Far ultraviolet imaging from the IMAGE spacecraft. 1. System design. Space Sci. Rev. **91**, 243–270 (2000a)

Mende, S.B., Heetderks, H., Frey, H.U., Lampton, M., Geller, S.P., Abiad, R., Siegmund, O.H.W., Tremsin, A.S., Spann, J., Dougani, H., Fuselier, S.A., Magoncelli, A.L., Bumala, M.B., Murphree, S., Trondsen, T.: Far ultraviolet imaging from the IMAGE spacecraft. 2. Wideband FUV imaging. Space Sci. Rev. **91**, 271–285 (2000b)

Meng, C.-I., Holzworth, R.H., Akasofu, S.-I.: Auroral circle: delineating the poleward boundary of the quiet auroral belt. J. Geophys. Res. **82**, 164 (1977)

Milan, S.E.: A simple model of the flux content of the distant magnetotail. J. Geophys. Res. **109**, A07210 (2004a). doi:10.1029/2004JA010397

Milan, S.E.: Dayside and nightside contributions to the cross polar cap potential: placing an upper limit on a viscous-like interaction. Ann. Geophys. **22**, 3771–3777 (2004b)

Milan, S.E.: Both solar wind-magnetosphere coupling and ring current intensity control of the size of the auroral oval. Geophys. Res. Lett. **36**, L18101 (2009). doi:10.1029/2009GL039997

Milan, S.E.: Modelling Birkeland currents in the expanding/contracting polar cap paradigm. J. Geophys. Res. **118**, 5532–5542 (2013). doi:10.1002/jgra.50393

Milan, S.E., Slavin, J.A.: An assessment of the length and variability of Mercury's magnetotail. Planet. Space Sci. **59**, 2048–2065 (2011). doi:10.1016/j.pss.2001.05.007

Milan, S.E., Lester, M., Cowley, S.W.H., Brittnacher, M.: Dayside convection and auroral morphology during an interval of northward interplanetary magnetic field. Ann. Geophys. **18**, 436–444 (2000)

Milan, S.E., Lester, M., Cowley, S.W.H., Oksavik, K., Brittnacher, M., Greenwald, R.A., Sofko, G., Villain, J.-P.: Variations in polar cap area during two substorm cycles. Ann. Geophys. **21**, 1121–1140 (2003)

Milan, S.E., Cowley, S.W.H., Lester, M., Wright, D.M., Slavin, J.A., Fillingim, M., Singer, H.J.: Response of the magnetotail to changes in the open flux content of the magnetosphere. J. Geophys. Res. **109**, A04220 (2004). doi:10.1029/2003JA010350

Milan, S.E., Bunce, E.J., Cowley, S.W.H., Jackman, C.M.: Implications of rapid planetary rotation for the Dungey magnetotail of Saturn. J. Geophys. Res. **110**, A03209 (2005a). doi:10.1029/2004JA010716

Milan, S.E., Hubert, B., Grocott, A.: Formation and motion of a transpolar arc in response to dayside and nightside reconnection. J. Geophys. Res. **110**, A01212 (2005b). doi:10.1029/2004JA010835

Milan, S.E., Wild, J.A., Hubert, B., Carr, C.M., Lucek, E.A., Bosqued, J.M., Watermann, J.F., Slavin, J.A.: Flux closure during a substorm observed by Cluster, Double Star, IMAGE FUV, SuperDARN, and Greenland magnetometers. Ann. Geophys. **24**, 751–767 (2006)

Milan, S.E., Provan, G., Hubert, B.: Magnetic flux transport in the Dungey cycle: a survey of dayside and nightside reconnection rates. J. Geophys. Res. **112**, A01209 (2007). doi:10.1029/2006JA011642

Milan, S.E., Boakes, P.D., Hubert, B.: Response of the expanding/contracting polar cap to weak and strong solar wind driving: implications for substorm onset. J. Geophys. Res. **113**, A09215 (2008). doi:10.1029/2008JA013340

Milan, S.E., Grocott, A., Forsyth, C., Imber, S.M., Boakes, P.D., Hubert, B.: A superposed epoch analysis of auroral evolution during substorm growth, onset and recovery: open magnetic flux control of substorm intensity. Ann. Geophys. **27**, 659–668 (2009a)

Milan, S.E., Hutchinson, J., Boakes, P.D., Hubert, B.: Influences on the radius of the auroral oval. Ann. Geophys. **27**, 2913–2924 (2009b)

Milan, S.E., Gosling, J.S., Hubert, B.: Relationship between interplanetary parameters and the magnetopause reconnection rate quantified from observations of the expanding polar cap. J. Geophys. Res. **117** (2012). doi:10.1029/2011JA017082

Morelli, J.P., Bunting, R.J., Cowley, S.W.H., Farrugia, C.J., Freeman, M.P., Friis-Christensen, E., Jones, G.O.L., Lester, M., Lewis, R.V., Luhr, H., Orr, D., Pinnock, M., Reeves, G.D., Williams, P.J.S., Yeoman, T.K.: Radar observations of auroral zone flows during a multiple-onset substorm. Ann. Geophys. **13**, 1144–1163 (1995)

Morley, S.K., Freeman, M.P.: On the association between northward turnings of the interplanetary magnetic field and substorm onsets. Geophys. Res. Lett. **34**, L08104 (2007). doi:10.1029/2006GL028891

Øieroset, M., Raeder, J., Phan, T.D., Wing, S., McFadden, J.P., Li, W., Fujimoto, M., Rème, H., Balogh, A.: Global cooling and densification of the plasma sheet during an extended period of purely Northward IMF on October 22–24, 2003. Geophys. Res. Lett. **32**, L12S07 (2005). doi:10.1029/2004GL021523

Ouellette, J.E., Brambles, O.J., Lyon, J.G., Lotko, W., Rogers, B.N.: Properties of outflow-driven sawtooth substorms. J. Geophys. Res. Space Phys. **118**, 3223–3232 (2013). doi:10.1002/jgra.50309

Partamies, N., Pulkkinen, T.I., McPherron, R.L., McWilliams, K., Bryant, C.R., Tanskanen, E., Singer, H.J., Reeves, G.D., Thomsen, M.F.: Statistical survey on sawtooth events, SMCs and isolated substorms. Adv. Space Res. **44**, 376 (2009). doi:10.1016/j.asr.2009.03.013

Petrinec, S.M., Russell, C.T.: Near-Earth magnetotail shape and size as determined from the magnetopause flaring angle. J. Geophys. Res. **101**, 137–152 (1996)

Reiff, P.H., Spiro, R.W., Hill, T.W.: Dependence of polar cap potential drop on interplanetary parameters. J. Geophys. Res. **86**, 7639–7648 (1981)

Rich, F.J., Hairston, M.: Large-scale convection patterns observed by DMSP. J. Geophys. Res. **99**, 3827–3844 (1994)

Rostoker, G., Akasofu, S.-I., Foster, J., Greenwald, R.A., Kamide, Y., Kawasaki, K., Lui, A.T.Y., McPherron, R.L., Russell, C.T.: Magnetospheric substorms—definition and signatures. J. Geophys. Res. **85**, 1663–1668 (1980)

Ruohoniemi, J.M., Greenwald, R.A.: Statistical patterns of high-latitude convection obtained from Goose Bay HF radar observations. J. Geophys. Res. **101**, 21743–21763 (1996)

Senior, C., Cerisier, J.-C., Rich, F., Lester, M., Parks, G.K.: Strong sunward propagating flow bursts in the night sector during quiet solar wind conditions, SuperDARN and satellite observations. Ann. Geophys. **20**, 771–779 (2002)

Sergeev, V.A.: On the state of the magnetosphere during prolonged periods of southward oriented IMF. Phys. Solariterr. Potsdam **5**, 39 (1977)

Shepherd, S.G.: Polar cap potential saturation: observations, theory, and modelling. J. Atmos. Sol. Terr. Phys. **69**, 234–248 (2006)

Shukhtina, M.A., Dmitrieva, N.P., Popova, N.G., Sergeev, V.A., Yahnin, A.G., Despirak, I.V.: Observational evidence of the loading-unloading substorm scheme. Geophys. Res. Lett. **32**, L17107 (2005). doi:10.1029/2005GL023779

Shukhtina, M.A., Sergeev, V.A., DeJong, A.D., Hubert, B.: Comparison of magnetotail magnetic flux estimates based on global auroral images and simultaneous solar wind-magnetotail measurements. J. Atmos. Sol. Terr. Phys. **72**, 1282–1291 (2010). doi:10.1016/j.jastp.2010.09.013

Siscoe, G.L., Huang, T.S.: Polar cap inflation and deflation. J. Geophys. Res. **90**, 543–547 (1985)

Siscoe, G.L., Crooker, N.U., Siebert, K.D.: Transpolar potential saturation: roles of region 1 current system and solar wind ram pressure. J. Geophys. Res. **107**, 1321 (2002). doi:10.1029/2001JA009176

Siscoe, G., Raeder, J., Ridley, A.J.: Transpolar potential saturation models compared. J. Geophys. Res. **109**, A09203 (2004). doi:10.1029/2003JA010318

Slavin, J.A., Fairfield, D.H., Lepping, R.P., Hesse, M., Ieda, A., Tanskanen, E., Østgaard, N., Mukai, T., Nagai, T., Singer, H.J., Sutcliffe, P.R.: Simultaneous observations of earthward flow bursts and plasmoid ejection during magnetospheric substorms. J. Geophys. Res. **107**(A7), 1106 (2002). doi:10.1029/2000JA003501

Slavin, J.A., Anderson, B.J., Baker, D.N., Benna, M., Boardsen, S.A., Gloeckler, G., Gold, R.E., Ho, G.C., Korth, H., Krimigis, S.M., McNutt Jr., R.L., Nittler, L.R., Raines, J.M., Sarantos, M., Schriver, D., Solomon, S.C., Starr, R.D., Trávníček, P.M., Zurbuchen, T.H.: MESSENGER observations of extreme loading and unloading of Mercury's magnetic tail. Science **329**, 655 (2010). doi:10.1126/science.1188067

Taylor, J.R., Yeoman, T.K., Lester, M., Emery, B.A., Knipp, D.J.: Variations in the polar cap area during intervals of substorm activity on 20–21 March 1990 deduced from AMIE convection maps. Ann. Geophys. **14**, 879–887 (1996)

Wild, J.A., Woodfield, E.E., Morley, S.K.: On the triggering of auroral substorms by northward turnings of the interplanetary magnetic field. Ann. Geophys. **27**, 3559–3570 (2009)

Willis, D.M., Lockwood, M., Cowley, S.W.H., Van Eyken, A.P., Bromage, B.J.I., Rishbeth, H., Smith, P.R., Crothers, S.R.: A survey of simultaneous observations of the high latitude ionosphere and interplanetary magnetic field with EISCAT and AMPTE-UKS. J. Atmos. Terr. Phys. **48**, 987–1008 (1986)

Woch, J., Lundin, R.: Magnetosheath plasma precipitation in the polar cusp and its control by the interplanetary magnetic field. J. Geophys. Res. **97**, 1421 (1992)

Zhu, L., Schunk, R.W., Sojka, J.J.: Polar cap arcs: a review. J. Atmos. Sol. Terr. Phys. **59**, 1087–1126 (1997)

Chapter 3
Triggered VLF Emissions-an On-Going Nonlinear Puzzle

David Nunn

Abstract The well-known phenomena of triggered VLF emissions and the related VLF chorus are due to strongly nonlinear wave particle interactions between cyclotron resonant keV/MeV electrons and a narrow band or band limited VLF wavefield, taking place in the equatorial region of the magnetosphere. Nearly all theoretical and numerical work in this field assumes the VLF wavefield is parallel propagating or 'ducted' along the field line by field aligned density irregularities. Although complex, the causative mechanisms are very amenable to both numerical modelling and to theoretical and even analytic analysis.

After presenting some new observational results we shall review the origins of theoretical research into this phenomenon dating back to the work of Jim Dungey and his team at Imperial College in the 1960s and 1970s. It will be shown that Dungey's original notion that proper application of Liouville's theorem is the key to understanding the non-linear theory, and we shall review the essence of trapping theory including sideband theory.

Numerical analysis may either take the Vlasov approach, which may be applied to band limited numerical simulation of triggered emissions, or alternatively use the classical PIC method for broadband numerical modelling, the former method being particularly relevant to chorus simulation.

Some recent numerical results for VLF emission triggering will be presented and critical issues and problem areas will be highlighted.

3.1 Introduction

Triggered VLF emissions (TEs) are a well known and fascinating radio phenomenon since their first observations in the 1960s. (Helliwell 1965). They are generated in the collision free plasma in the Earth's magnetosphere and result from

D. Nunn (✉)
School of Electronics and Computer Science, Southampton University, Southampton, Hampshire SO17 1BJ, UK
e-mail: dn@ecs.soton.ac.uk

D. Southwood et al. (eds.), *Magnetospheric Plasma Physics: The Impact of Jim Dungey's Research*, Astrophysics and Space Science Proceedings 41,
DOI 10.1007/978-3-319-18359-6_3

nonlinear electron cyclotron resonance between energetic radiation belt electrons of ~keV energy and a narrow band or quasi narrow band VLF wavefield. The essential feature of TE's is that a triggering signal, which may be quite weak, 'triggers' or initiates a self-sustaining long enduring VLF emission which may be of sizeable amplitude and usually has a sweeping frequency with sweep rates ~kHz/s. Emissions often cover wide frequency bands ~kHz and the sweep rate may remain amazingly constant over a wide frequency range. Generally rising emissions seem favoured though falling tones are often seen. Various more complex forms are observed as will be demonstrated in the data later in this paper—namely upward and downward hooks or even oscillating frequency tones.

The generation region of these emissions is believed to be in the equatorial region, with an L-shell range of $L = 3$–10. Not only is there ample observational evidence from the Cluster satellites (Santolik et al. 2003; Santolik and Gurnett 2003a) of equatorial generation, but simple theoretical arguments point to this. Cyclotron resonance energy increases rapidly away from the equator making off-equatorial generation not feasible. In the early years (Helliwell 1967) it was widely believed that emission amplitudes were low, ~1 pT, but recent satellite observation from Cluster, Geotail, and RBSP consistently report large amplitudes, ~5–200 pT, at least for the very closely related phenomenon of VLF chorus. Such amplitudes are vastly more than enough for the phenomenon of nonlinear resonant particle trapping.

Triggered VLF emissions may be very narrow band indeed, as little as 10 Hz bandwidth, which would strongly favour nonlinear wave particle interaction. However some emissions may have broader spectra (~100 Hz) or exhibit complex sideband structure. Most theoretical work assumes the triggered emission is narrow band consisting of a single sweeping frequency. The vast majority of theoretical and simulation work on TE's assumes propagation parallel or quasi-parallel to the ambient magnetic field. This must be the case for ground observations of TE's but Cluster observations confirm this for VLF chorus generation regions. In some cases though, wide propagation angles are seen for emissions and chorus, notably at the Gendrin angle or close to the resonance cone, particularly for falling tones.

The triggering signal for TEs may vary widely. A widely observed strong trigger source is that of VLF whistlers, where triggering takes place at the lower or upper frequency end of the whistler (Nunn and Smith 1996). Another common trigger source are Power Line Harmonic Radiation lines (PLHR) and Magnetospheric Lines (Nunn et al. 1999), which are arrays of monochromatic VLF spectral lines observed in the magnetosphere. In the case of PLHR these lines are believed to originate from VLF radiation from terrestrial power grid systems. In the early days TEs were observed triggered by VLF Morse pulses transmitted by high power USN VLF transmitters NAA and NPG (Helliwell 1965). Emissions may appear to have no trigger source or may arise from ambient hiss or very commonly from the top of a hiss band. In the 1980s the famous Siple VLF experiment was set up by the Stanford University group (Helliwell 1983; Helliwell and Katsufrakis 1974) in which a horizontal antenna on the Antarctic polar plateau transmitted a wide variety of transmission forms into the magnetosphere. The corresponding triggered VLF emissions were picked up at Roberval, Quebec which is the conjugate point. This represents a superb data base which may be further exploited. More recently the HAARP experiment has been set up (Golkowski et al. 2010), in which UHF beams interact nonlinearly in the ionosphere injecting VLF signals into the magnetosphere. These have been observed to trigger emissions but the data base is still small compared to Siple. There are also sizeable data bases of discrete emissions and emissions triggered by hiss bands and whistlers obtained at the BAS station at Halley Bay Antarctica and also from Sodankyla Geophysical Observatory (SGO) in Finnish Lapland. We might note at this point that triggered VLF emissions and chorus are not confined to Earth. Missions to Jupiter and Saturn have observed VLF chorus in the magnetospheres of these planets.

In this paper we shall first show how early work done at Imperial College in Jim Dungey's group established the basic mechanism of triggered emission production, based upon proper application of Liouville's theorem (Dungey 1961). Some recent ground observations of triggered emissions will be presented followed by a description of the fundamental plasma theoretical mechanism involved. Numerical simulation of TEs and various approaches to it will be discussed and recent results for triggered emission simulation presented.

3.2 Ground Observations of Triggered VLF Emissions

It is of relevance to present some interesting ground-based observations—not previously published—of triggered VF emissions with rather interesting characteristics. Figure 3.1 shows a spectrogram of a triggered VLF emission obtained in Finnish Lapland in 1997. The emission appears triggered by a tone of constant frequency, and extends over a huge frequency range from 1,500 to 6,500 Hz. Astonishingly sufficient growth for self-sustaining emission prevails over this whole range and the df/dt value for this emission is held at an almost constant value throughout. This implies that the frequency sweep rate, once established, is a property of the emission generation region and may not be readily changed.

Figure 3.2 is a ground-based observation from the same location and exhibits a remarkable discrete emission apparently triggered from a magnetospheric line or power line harmonic line. A steep riser with sweep rate ~2 kHz/s abruptly changes its configuration to a steep faller, possibly due to interaction with another magnetospheric line. Note that the riser appears steeper due to dispersion between the equator and the ground. Again on reaching the triggering line there is another abrupt change to a steep riser, presumably due to nonlinear wave-wave interaction. The entire N-shaped emission appears to twice undergo a 2-hop bounce with spectral broadening and dispersion but no evidence of further wave particle interaction.

Figure 3.3 shows a spectrogram obtained from a SGO ground-based VLF campaign in Kannuslehto in Finland in 2008. There are several risers triggered from the obvious narrow hiss band with sweep rates ~1.5 kHz s^{-1}, though the sweep rate is not held so constant in this case. At the high frequency end downward hooks

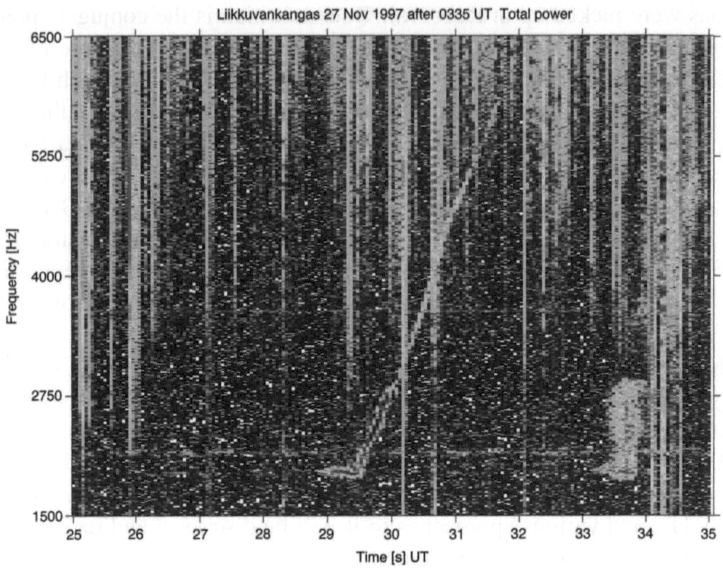

Fig. 3.1 Spectrogram of a ground-based observation of a very long enduring riser, recorded in Finnish Lapland

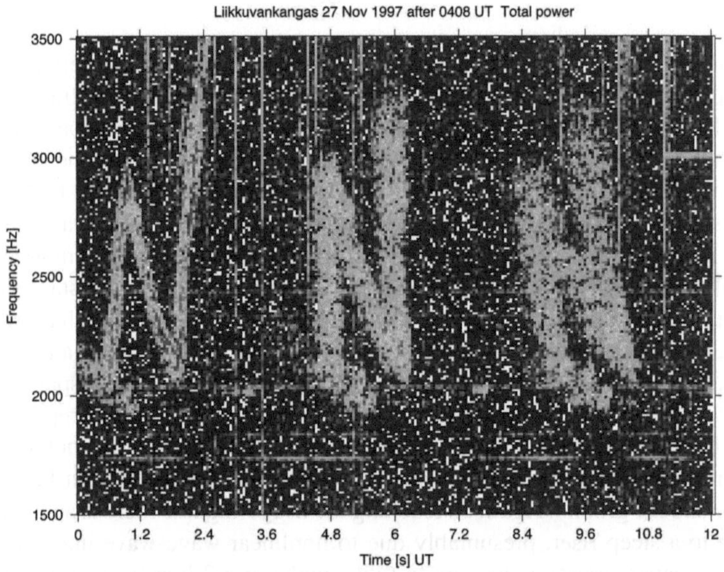

Fig. 3.2 Spectrogram of a distinctive triggered VLF emission with an N shaped spectral profile. The emission undergoes two 2-hop reflections. Each time there is further dispersion and spectral broadening, but no evidence of further non-linear wave particle interaction effects

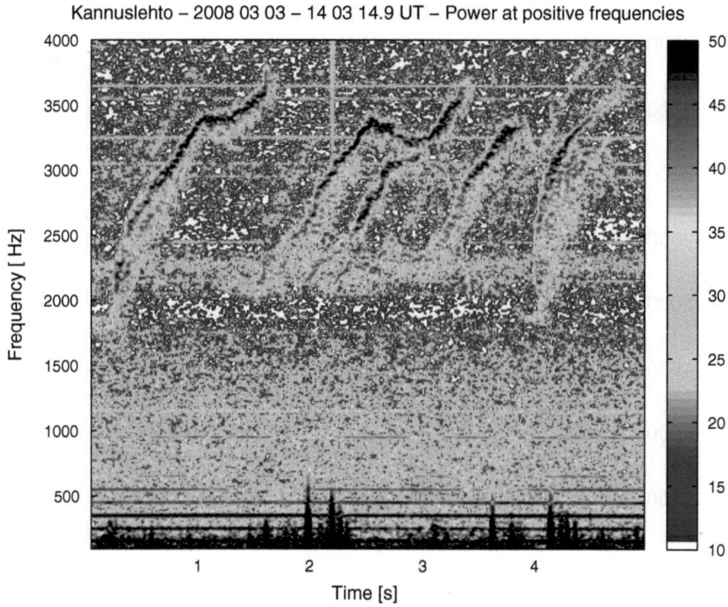

Fig. 3.3 Terrestrial observation of risers triggered by a VLF hiss band observed in Finnish Lapland

or downward and then upward hooks are seen. Figure 3.4 is a spectrogram from the same campaign and shows a succession of very pronounced downward hooks triggered from a lower hiss band and risers triggered from the upper hiss band.

An excellent VLF database has been accumulated from the British Antarctic Survey station at Halley Bay over the past three decades. Triggered emissions occur commonly, particularly emissions triggered by lightning whistlers and at the top of broad hiss bands. Figure 3.5 is from the 1995 data set and shows multiple risers arising from the top of the low frequency hiss band. Finite width hiss bands can produce, by quasi-linear diffusion, a step in the distribution at the equatorial parallel velocity corresponding to the cyclotron resonance velocity at the top frequency (Trakhtengerts 1995, 1999). This will give enhanced growth rates at this frequency, encouraging rapid growth and non-linear triggering. In some examples the triggering from the upper edge of a hiss band can be quite violent with a mass of rising emissions. Examples of the triggering of fallers from the lower edge of a hiss band have been noted but this is a relatively rare phenomenon. Figure 3.6 from the 1989 data set is a good example of triggering from whistlers. The tendency is for triggering at either the top upper frequency end of the wave train or at the end of the lower frequency branch—essentially *termination triggering*. We see triggering of a powerful riser from the lower end from the right hand whistler. The first whistler triggers risers and also an oscillating tone.

At this point it is well worth introducing the currently hot topic of VLF chorus, closely related to that of triggered VLF emissions. Chorus occurs most widely

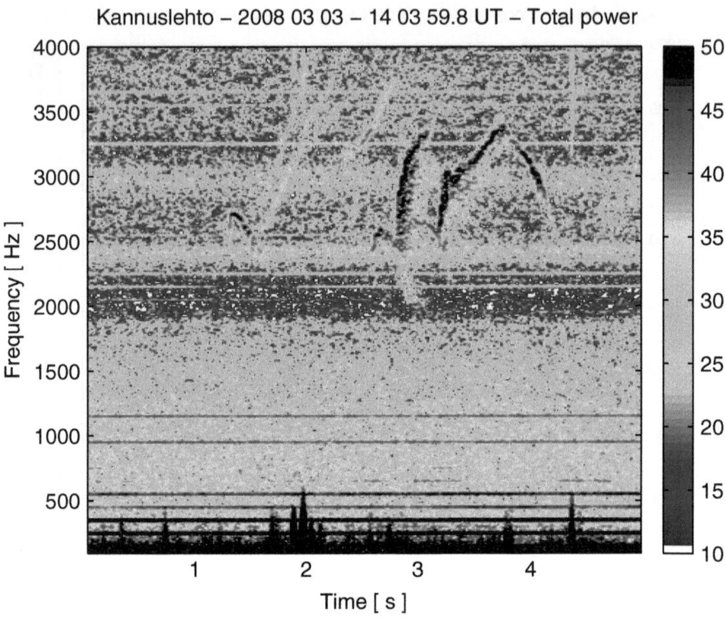

Fig. 3.4 Triggered downward hooks and risers observed in a ground-based campaign in Finland. The trigger source appears to be a narrow band of VLF hiss

Fig. 3.5 Risers triggered from the top of a hiss band observed at Halley Bay Antarctica

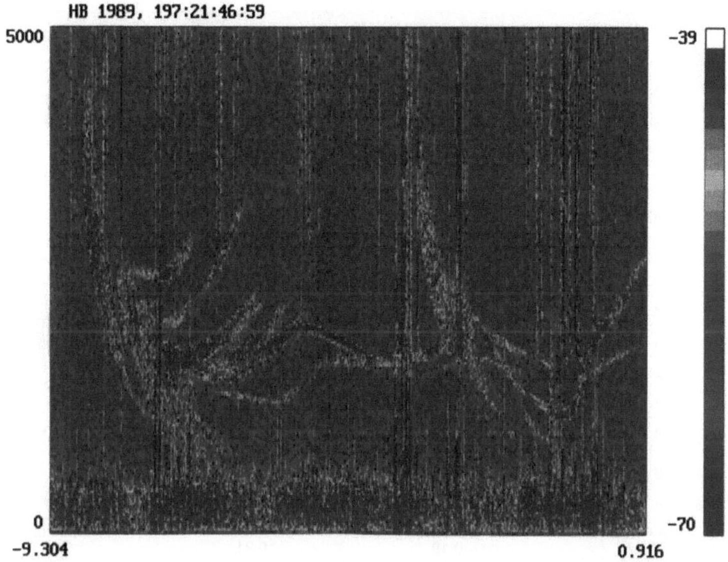

Fig. 3.6 Risers triggered by lightning whistlers observed at Halley Bay. Triggering takes place at the termination of the upper and lower frequency branches. Emission spectra are either risers or oscillating tones

outside the plasmapause and consists of sequences of VLF emissions, with no obvious trigger except that of the previous element. Rising frequency chorus rather predominates. Chorus is observed to be either upper band (above half the local equatorial gyrofrequency) or lower band (below half the local equatorial gyrofrequency). Generally lower band chorus is more common and has more spectral structure. These individual elements may be widely separated with random spacings in time, as for example observed on the Geotail satellite at $L = 10$ (Nunn et al. 1997), in which case one would suppose the generation mechanism is similar to that of triggered emissions. Some chorus, however, consists of tightly packed sequences of elements or emissions, in which case highly complex physics, probably not at present fully understood, determines the element spacing and how one element controls the next. In some cases the chorus elements are quite broad in spectrum and the 'swishy' chorus can only be properly investigated with a broadband simulation code (Omura et al. 2009). Figure 3.7 presents an example spectrogram of rising chorus observed on one of the Cluster satellites (Santolik et al. 2003). This lower band chorus is spectrally well formed but on a background of hiss, suggesting the elements arise from a highly non-linear amplification of the hiss.

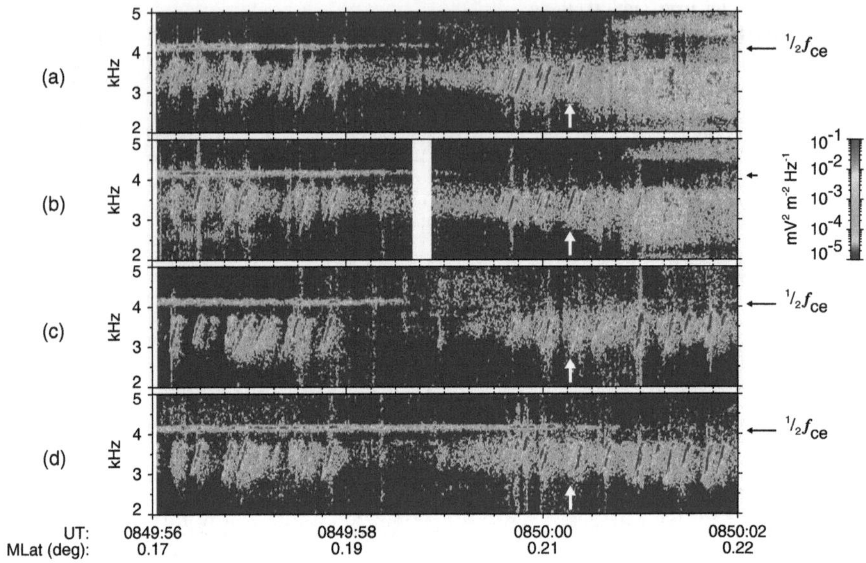

Fig. 3.7 Satellite (Cluster) observation of lower band rising chorus. Note the background low level hiss. Individual elements are rather unevenly and well-spaced, and may arguably be regarded as single triggered emissions

3.3 Non-linear Resonant Particle Dynamics

The first theory of triggered VLF emissions was due to Helliwell (1967), who considered the wave particle interaction process would be concentrated at the point where inhomogeneity factor S = 0, which is synonymous with the point of second order resonance and is where a resonant electron sees zero first order time derivative of cyclotron resonance velocity. This point is downstream from the equator for a faller and upstream for a riser. However since these early pioneering days our understanding has moved on. The optimum power input occurs for |S| ~ 0.4 for a constant S scenario, and not S = 0, as was demonstrated by Omura et al. (2009) and Hikishima et al. (2009). Further the generation region is not a point object but has finite size, since the essence of non-linearity is long-time interaction. This concept of Helliwell still has relevance today, particularly for steep sweep rates, where the 'Helliwell point' will roughly position the generation region.

It was realised by Jim Dungey in the 1960s (Dungey 1963a, 1966, 1968, 1969) and also by researchers in the Soviet Union (Karpman and Shklyar 1972) that the fundamental plasma physical process underlying the generation mechanism of triggered VLF emissions and VLF chorus was the non-linear trapping of cyclotron-resonant electrons in an inhomogeneous medium (Nunn 1970). We should point out at once that nearly all research in this area assumes a VLF wavefield propagating parallel to the ambient magnetic field (in the 'z' direction), and that the wavefield is narrow band. Research at Imperial College first studied non-linear trapping of Landau resonant electrons in an electrostatic wave in the presence of an inhomogeneity (Nunn 1970). Essentially if ψ is the wave phase seen by a Landau-resonant electron then it obeys the trapping equation

$$d^2\psi/dt^2 + \omega_{tr}{}^2(cos\,\psi + S(z,t)) = 0 \tag{3.1}$$

where

$$\omega_{tr}{}^2 = ek_z|Ez|\,/m \tag{3.2}$$

is the square of the trapping frequency and

$$S(z,t) = [-d/dt V_{res} + F/m]\,k_z/\omega_{tr}{}^2 \tag{3.3}$$

is the *collective inhomogeneity factor*. The first term represents the rate of change of the Landau resonance velocity in the frame of the resonant particle, such as might be due to a gradient of wave number *k* for example. In the second term *F* represents an external DC force such as might be due to a DC electric field or more realistically the mirror force in a magnetic field dipole geometry. It was shown that the condition for trapping is $|S| < 1$, whence stably trapped particles remain phase locked and their velocity remains close to the changing resonance velocity. Their phase oscillates about the phase locking angle $cos^{-1}S$. If $|S| > 1$ trapping is not allowed and electrons are swept through resonance. Trapped particles in an inhomogeneous medium undergo relatively large changes in V_z . By application of Liouville's theorem the distribution function of trapped particles will be that appertaining to when they were first trapped, which will give a 'hill' or 'hole' in distribution function in the region of phase space corresponding to trapping, depending on the sign of S and also on the zero order distribution function. Jim Dungey always emphasised the importance of the strict application of Liouville's theorem in obtaining a proper plasma theoretic description of this nonlinear plasma theory (Dungey 1961, 1963a, 1969).

The concept of non-linear trapping in an inhomogeneous medium was soon applied to the triggered VLF emission problem, whence cyclotron resonant electrons were found to obey the same trapping equation (Dungey 1963a, 1969; Nunn 1973, 1974b; Nunn and Rycroft 2005; Karpman et al. 1974; Istomin et al. 1976), although now S is a function of pitch angle and in general also of z and t. In the whistler case the inhomogeneity factor S derives from the gradient of the ambient magnetic field in a dipole geometry, and also from the sweeping frequency of the emission itself. Omura and co-workers have developed the fully relativistic problem and shown resonant particles obey the same trapping equation (Omura et al. 2007, 2008, 2009; Hikishima et al. 2009) where ψ is now the phase angle between the perpendicular velocity vector and the wave magnetic field vector and inhomogeneity factor S may be simply expressed in the form

$$S(z,t,V_{perp}) = \{A\,dB_0/dz + B\,df/dt\}/|B_w| \tag{3.4}$$

where df/dt is the sweep rate in the frame of the resonant particle, and $|B_w|$ is the local wave amplitude. For each pitch angle, trapped particles occupy a region in V_z—ψ space centred on the stable phase locking angle $cos^{-1} S$ (Nunn 1974a; Omura et al. 2009, 2008, 2007). It should be pointed out that simple analyses of trapping theory often assume S to be constant in the particle frame. In a parabolic field inhomogeneity S is not constant and simple trapping dynamics is only a first approximation. Where $|S|$ is decreasing, the trap enlarges and new particles will become trapped, and when $|S|$ is increasing trapped particles will be forced out of resonance. We may remark here on the conditions required for nonlinear trapping. We require $|S| < 1$ of course but this condition must be sustained in the particle frame and the geometry of the problem must be such that trapped particles are able to execute at least one trapping oscillation, or trapping time $> 1/\omega_{tr}$.

Some early work has examined the nonlinear resonant particle dynamics where one or more sideband waves are present in the wavefield (Nunn 1973, 1974a, 1985). Where the sideband is weak and frequency separation is of the order of the trapping frequency, a resonance interaction with the trapping oscillation occurs, trapped particles undergo increasing amplitude of oscillation, eventually being detrapped. The effect of spectral broadening and of sidebands is to decrease the level of nonlinearity and limit trapping times. It was found that where sidebands have magnitudes of order of the main wave, resonant particle motions become very complex and chaotic, but nonlinearity still remained.

3.4 Resonant Particle Distribution Function

The correct way to calculate the resonant particle distribution $F_{res}(z,\underline{V},t)$ appropriate to electrons at or near cyclotron resonance was outlined by Jim Dungey in the 1960s (Dungey 1961, 1963a, b, 1969), and involves the strict application of Liouville's theorem. Starting from the phase space point in question the resonant particle trajectory is followed *backwards* until there is no further significant interaction with the wave field, and certainly before trapping occurred. From Liouville's theorem we then have

$$F_{res}(\mu, W) = F_0(\mu - \Delta\mu, W - \Delta W) \tag{3.5}$$

where F_0 is the time-independent zero order distribution function taken to be a function of invariants energy W and magnetic moment μ, and where $\Delta\mu$ is the integrated change in magnetic moment that electron has undergone since starting to interact with the wave field and ΔW is the integrated energy change. The technique of backward trajectory integration is particularly useful when calculating the resonant particle distribution function, resonant particle current and non-linear growth rates in the non-self-consistent case where the wavefield is specified in advance.

What implications does non-linear trapping in an inhomogeneous case have for the nature of the resonant particle distribution function? Trapped particles are essentially dragged through invariant energy and magnetic moment space by the trapping process and they will normally have a distribution function that is radically different from that of the surrounding untrapped particles, giving rise to a hole or hill in distribution function in velocity space, at least provided the trapping time is greater than the trapping period. (Nunn 1974a). The reader will find an elegant description of trapping and the geometry of the trap in velocity space for the fully relativistic case in Omura et al. (2007, 2008, 2009). A full mathematical development will be found in (Nunn 1990), but for negative inhomogeneity, $S < 0$, trapped particles undergo an *increase* in both energy and magnetic moment. These changes may be substantial for strong VLF waves ~50 pT say, of order keV in energy and 20° in equatorial pitch angle, and so for $S < 0$ trapping is an important process for electron heating but not significant as far as particle precipitation is concerned. Conversely for positive inhomogeneity $S > 0$ trapped particles have a decrease in energy and magnetic moment. It is found that for a constant frequency narrow band (CW) wavefield the region downstream from the equator (receiver side) has negative inhomogeneity and vice versa. For the wavefield appropriate to a rising frequency VLF emission or rising chorus the generation region will have predominantly negative inhomogeneity (Omura and Nunn 2011). If one assumes the zero order distribution function is anisotropic and linearly unstable then negative inhomogeneity will give a 'hole' in the distribution function at the location of the trap (Nunn 1974a; Omura et al. 2007, 2008, 2009; Hikishima et al. 2009).

Clearly the depth of the hole will be roughly proportional to *dF* and thus trapping time, though for strong VLF waves saturation effects will kick in as the hole becomes close to a void. Positive inhomogeneity will occur upstream from the equator in a CW wave field, and it was shown by Nunn and Omura (2012) that falling tone VLF emissions will have generation regions characterised by $S > 0$. In this case a hill in the distribution function will be found at the location of the trap and trapped particles are de-energised and have their equatorial pitch angle reduced. A popular misconception is that because trapped particles have large *dF* they dominate the power transfer between particles and the wave field. However the passing untrapped particles with smaller *dF* make a vital non-negligible contribution to power transfer as they are actually far more numerous than the trapped particles.

Recalling how the distribution function is calculated using Liouville's theorem it is apparent that the depth of distribution function holes/hills will depend absolutely on F_0. For example a loss cone distribution function will produce a deeper hole than a bi-Maxwellian. Further, in the event that the zero order distribution function is stable and isotropic, negative inhomogeneity S will result in a hill at the trap location, and a positive inhomogeneity will give a hole. Such a scenario is improbable, however, since in a plasma with linear damping non-linear amplitudes are unlikely to be reached, unless a very strong signal is injected into the plasma.

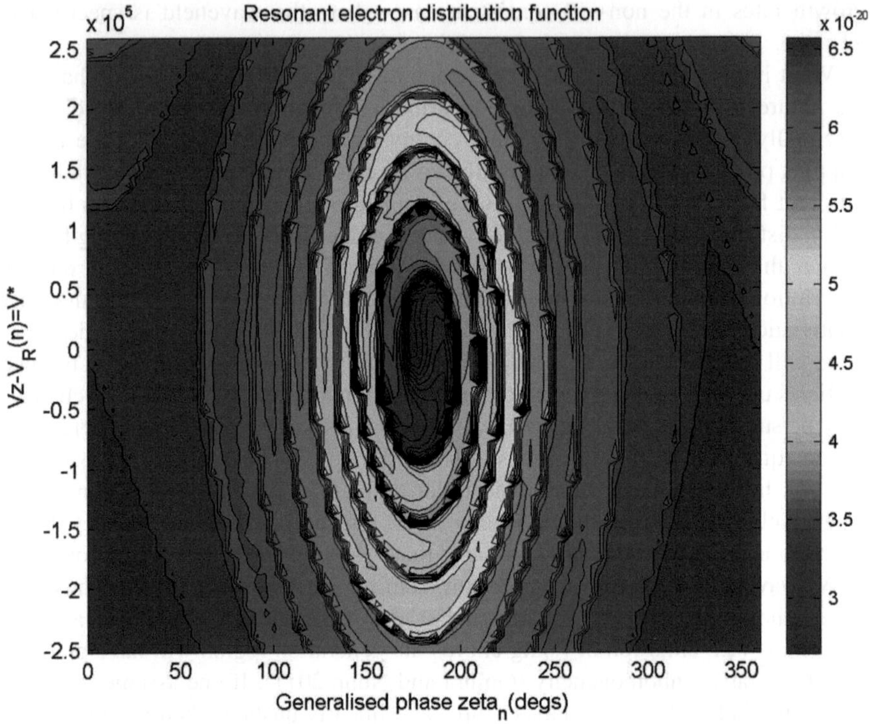

Fig. 3.8 View of distribution function F_{res} at the equator for a pitch angle of 50°, in the parallel velocity/gyrophase plane, where gyrophase is measured relative to the wave magnetic field vector

By way of illustration we present in Fig. 3.8 a non self-consistent computation of the resonant electron distribution at the equator for a pitch angle of 50°, shown as a function of parallel velocity about resonance $\{Vz - V_{res}\}$ and gyrophase relative to that of the wave magnetic field vector. The wavefield is assumed to be CW with an amplitude of 20 pT, L = 4.2, the cold plasma density being 100 cm^{-3}, frequency 4 kHz, and electron gyrofrequency 10 kHz. The zero order distribution function F_0 is bi-Maxwellian with anisotropy A = 2, plus a loss cone factor. Trapped particles arriving at the equator have come from the negative inhomogeneity region downstream and we see a deep hole in the distribution function of order 40 % of the zero order value at the trap centre, giving a Q factor as defined by *Omura* of $Q \sim 0.55$. The maximum historical integrated energy change at the trap centre is ~+0.5 keV, and the integrated equatorial pitch angle change there is +18°.

3.5 Resonant Particle Current

The matter of how to calculate the resonant particle distribution function was a subject of considerable dispute in the late 1960s. Jim Dungey emphasised the classical approach of integrating $-e\,\underline{v}\,F_{res}$ over velocity space in the neighbourhood of the cyclotron resonance velocity, as in the familiar expression in Eq. 3.4, the distribution function being derived by backward trajectory integration and application of Liouville's theorem (Dungey 1961, 1963a, 1969).

$$J_R(z,t,\psi) = -e\iiint \mathbf{v} F_{res}|v_\perp|dv_z d|v_\perp|d\phi \tag{3.6}$$

However Helliwell and co-workers (Helliwell 1967; Helliwell and Crystal 1970) postulated the 'phase bunching current', derived by following the phase progression in time of some 16 particles. However, such a current does not provide any guide as to what the current expression in Eq. 3.6 is. Surprisingly, even contemporary works refer to particle bunching. However as far as non-linear wave particle interactions are concerned particle trapping is the only relevant process.

We note here that inhomogeneous non-linear trapping is actually a very simple process compared to the homogeneous case, and provided we have a very narrow band wavefield and S is constant or slowly varying, simple models for non-linear current are available (Nunn 1974a; Omura et al. 2009). Under these simple conditions the resonant particle trap will coincide with a pronounced hill or a hole in the distribution function. This will give a dominant contribution to the resonant particle current with a phase equal to the trapping phase $cos^{-1} S$ relative to the wave magnetic field vector when we have a hole and $S < 0$. If $S > 0$ and we have a 'hill' then the current phase will be $cos^{-1} S + \pi$. It should be noted that the 'hole' depth or 'hill' height and thus the current at any position and time is a function of the entire trapping history there, and thus a function of the whole wavefield and its history. This is in sharp contrast to the *phase* of the non-linear current which is closely controlled by the local value of inhomogeneity S. It must be re-emphasised that the current is derived from an integration over pitch angle/perpendicular velocity, and that S and thus trapping conditions are functions of pitch angle. For reasonable choices of F_0 the dominant contribution to the current will come from the pitch angle range 45–60°, and indeed very low and high pitch angles may well have $|S| > 1$ and be linear in behaviour. The contribution from each pitch angle is weighted by the appropriate gradient term in F_0, which is the same factor which weights the contribution to the linear current as a function of pitch angle.

3.6 Time Development of the Wavefield

The resonant particle current continuously modifies the wavefield in a self-consistent fashion, which leads to the formation of the triggered emission. A useful view of this process was given in Helliwell (1967) who perceived the resonant particle current as acting like an antenna tuned to the ambient plasma and radiating new fields which are summed with the current field in a continuous fashion.

In a broadband view of the problem, as in the PIC codes of Omura and co-workers (Omura et al. 2009), the wave field is pushed using Maxwell's equations directly. However under the assumption of a quasi- narrow band wavefield, wave equations for pushing the wavefield may be derived and these provide much insight into the particle interaction process (Nunn 1974a; Omura et al. 2009). Firstly we have the equation for evolution of wave phase

$$\partial\varphi/\partial t + V_g \partial\varphi/\partial z = \omega_1 \tag{3.7}$$

where

$$\omega_1 = -\mu_0 V_g J_B/(2B_w). \tag{3.8}$$

Here φ is the additional wave phase over and above the base phase (prior to modification by the current), J_B is the component of current parallel to wave magnetic field, B_w is the wavefield magnitude, and V_g is the group velocity. Clearly component J_B directly modifies the wave phase and must be responsible for producing the sweeping frequency of the emission. Secondly we have the equation for evolution of wavefield amplitude B_w

$$\partial B_w/\partial t + V_g \partial B_w/\partial z = \omega_2. \tag{3.9}$$

where

$$\omega_2 = -\mu_0 V_g J_E/2. \tag{3.10}$$

and J_E is the component of current parallel to the wave electric field and provides the non-linear growth allowing the formation of a self-sustaining generation region.

A major problem has been to understand how the sweeping frequency of an emission comes about. Some very elegant manipulation by Katoh and Omura (2006, 2007) results in the following expression for wave frequency ω

$$\partial\omega/\partial t + V_g \partial\omega/\partial z = -\mu_0 \partial/\partial t(V_g J_B/B_W) \sim 0 \tag{3.11}$$

Since the term on the right must be small and cannot have a significant constant component, then a frequency sweep at a fixed point can only result from the establishment of a spatial gradient of frequency in the wavefield. The question then remains, how such a gradient is set up. It was shown by Nunn and Omura (2012) and Omura and Nunn (2011) that the current field $J_B(z,t)$ is able to set up such a gradient during the triggering phase.

Another topic of importance is that of sideband stability. An analytic analysis was performed by Nunn (1985) which considered the linear perturbation of trapped particles by weak resonant sidebands. It was shown that for a rising tone the inhomogeneity $S < 0$, and the resonant upper sideband is unstable and the lower sideband is damped. This is consistent with a rising tone, but describing the rising frequency as being due to successive transfer of power to higher sidebands is another way of looking at the problem. In the case of a falling tone inhomogeneity $S > 0$, and it was shown that the lower sideband is unstable in that case and the upper sideband damped. However, the Siple VLF active experiment (Helliwell and Katsufrakis 1974; Helliwell 1983) has probed sideband behaviour which was found to be in agreement with the above theory.

3.7 Later Work-Numerical Simulation of Triggered VLF Emissions

The actual scenario in which VLF emissions are triggered is rather more complex than may be described by simple trapping theory, which assumes constant S and a monochromatic wavefield. The dominant inhomogeneity is due to the parabolic dependence of the ambient magnetic field on distance from the equator z, and thus S will have an approximately linear z dependence, as well as being a function of pitch angle. Further, in any simulation the wavefield will have a finite bandwidth and wavefield amplitude, and additional phase will be functions of z and t.

To fully understand the process of emission triggering it is necessary to resort to a full numerical simulation. Two approaches are available. The first is a narrow band code which assumes the wavefield has a limited bandwidth <100 Hz, the field being pushed by the narrow band field equations above. The Vlasov VHS code described by Nunn (1990, 1993); and Nunn et al. (2009) uses a special technique that follows particles continuously forwards in time. Based upon the early philosophy of Dungey it uses Liouville's theorem to numerically solve the Vlasov equation. From Liouville the distribution function appertaining to each simulation particle is known. At each time step therefore the distribution function is interpolated using a low order interpolator onto the regular phase space grid, whence the resonant current field is readily calculated.

Figure 3.9 shows the spectrogram of a triggered VLF emission with a hook form, obtained using the VHS code. This is for a simulation inside the plasmapause at $L = 4.2$, cold plasma density 400 cm^{-3} with a bi-Maxwellian ambient hot plasma with anisotropy $A = 2$. The trigger is a CW pulse of length 100 ms. The linear equatorial growth rate at the start frequency is 120 dB s^{-1}. Figure 3.10 shows the wave field amplitude in pT for the entire event as a function of z and t. What we see

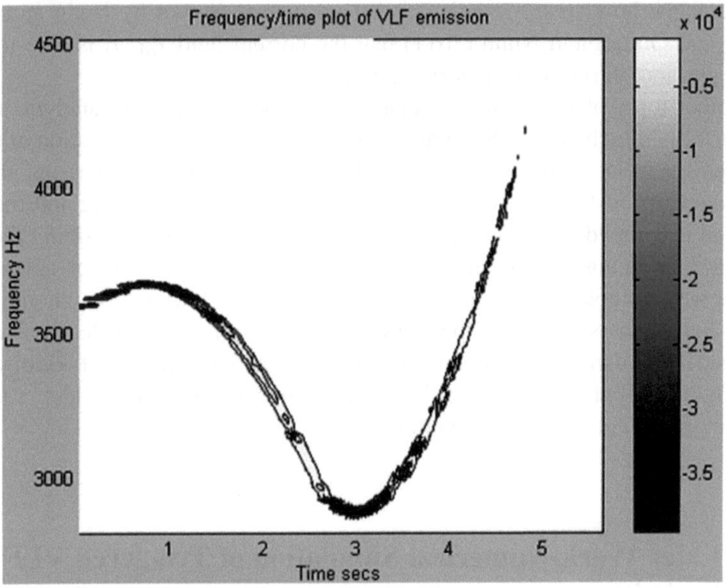

Fig. 3.9 Numerical simulation of triggered upward hook produced by band limited (100 Hz) Vlasov Hybrid Simulation (VHS) code

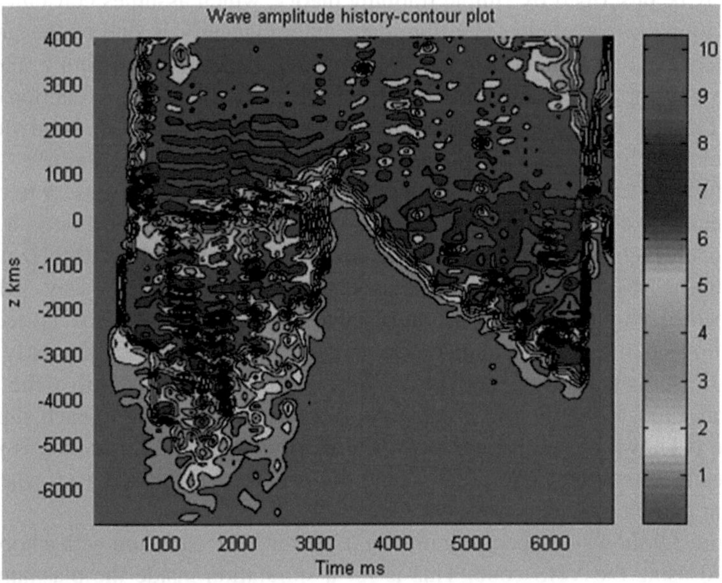

Fig. 3.10 For the simulation in Fig. 3.9, plot of wave amplitude in pT in the z t plane, where z is the distance from the equator along the magnetic field line. The quasi-static generation region of the falling frequency segment is seen to be located upstream of the equator where inhomogeneity factor *S* is positive. The riser segment has the generation region mainly downstream of the equator where inhomogeneity S is negative. In both cases the sweeping frequency is due to the spatial gradient of frequency set up during the triggering phase

is the formation of a generation region, which is a fully nonlinear stable quasi static construct with a frequency gradient across it. The first part of the emission is a faller and we see that the wave envelope extends upstream from the equator where inhomogeneity $S > 0$. The second part of the emission is a riser and the generation is seen to be in the $S < 0$ region downstream from the equator. The role of the current component anti-parallel to the wave electric field $-J_E$ provides the non-linear growth to sustain the wave profile in a constant position. The role of J_B, the current component parallel to the wave B field is to set up the spatial frequency gradient during the triggering phase (Omura and Nunn 2011; Omura et al. 2009).

The second approach is that due to Omura and co-workers (Omura et al. 2008, 2009, 2012), and employs broadband simulations using the classical PIC method. The simulations are very numerically intensive and have succeeded in simulating triggered VLF emissions (Hikishima et al. 2010a) and remarkably VLF chorus consisting of separated narrow band rising frequency elements starting from broadband noise (Hikishima et al. 2009). These codes have the advantage of avoiding certain narrow band approximations and are thus quite realistic. The PIC approach is particularly useful when simulating heating and particle precipitation phenomena (Hikishima et al. 2010b).

Mention should also be made of some theoretical and analytic studies which are current. Omura has developed chorus equations (Omura et al. 2008, 2009, 2012) that predict frequency sweep rates of chorus elements. Since resonant particle fluxes only have significant values in the equatorial region, it is argued that nonlinear trapping must occur there. In a scenario where $S = \text{constant}$ maximum non-linear growth occurs for $|S| = 0.4$, so by applying this criterion at the equator for a median value of pitch angle the series of equations governing VLF chorus properties are derived.

An ingenious model, namely that of the backward wave oscillator (BWO) has been developed by Trakhtengerts and co-workers (Trakhtengerts 1995, 1999). Interesting simulations have been performed using the BWO equations which are very realistic and bear good relation to observations of chorus on Cluster.

3.8 Conclusions

The basic plasma theory underlying the phenomenon of triggered VLF emissions and the related problem of VLF chorus was established in the 1960s using the theoretical approach pioneered by Jim Dungey. It became apparent that the key process was non-linear phase trapping of cyclotron resonant keV/MeV electrons in the equatorial zone in a parabolic inhomogeneity. This process is really quite straightforward and easy to understand. In the presence of inhomogeneity—or spatial/time variation of resonance velocity—it is easily shown that trapped electrons undergo relatively large changes in energy and magnetic moment. Proper application of Liouville's theorem gives the result that the distribution function at the site of the resonant particle trap will have a depressed value—giving a hole—or an elevated value giving a hill. This will give a readily modelled contribution to resonant current whose phase is that of the phase trapping angle (plus 180° for $S > 0$) which is controlled by the local inhomogeneity factor S. Further the sizeable component of current anti parallel to the wave electric field furnishes the non-linear growth. The component of current J_b parallel to wave magnetic field has a non-linear dependence on z and t, and it is this that causes changes to the wave phase and frequency. Note that the current produced by the linear cold plasma is also in the B direction, but a linear J_B merely contributes to the linear dispersion relation and will not cause frequency changes.

Because of variation in S due to the parabolic inhomogeneity, finite field bandwidth and the presence of sidebands, plus the pitch angle dependence of the inhomogeneity factor, the actual VLF emission triggering problem becomes rather complex in practice. Narrow band Vlasov simulations and broadband PIC simulations have both been successful in simulating triggered VLF emissions and VLF chorus, and have identified the structure of riser and faller generation regions and confirmed that frequency shift is caused by spatial gradients of frequency set up in the triggering phase. However it should be realised that there are many problems remaining and much further challenging work to do. One such problem is the slow exponential growth of received wave amplitude characteristic of Siple experiments with a key-down transmission format. This must be non-linear but simulations fail to reproduce this. An important subject is how emissions and chorus can arise from broadband noise and why and how the emissions remains sharply narrow band in character. The PIC simulations of Omura and colleagues have amazingly reproduced this behaviour, but the exact details are not quite well understood.

Of necessity simulations up to now have assumed parallel propagation with one spatial dimension and three velocity dimensions. However, physical reality is rather different with 3D propagation and oblique propagation. At present full 3D simulations are not attainable, though there are many non self-consistent theoretical studies of non-linear wave particle interactions in oblique whistlers, notably the works of Bell (1986). The way ahead is to progress to full 2D simulations, though the 1D simulations could be considerably improved by use of greater computer power and more precision. All these developments require the use of ever more powerful super computers, which certainly exist!

Acknowledgements For VLF ground data from Sodankyla Geophysical Observatory the author acknowledges the support of the European Community—Research Infrastructure Action under the FP6 "Structuring the European Research Area" Programme, LAPBIAT (RITA-CT-2006-025969). The author thanks A.J. Smith and the British Antarctic Survey for the VLF Halley Bay data, and O. Santolik for use of Cluster VLF data.

References

Bell, T.F.: The wave magnetic field amplitude threshold for nonlinear trapping of energetic gyroresonant electrons and landau resonant electrons by nonducted VLF waves in the magnetosphere. J. Geophys. Res. **91**(A4), 4365–4379 (1986)

Dungey, J.W.: The action of Vlasov waves on the velocity distribution in a plasma. J. Fluid Mech. **10**(3), 473–479 (1961). doi:http://dx.doi.org /10.1017/ S0022112061001074

Dungey, J.W.: Resonant effect of plasma waves on charged particles in a magnetic field. J. Fluid Mech. **15**(1), 74–82 (1963a). doi:10.1017/S0022112063000069

Dungey, J.W.: Loss of Van Allen electrons due to whistlers. Planet. Space Sci. **11**, 6591–6595 (1963b)

Dungey, J.W.: The motion of a charged particle in the field of a whistler mode wave packet. In: Thomas, J.O., Landmark, B.J. (eds.) Plasma Waves in Space and in the Laboratory, p. 407. American Elsevier, New York (1969)

Dungey, J.W.: Survey of acceleration and diffusion. In: Radiation Trapped in the Earth's Magnetic Field, pp. 389–397. Astrophysics and Space Science Library, Springer, London (1966)

Dungey, J.W.: Waves and particles in the magnetosphere. In: Physics of the Magnetosphere, pp. 218–259. Astrophysics and Space Science Library, Springer, London (1968)

Golkowski, M., Inan, U.S., Cohen, M.B., Gibby, A.R.: Amplitude and phase of nonlinear magnetospheric wave growth excited by the HAARP HF heater. J. Geophys. Res. **115**, A00F04 (2010). doi:10.1029/2009JA014610

Helliwell, R.A.: Whistlers and related ionospheric phenomena. Stanford University Press, Stanford, CA (1965)

Helliwell, R.A.: A theory of discrete VLF emissions from the magnetosphere. J. Geophys. Res. **72**, 4773–4790 (1967)

Helliwell, R.A.: Controlled stimulation of VLF emissions from Siple station Antarctica. Radio Sci. **18**, 801–814 (1983)

Helliwell, R.A., Crystal, T.L.: A feedback model of cyclotron interaction between whistler mode and energetic electrons in the magnetosphere. J. Geophys. Res. **78**, 7357–7371 (1970)

Helliwell, R.A., Katsufrakis, J.P.: VLF wave injection into the magnetosphere from the Siple station, Antarctica. J. Geophys. Res. **79**, 2511–2518 (1974)

Hikishima, M., Yagitani, S., Omura, Y., Nagano, I.: Full particle simulation of whistler-mode rising chorus emissions in the magnetosphere. J. Geophys. Res. **114**, A01203 (2009). doi:10.1029/2008JA013625

Hikishima, M., Omura, Y., Summers, D.: Self-consistent particle simulation of whistler-mode triggered emissions. J. Geophys. Res. **115**, A12246 (2010a). doi:10.1029/2010JA015860

Hikishima, M., Omura, Y., Summers, D.: Microburst precipitation of energetic electrons associated with chorus wave generation. Geophys. Res. Lett. **37**, L07103 (2010b). doi:10.1029/2010GL042678

Istomin, Y.N., Karpman, V.I., Shklyar, D.R.: Contribution to the theory of triggered emissions. Geomag. Aeron. **16**, 67–69 (1976)

Karpman, V.I., Shklyar, D.: Nonlinear damping of potential monochromatic waves in an inhomogeneous plasma. Sov. Phys. JETP **35**(3), 500 (1972)

Karpman, V.I., Istomin, J.N., Shklyar, D.R.: Nonlinear theory of a quasi monochromatic whistler mode wave packet in inhomogeneous plasma. Plasma Phys. **16**, 685 (1974)

Katoh, Y., Omura, Y.: A study of generation mechanism of VLF triggered emission by self-consistent particle code. J. Geophys. Res. **111**, A12207 (2006). doi:10.1029/2006JA011704

Katoh, Y., Omura, Y.: Computer simulation of chorus wave generation in the Earth's inner magnetosphere. Geophys. Res. Lett. **34**, L03102 (2007). doi:10.1029/2006GL028594

Nunn, D.: Wave particle interactions in electrostatic waves in an inhomogeneous medium. J. Plasma Phys. **6**(2), 291–307 (1970)

Nunn, D.: The sideband instability of electrostatic waves in an inhomogeneous medium. Planet. Space Sci. **21**, 67–88 (1973)

Nunn, D.: A theoretical investigation of banded chorus. J. Plasma Phys. **11**(2), 189–212 (1974a)
Nunn, D.: A self-consistent theory of triggered VLF emissions. Planet. Space Sci. **22**, 349–378 (1974b)
Nunn, D.: A non-linear theory of sideband stability of ducted whistler mode waves. Planet. Space Sci. **34**(5), 429–451 (1985)
Nunn, D.: The numerical simulation of non-linear VLF wave-particle interactions using the Vlasov hybrid simulation technique. Comput. Phys. Commun. **60**, 1–25 (1990)
Nunn, D.: Vlasov hybrid simulation—a novel method for the numerical simulation of hot collision-free plasmas. J. Comput. Phys. **108**(1), 180–196 (1993)
Nunn, D., Omura, Y.: A computational and theoretical analysis of falling frequency VLF emissions. J. Geophys. Res. **117**, A08228 (2012). doi:10.1029/2012JA017557
Nunn, D., Rycroft, M.J.: A parametric study of triggered VLF emissions and the control of radio emission frequency sweep rate. Ann. Geophys. **23**, 1–12 (2005)
Nunn, D., Smith, A.J.: Numerical simulation of whistler-triggered VLF emissions observed in Antarctica. J. Geophys. Res. **101**(A3), 5261–5277 (1996)
Nunn, D., Omura, Y., Matsumoto, H., Nagano, I., Yagitani, S.: The numerical simulation of VLF chorus and discrete emissions observed on the Geotail satellite. J. Geophys. Res. **102**(A12), 27083–27097 (1997)
Nunn, D., Manninen, J., Turunen, T., Trakhtengerts, V., Erokhin, N.: On the non-linear triggering of VLF emissions by power line harmonic radiation. Ann. Geophys. **17**, 79–94 (1999)
Nunn, D., Santolik, O., Rycroft, M., Trakhtengerts, V.: On the numerical modelling of VLF chorus dynamical spectra. Ann. Geophys. **27**, 1–19 (2009)
Omura, Y., Nunn, D.: Triggering process of whistler mode chorus emissions in the magnetosphere. J. Geophys. Res. **116**, A05205 (2011). doi:10.1029/2010JA016280
Omura, Y., Furuya, N., Summers, D.: Relativistic turning acceleration of resonant electrons by coherent whistler-mode waves in a dipole magnetic field. J. Geophys. Res. **112**, A06236 (2007). doi:10.1029/2006JA012243
Omura, Y., Katoh, Y., Summers, D.: Theory and simulation of the generation of whistler-mode chorus. J. Geophys. Res. **113**, A04223 (2008). doi:10.1029/2007JA012622
Omura, Y., Hikishima, M., Katoh, Y., Summers, D., Yagitani, S.: Nonlinear mechanisms of lower-band and upper-band VLF chorus emissions in the magnetosphere. J. Geophys. Res. **114**, A07217 (2009). doi:10.1029/2009JA014206
Omura, Y., Nunn, D., Summers, D.: Generation processes of whistler mode chorus emissions: current status of nonlinear wave growth theory. In: Summers, D., Mann, I.R., Baker, D.N., Schulz, M. (eds.) Dynamics of the Earth's Radiation Belts and Inner Magnetosphere, Geophysical Monograph Series, vol. 199, pp. 243–254. AGU, Washington, DC (2012). doi:10.1029/2012GM001347
Santolik, O., Gurnett, D.A.: Transverse dimensions of chorus in the source region. Geophys. Res. Lett. **30**(2), 1031 (2003). doi:10.1029/2002GL016178
Santolik, O., Gurnett, D.A., Pickett, J.S.: Spatio-temporal structure of storm time chorus. J. Geophys. Res. **108**(A7), 1278 (2003). doi:10.1029/2002JA009791
Trakhtengerts, V.Y.: Magnetosphere cyclotron maser: BWO generator regime. J. Geophys. Res. **100**, 17205–17210 (1995)
Trakhtengerts, V.Y.: A Generation mechanism for chorus emission. Ann. Geophys. **17**, 95–100 (1999)

Chapter 4
Auroral Kilometric Radiation as a Consequence of Magnetosphere-Ionosphere Coupling

Robert J. Strangeway

Abstract Prof. Jim Dungey's Masters degree student Stephen Knight published a paper (Planet. Space Sci. 21:741–750, 1973) that explained how auroral flux tubes could carry enhanced upward currents through the acceleration of magnetospheric electrons by a parallel electric field. While the model was developed to explain field-aligned currents, the model was later found to be essential in understanding the generation of Auroral kilometric radiation (AKR), which is an intense radio emission generated above Earth's auroral zone. AKR is generated in the region of upward field-aligned current, and the field-aligned currents are required in order to transmit the stresses imposed by magnetospheric convection to the ionosphere. The accelerating potential, in concert with the magnetic mirror force, introduces perpendicular gradients in the electron distribution function, which has a horseshoe-like appearance. This horseshoe distribution is unstable to X-mode waves near the electron gyro-frequency, thereby generating AKR. Given the ubiquity of AKR-like radio emissions, it is likely that these emissions are a manifestation of coupling between different plasmas through field-aligned currents that require accelerating potentials to provide the current. Many of these emissions have fine structure, and this fine structure provides clues as to microphysical processes occurring in the source region of the emission.

R.J. Strangeway (✉)
Institute of Geophysics and Planetary Physics, University of California, Los Angeles, Los Angeles, CA, USA

Department of Earth, Planetary, and Space Sciences, University of California, Los Angeles, Los Angeles, CA 90095, USA
e-mail: strangeway@igpp.ucla.edu

D. Southwood et al. (eds.), *Magnetospheric Plasma Physics: The Impact of Jim Dungey's Research*, Astrophysics and Space Science Proceedings 41,
DOI 10.1007/978-3-319-18359-6_4

4.1 Introduction

In 1974 I joined Prof. Jim Dungey's group as a graduate student. He assigned to me the topic of investigating electron beam instabilities for a spatially confined beam moving along a uniform magnetic field. This work followed on from earlier work by another of Jim's students, David Elliott (1975). We investigated how a beam of finite width can support wave modes in addition to the standard cold plasma modes present because of the change in plasma properties inside and outside of the beam (Dungey and Strangeway 1976; Strangeway 1977). Because the beam density was large, of order 10 % of the background plasma, it was felt that the results better applied to laboratory experiments such as those carried out in a large vacuum facility by Bernstein et al. (1975). We did speculate that the results could apply to the auroral zone, possible even being responsible for the recently discovered kilometric radiation at the Earth (Gurnett 1974). Because this radiation was found to come from the Earth's auroral zone (Kurth et al. 1975) it became known as auroral kilometric radiation (AKR).

Auroral processes have long been an interest of Jim. Indeed one of the conclusions of Jim's seminal work on reconnection (Dungey 1961) was that reconnection could generate aurora. The aurora are associated with particles carrying field-aligned currents, and Jim Dungey's Masters student, Stephen Knight, developed a theory to explain how a parallel electric field could enhance the ability of a flux-tube to carry field-aligned current (Knight 1973). This work is the foundation of much of what we now know about auroral particle acceleration, and ties together the role of field-aligned currents in magnetosphere-ionosphere coupling and the generation of AKR.

While my work at Imperial was not specifically related to the generation of AKR, this was a topic of conversation among the group. One of the most intriguing facts about AKR is that it is observed to be propagating in the cold plasma R-X mode (Kaiser et al. 1978). The R-X mode is a wave that has phase speed faster than light, and also has no upper frequency cut-off (Boyd and Sanderson 1969). Landau resonance and the type of beam instabilities I was investigating as part of my thesis generate waves with phase velocities less than the speed of light, and it was by no means obvious how AKR could be generated. There were several ingenious theories that invoked mode conversion, some of which I will discuss later in this paper. But these were inherently inefficient. There appeared to be no satisfactory generation mechanism for AKR.

At this time we had an academic visitor from the University of Maryland, C. S. Wu. After his return to the United States he published a paper with L.C. Lee (Wu and Lee 1979) that essentially solved the AKR problem. Wu and Lee noted that in a low density plasma the gyro-resonance condition has to be modified to include relativistic effects. I remember discussing gyro-resonance as a source for AKR with the group at Imperial, but without the relativistic correction, growth is in general not possible for high phase velocity waves. I will show how we came to that conclusion before I discuss how Wu and Lee (1979) solved the problem.

The outline of the rest of the paper is as follows. In the next section I will show why field-aligned currents are an essential part of magnetosphere-ionosphere coupling, and then present the derivation the "Knight" relation (Knight 1973). I will also show that in situ observations of accelerated auroral electrons are consistent with the acceleration by a parallel electric field, assumed by Knight (1973). In Sect. 4.3 I will discuss wave propagation in a plasma, and why in the early 1970s we had difficulty in understanding how AKR could be generated directly. This section will also show the insight of Wu and Lee (1979) that gyro-resonance, which allows for direct coupling to the R-X mode, must include relativistic effects. Evidence that confirms auroral electrons are the source of AKR is presented in Sect. 4.4, where I also discuss an important modification to the Wu and Lee mechanism. Because the plasma density is very low in the auroral acceleration region, the wave dispersion itself must also include relativistic corrections. This allows more of the electron distribution to provide energy to AKR. In Sect. 4.5 I briefly discuss some of the remaining issues with the AKR generation process, and I provide concluding remarks in Sect. 4.6.

4.2 Magnetosphere-Ionosphere Coupling and the Knight Relation

Dungey (1961) first postulated that magnetospheric convection was driven by the reconnection of interplanetary and magnetospheric field lines. These reconnected field lines are connected to the solar wind and the ionosphere and are being transported anti-sunward by the solar wind [here we are making use of the frozen-in theorem that states that the magnetic field is frozen to the plasma (Alfvén 1943)]. As a consequence, the field lines that thread the ionosphere are also pulled anti-sunward by the solar wind flow. These field lines are hence transported over the polar cap to the nightside where they again reconnect. The newly closed field lines are in turn transported sunward to replace the magnetic flux eroded at the dayside by the reconnection there. Because the ionosphere is collisional, and has finite conductivity, if the ionosphere is made to move relative to the neutrals, then forces are required to overcome the friction. As shown below, the force in the ionosphere is provided by perpendicular currents, but since the ionosphere is dissipative, this requires the presence of field-aligned currents to connect currents flowing in the ionosphere to currents at the magnetopause and in the magnetosphere.

Parker (1996) and Vasyliūnas (2001) both emphasize that in order to understand dynamically coupled systems it is better to consider the forces and flows within a plasma (the "B, v" paradigm). This makes the plasma momentum equation the basis for understanding the system, and Strangeway and Raeder (2001) present a form of the plasma momentum equation that includes collisions:

$$n(m_i\nu_{in} + m_e\nu_{en})(\mathbf{U}_i - \mathbf{U}_n) - m_e\nu_{en}\frac{\mathbf{j}}{e} = \mathbf{F} + \mathbf{j}\times\mathbf{B}, \tag{4.1}$$

where n is the plasma density and we have assumed quasi-neutrality and only one ion species, m_i is the ion mass, ν_{in} is the ion-neutral collision frequency, m_e is the electron mass, ν_{en} is the electron-neutral collision frequency, $\mathbf{U}_i$ is the ion fluid flow velocity, $\mathbf{U}_n$ is the neutral gas flow velocity, $\mathbf{j}$ is the current density, e is the magnitude of the electron charge, and $\mathbf{B}$ is the magnetic field.

At first sight Eq. (4.1) appears to not have any time derivatives, and hence be steady state, but the term $\mathbf{F}$ includes the plasma inertia and non-electromagnetic forces. In order to simplify our discussion we shall ignore gravity, and assume isotropic plasma pressure, in which case

$$\mathbf{F} = -\nabla P - nm_i\, d\mathbf{U}_i/dt_i - nm_e\, d\mathbf{U}_e/dt_e, \tag{4.2}$$

where P is plasma pressure, and d/dt_i and d/dt_e are the total time derivatives with respect to the ion and electron fluids, respectively. It is these derivatives that allow for time variation in Eq. (4.1). We can further simplify the discussion by reducing Eq. (4.2) to a single fluid form. To do this we define the mass density as $\rho = n(m_i + m_e) \approx nm_i$, and the center of mass velocity as $\mathbf{U} = n(m_i\mathbf{U}_i + m_e\mathbf{U}_e)/\rho \approx \mathbf{U}_i$, where we have used the condition $m_e << m_i$ in the approximation. In addition, the current density is given by $\mathbf{j} = ne(\mathbf{U}_i - \mathbf{U}_e) \approx ne(\mathbf{U} - \mathbf{U}_e)$. In that case $d/dt_i \approx d/dt = \partial/\partial t + \mathbf{U}\cdot\nabla$ and $d/dt_e \approx d/dt - (\mathbf{j}/ne)\cdot\nabla$. The current density dependent term in the total time derivative for the electron fluid can be ignored since it is $O(m_e/m_i)$ smaller than the corresponding derivative for the ion fluid. We then rewrite $\mathbf{F}$ as

$$\mathbf{F} = -\nabla P - \rho d\mathbf{U}/dt. \tag{4.3}$$

Some simplifying assumptions were made by Strangeway (2012) in order to elucidate why field-aligned currents are a necessary component of magnetosphere-ionosphere coupling. First, the generalized Ohm's law [see, for example, Eq. (4.9) in Strangeway and Raeder (2001)] can be simplified by assuming that the electron fluid is frozen in, that is

$$\mathbf{E} + \mathbf{U}_e\times\mathbf{B} = \mathbf{E} + \mathbf{U}\times\mathbf{B} - \mathbf{j}\times\mathbf{B}/ne \approx 0, \tag{4.4}$$

where $\mathbf{E}$ is the electric field. Second, for the magnetosphere it is assumed that collisions can be ignored in the momentum equation, and

$$\rho d\mathbf{U}/dt = \mathbf{j}\times\mathbf{B} - \nabla P. \tag{4.5}$$

For the ionosphere, electron collisions are ignored, which also follows from assuming the electron frozen-in condition. Furthermore, the terms represented by $\mathbf{F}$ in Eq. (4.1) are assumed small. Hence

$$\rho\nu_{in}(\mathbf{U}_i - \mathbf{U}_n) = \mathbf{j}\times\mathbf{B}. \tag{4.6}$$

Both Eqs. (4.5) and (4.6) include the term $\mathbf{j}\times\mathbf{B}$, which only includes the current perpendicular to $\mathbf{B}$ since $\mathbf{j}\times\mathbf{B} = \mathbf{j}_\perp\times\mathbf{B}$. It is not immediately obvious how these two equations require field-aligned currents, but

$$\nabla\times(\mathbf{j}\times\mathbf{B}) = (\mathbf{B}\cdot\nabla)\mathbf{j} - (\mathbf{j}\cdot\nabla)\mathbf{B} \tag{4.7}$$

under the assumption $\nabla\cdot\mathbf{j} = 0$, which is a very good approximation for time-scales longer than electron plasma periods. The first term on the right-hand side of Eq. (4.7) includes the gradient in the field-aligned current. Thus the field-aligned current can be derived by taking the dot product of the curl of Eqs. (4.5) and (4.6) with $\mathbf{B}$, as done, for example in Strangeway (2012) (see also, Hasegawa and Sato 1979; Haerendel 2011). For the magnetosphere

$$(\mathbf{B}\cdot\nabla)\left(\frac{\mathbf{j}\cdot\mathbf{B}}{B^2}\right) = \frac{\mathbf{B}}{B^2}\cdot\left[2\left(\nabla P + \rho\frac{D\mathbf{u}}{Dt}\right)\times\frac{\nabla B}{B} + \nabla\times\left(\rho\frac{D\mathbf{u}}{Dt}\right)\right], \tag{4.8}$$

while for the ionosphere

$$\mathbf{B}\cdot(\mathbf{B}\cdot\nabla)\mathbf{j} - B\mathbf{j}\cdot\nabla B = \rho\nu_{in}\mathbf{B}\cdot(\boldsymbol{\omega}_i - \boldsymbol{\omega}_n) - \mathbf{B}\cdot[(\mathbf{u}_i - \mathbf{u}_n)\times\nabla\rho\nu_{in}], \tag{4.9}$$

where $\boldsymbol{\omega} = \nabla\times\mathbf{u}$ is the vorticity.

In order to determine the net field-aligned current on a flux-tube Eqs. (4.8) and (4.9) must be integrated over the regions where the gradient in the field-aligned current is non-zero. Thus Eq. (4.8) should be integrated along the flux tube down to the ionosphere, while Eq. (4.9) must be integrated through the ionosphere. Not addressed here is how the resultant field-aligned currents are arranged so that $\nabla\cdot\mathbf{j} = 0$ everywhere. Presumably as magnetospheric flow patterns change, for example, magnetohydrodynamics waves transfer information about these changes to the ionosphere, and the flows change in the ionosphere, but it is also likely that the magnetospheric flows have to in turn adapt to the ionospheric flows until a new equilibrium is reached.

Nevertheless, Eqs. (4.8) and (4.9) can be used to provide physical insight into the cause of field-aligned currents such as the classical Region-1 and -2 current system (Iijima and Potemra 1978). The higher-latitude Region-1 currents flow along field lines that map to the outer magnetosphere and the magnetopause. Here plasma flows are likely to dominate, especially for reconnected field-lines that are being dragged tailward by the solar wind, and the flow-velocity dependent terms in Eq. (4.8) dominate. The lower-latitude Region-2 currents map to the inner magnetosphere, and the thermal pressure terms in Eq. (4.8) are more probably responsible for maintaining field-aligned currents. In terms of the ionosphere, Region-1 currents are close to, or even within, the polar cap. Thus they are bounded by anti-sunward flows at higher latitudes, and return (sunward) flows at lower latitude. This flow shear has vorticity. By the same token, the Region-2 currents tend to confine the return flows to higher latitudes. Again we have a flow shear, but with the opposite sign of vorticity, and the Region-2 currents are of opposite sign to the higher latitude Region-1 currents for the same local time sector. The Region-1 currents are downward at dawn and upward at dusk, with Region-2 having the opposite polarity.

It is the presence of upward currents that leads to the need for the Knight relation. Field-aligned currents are typically carried by electrons, since the ions are much more massive. Upward current is carried by precipitating electrons. These electrons must come from the magnetosphere, where the plasma is relatively rarefied in comparison to the ionosphere. If we consider an isotropic Maxwellian, then the upward current carried by downward traveling electrons is

$$j_0 = n_0 e v_T / 2\pi^{1/2}, \tag{4.10}$$

where n_0 is the electron density for an isotropic Maxwellian occupying 4π steradians (at this stage we are not considering the effects of the loss-cone), v_T is the thermal velocity defined as $\frac{1}{2} m_e v_T^2 = k_B T$, m_e is the electron mass, and T is the temperature. For a density of 1 cm^{-3}, and a temperature of 1 keV, $j_0 \approx 0.85$ μA/m^2. If the field-aligned current required is greater than j_0, then some process must increase the flux into the atmosphere. In his paper Knight (1973) assumed that this was done through the presence of a parallel electrostatic potential that accelerated the precipitating electrons.

Knight (1973) used Liouville's theorem to determine how the particle distribution function is mapped along the magnetic field in the presence of an accelerating potential. Liouville's theorem states that the distribution function is constant following a particle trajectory in phase space. That trajectory is determined by the constants of the motion. For the case considered here, where the only forces are from the magnetic field and the potential, then the constants of the motion are the total energy and the magnetic moment. For electrons the total energy is given by $W - e\phi = W_0$, where W is the electron kinetic energy, ϕ is the potential, and W_0 is the kinetic energy at zero potential, chosen to be the top of the acceleration region. The magnetic moment is given by $\mu = W_\perp / B$, where $W_\perp$ is the perpendicular kinetic energy of the electron, and B is the magnitude of the magnetic field.

The constants of the motion also specify the trajectory in velocity space for the electrons. For precipitating electrons the parallel and perpendicular energy are related by

$$W_{||} + W_\perp (1 - B_0/B) = W_{||0} + e\phi, \tag{4.11}$$

which is an ellipse in the two-dimensional velocity space defined by $v_{||}$, $v_\perp$ (the use of a two-dimensional velocity space implicitly assumes that the electrons are gyrotropic, which is a reasonable assumption for auroral flux tubes where the acceleration occurs). The subscript "0" defines the values at the top of the acceleration region. Setting $W_{||0} = 0$ defines an "acceleration ellipse," and all electrons present from above the acceleration region must lie outside this ellipse. This is also the case for the electrons that have been reflected at lower altitudes by the magnetic mirror force. In addition, any upgoing ionospheric and backscattered electrons that lie within this ellipse will be reflected before they reach the top of the acceleration region.

We can also relate the local energy to the parallel energy and magnetic field at the ionosphere

$$W_{\|} + W_{\perp}(1 - B_I/B) = W_{\|I} + e(\phi - \phi_I), \tag{4.12}$$

where the subscript "I" refers to the ionosphere. This corresponds to a hyperbola in the 2-D velocity space, and the limiting case is found by setting $W_{\|I} = 0$. Electrons that lie outside of this limiting curve will be (or have been) reflected before they reach the ionosphere.

The division of velocity space into different regions implied by these results was depicted by Chiu and Schulz (1978), from which Fig. 4.1 is adapted. The red portion of velocity space includes those electrons that will precipitate into the ionosphere, while the magenta area corresponds to magnetospheric electrons that mirror before reaching the ionosphere. The light green area includes backscattered

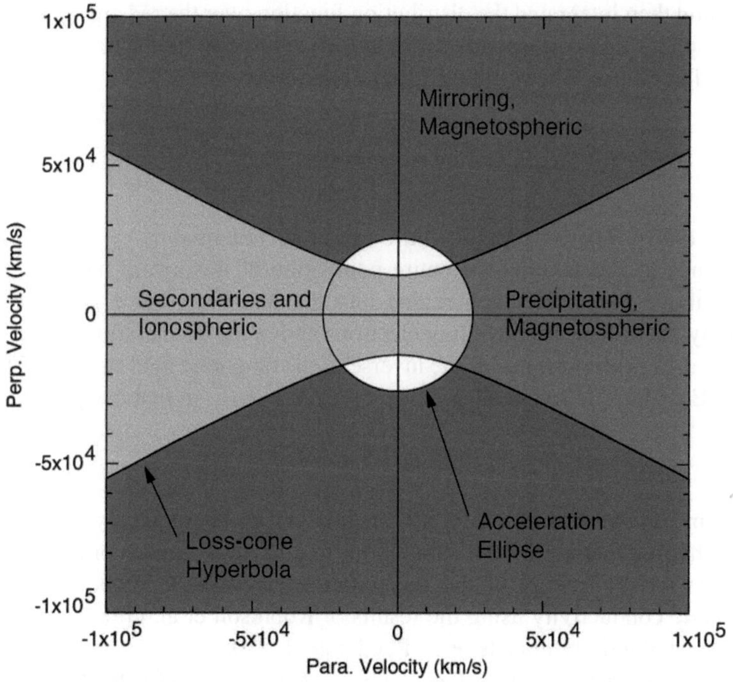

Fig. 4.1 The different regions accessible to electrons of either magnetospheric or ionospheric origin (after Chiu and Schulz 1978). In this figure positive parallel velocities correspond to downward moving electrons

secondaries and electrons of ionospheric origin, some of which are reflected above the altitude corresponding to the magnetic field value B, as used in Eqs. (4.11) and (4.12). The white region in the figure can contain trapped electrons, which may be present because of wave-particle interactions. But this region of velocity space is not directly accessible for particles coming from above the acceleration region or from the ionosphere.

The presence of the trapped region indicates an assumption has been made about how the potential ϕ varies as a function of B. In particular, what happens if the acceleration ellipse lies entirely within the loss-cone hyperbola at any position on the flux tube? In that case we would infer that the precipitating electron would be lost, as it is inside the loss-cone. But the condition for reflection is $W_{||} = 0$. Because these electrons are reflected, additional holes are created in the distribution function, beyond those implied by the boundaries shown in Fig. 4.1. This is equivalent to a statement that for any altitude along the flux tube, a point in velocity space that is identified as being accessible from either above or below is accessible everywhere along the flux tube. This reduces to the condition that $d\phi/dB > 0$ and $d^2\phi/dB^2 \leq 0$ (Chiu and Schulz 1978).

The separation of velocity space into different regions, and also the accessibility condition given above was implicit in Knight (1973). He used Liouville's theorem to map the distribution function from above the acceleration region to the ionosphere, and then integrated the distribution function over the red area in Fig. 4.1. At the ionosphere this corresponds to the positive velocity half-space outside of the acceleration ellipse. The resulting Knight relation is

$$j = j_0 \frac{B_I}{B_0} \left\{ 1 - \left(1 - \frac{B_0}{B_I} \right) \exp \left[\frac{-e\phi}{k_B T (B_I/B_0 - 1)} \right] \right\} \tag{4.13}$$

(Knight 1973). The asymptotic limit for large potential is $j = j_0 B_I / B_0$. This corresponds to accelerating the entire population of downward moving electrons present above the acceleration region into the loss-cone, since j_0 is the current carried by the downward travelling electrons above the acceleration region and the flux tube area is proportional to the inverse of the magnetic field strength. The other limit of Eq. (4.13) is for small ϕ, given by $e\phi << k_B T$. In that case

$$j \approx j_0 (1 + e\phi/k_B T). \tag{4.14}$$

This form, or the part that depends linearly on ϕ, is frequently used in global magnetohydrodynamic (MHD) simulations to provide estimates of the energy flux and characteristic energy of the precipitating electrons in order to modify the ionospheric conductivity using the results of Robinson et al. (1987), or using first principle conductivity models (e.g., Raeder et al. 2008).

Taking into account how the electron energization affects the ionospheric conductivity within global MHD codes is an important step in improving the models and how they address magnetosphere-ionosphere coupling, but there is a second aspect of the acceleration that is harder to include. This is the effect of the parallel potential, which means that the flows in the ionosphere are partially decoupled from the flows in the magnetosphere. Thus while Eqs. (4.8) and (4.9) can provide a context for understanding why the magnetosphere-ionosphere requires field-aligned currents, making quantitative statements about, say, the size of the field-aligned currents is more complicated, and almost certainly requires the inclusion of processes not encompassed in the MHD formalism.

Figure 4.2 shows particles and fields data acquired by the Fast Auroral Snapshot (FAST) Small Explorer (see Carlson et al. (1998) and Pfaff et al. (2001) for an overview of the FAST mission objectives and instrumentation). From 06:43:50 to 06:44:40, and 06:45:00 to 06:45:20, the electron energy spectra (middle panel) and the ion spectra (next panel) both show relatively narrow bands of enhanced differential energy flux. For the electrons this is around 4 keV, and for the ions initially near 1 keV. The width of the ion peak is much narrower than the peak for the electrons. While we have not shown the pitch angle data, the ion fluxes are confined to a very narrow beam along the magnetic field. Thus FAST is within the acceleration region, with electrons being accelerated into the ionosphere, and ions out of the ionosphere. Between these two intervals FAST passes below the

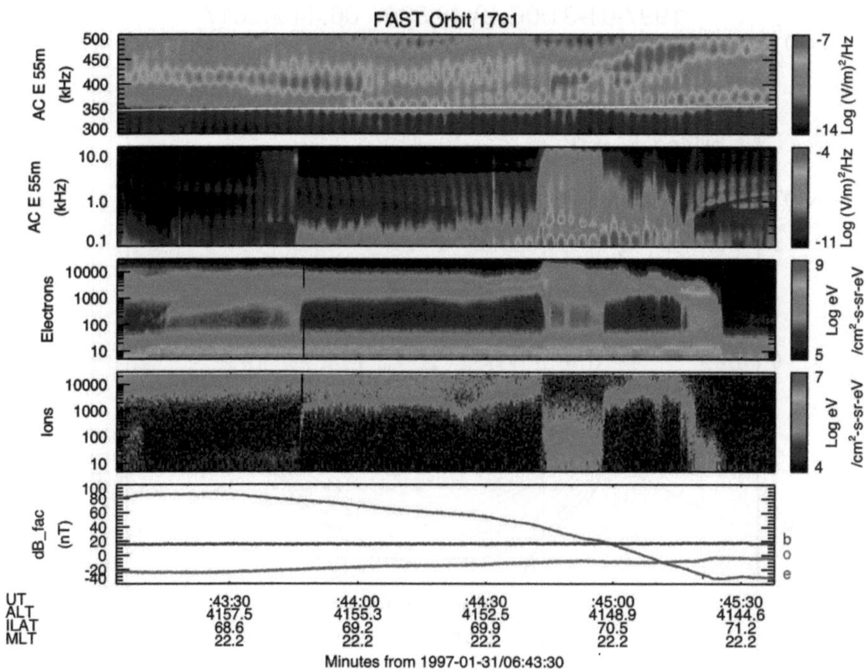

Fig. 4.2 Particle and fields data observed in the auroral acceleration region by FAST (after Strangeway et al. 1998). From *top* to *bottom* the *panels* show spin plane electric field data in AKR frequency range, spin plane VLF electric field data, electron differential energy flux spectra, ion differential energy flux spectra, and deviation of the magnetic field, with the *green* trace showing the eastward component. For FAST the spin-plane is near the spacecraft orbit plane, and at high latitudes the ambient magnetic field is close to this plane

acceleration region, as evidenced by the enhanced low energy electron fluxes, corresponding to backscattered secondaries. The ions have a conic distribution in pitch angle (not shown), and the fluxes are lower in energy, as expected for a transversely accelerated ion distribution.

The bottom panel in Fig. 4.2 shows the deviations of the magnetic field with respect to the International Geomagnetic Reference Field—Version 11 (*International Association of Geomagnetism and Aeronomy, Working Group V-MOD* 2010). The data have been cast into a field-aligned coordinate system with "b" (blue) along the model magnetic field, "o" (red) perpendicular to the background field in the meridian containing this field and the radius vector to the spacecraft, and "e" (green) eastward, completing the right-handed triad "o-e-b." The eastward magnetic field decreases as the spacecraft moves to higher latitudes, indicating an upward current. This is as expected for the electron acceleration region.

Figure 4.3 shows the phase space distribution for the electrons observed within the acceleration region, for the time interval shown in the title to the figure. This

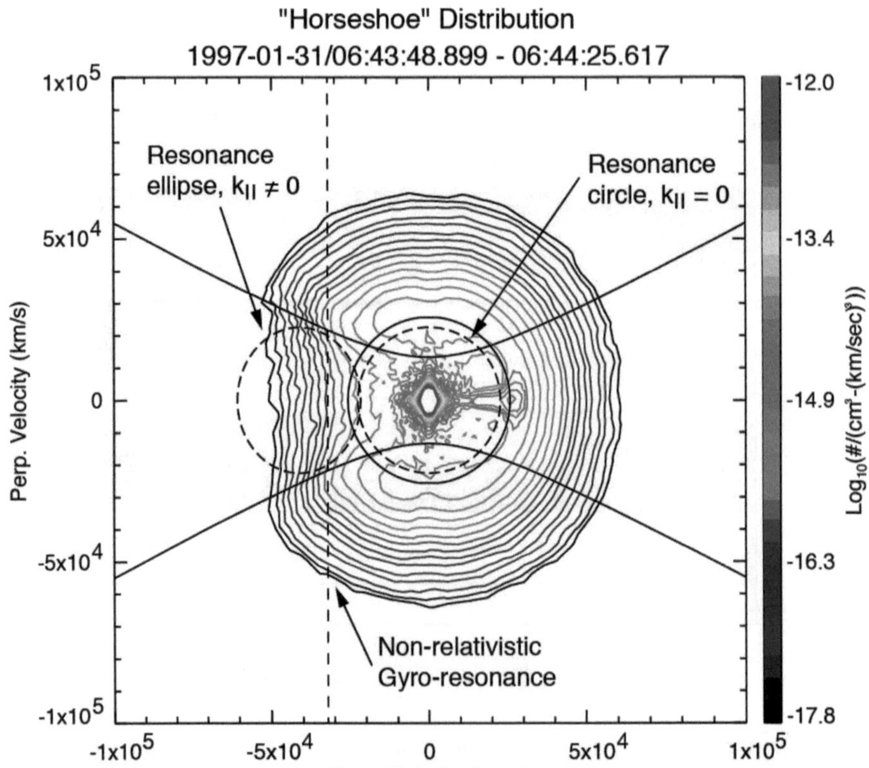

Fig. 4.3 Phase space density contours for the electrons observed in the auroral acceleration region shown in Fig. 4.2 (after Strangeway et al. 2001). The *solid lines* give the acceleration ellipse and loss-cone hyperbola as shown in Fig. 4.1. The *dashed lines* show different integration contours that are discussed further in Sect. 4.3

interval was chosen as the peak energy of the electrons was roughly constant over that interval. The acceleration ellipse and the loss-cone hyperbola are the same as shown in Fig. 4.1. In fact these curves were drawn based on the distribution shown in Fig. 4.3. The acceleration ellipse is close to a circle, indicating that the top of the acceleration region is at quite high altitudes such that $B_0/B << 1$.

While there are some additional features in the distribution, it is clear that it contains features that conform to the model used by Knight (1973). The distribution is referred to as a horseshoe distribution because of its shape in 2-D velocity space. The resultant gradients in the phase space distribution constitute a source of free energy for wave growth, be they parallel gradients that would generate waves through Landau resonance, or perpendicular gradients that would generate waves through gyro-resonance. The integration contours for gyro-resonance under different assumptions are indicated by the dashed lines. These will be discussed further in the next section.

4.3 The Auroral Kilometric Radiation Generation Mechanism

As noted in the introduction, AKR is an intense radio emission generated in the Earth's auroral zone that can escape from the Earth's magnetosphere. AKR has planetary counterparts, such as Jovian decametric radiation (DAM) (e.g., Warwick 1967), and Saturnian kilometric radiation (SKR) (e.g., Kaiser et al. 1980). Thus understanding the generation mechanism for AKR provides insight into the processes responsible for DAM and SKR.

That AKR and the planetary counterparts are observed as freely escaping wave modes raises questions about how the waves are generated. Figure 4.4 shows cold plasma dispersion curves for different propagation angles, θ, relative to the background field as a function of normalized wave number kc/Ω_e and wave frequency ω/Ω_e, where $\Omega_e = eB/m_e$ is the non-relativistic electron gyro-frequency, c is the speed of light, and m_e is the electron rest mass. The different cold-plasma wave modes are indicated with their usual names: the two faster than light modes that are unbounded (R-X and L-O), the L-X mode, the Z-mode (here we have made a distinction between the faster than light L-X mode and the slower Z-mode), and the whistler mode. For simplicity we have ignored the effect of ions at low frequencies. To make the mode identification easier we have assumed $\omega_{pe}/\Omega_e = 0.5$, where $\omega_{pe} = \sqrt{4\pi n_o e^2/\varepsilon_0 m_e}$ is the electron plasma frequency and ε_0 is the permittivity of free space. In the AKR source region this ratio can be much smaller.

The speed of light corresponds to the line of unit slope, and the other lines give the Landau resonance and gyro-resonance condition for an electron with a speed of $0.5c$. The general resonance condition is

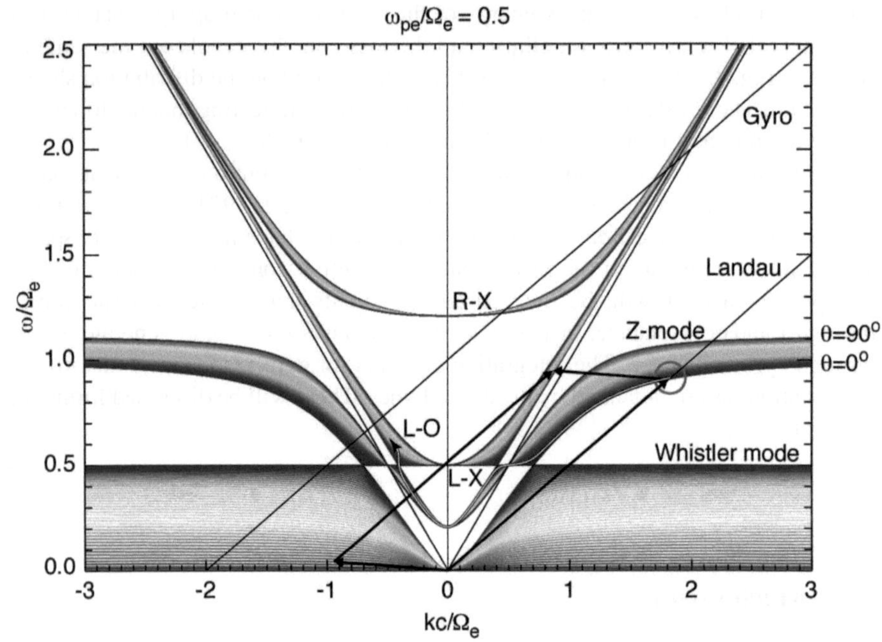

Fig. 4.4 Dispersion curves and associated wave modes for a cold plasma assuming $\omega_{pe}/\Omega_e = 0.5$. The wave number and wave frequency are normalized such that the speed of light corresponds to a line of unit slope. The *color-coding* shows the propagation angle, with *blue* corresponding to 0°, and *red* to 90°. The *curves* are plotted at two-degree intervals. The freely escaping R-X and L-O modes have phase speeds faster than light. The lines labeled "Landau" and "Gyro" show the respective resonance conditions for an electron moving at half the speed of light. The *grey circle* indicates where a parallel-propagating Z-mode wave is in Landau resonance, and the white and *black lines* with *arrowheads* show different mechanisms for coupling Landau-resonance generated waves to faster than light modes, as discussed in the text

$$\omega - k_{||}v_{||} = n\Omega_e/\gamma, \tag{4.15}$$

where $\gamma = 1/\sqrt{1 - v^2/c^2}$ is the Lorentz factor. In Eq. (4.15) $n = 0$ corresponds to Landau resonance, and $n = 1$ to gyro-resonance. In drawing the gyro-resonance line we have assumed $\gamma = 1$, for simplicity. We will return to the relativistic form of the gyro-resonance later.

In early work to explain AKR and DAM it was assumed that the auroral electrons were truly a beam of particles, and it was expected that the waves would be generated through Landau resonance. Figure 4.4 shows the difficulty with this, as electrons cannot be in Landau resonance with faster-than-light waves. Mode conversion is therefore required.

One approach was linear mode conversion (Oya 1974; Jones 1977), represented by the white line in the figure. The wave is first generated were the line giving the Landau resonance crosses the 0° Z-mode dispersion curve (for simplicity we are only considering parallel propagation for the resonant waves), marked by the grey circle in Fig. 4.4. The wave initially starts at this location and then follows the white line, propagating into regions of increasing magnetic field strength, and hence to lower ω/Ω_e, eventually passing through the speed of light to the L-X mode. As the wave continues to propagate, the wave encounters the L-X mode cut-off and is reflected. Assuming the reflected path is not the same as the incident path, it is possible that the wave could approach the L-X to L-O mode transition, and therefore passes through the Ellis window (Ellis 1956). (In his paper Ellis refers to the "second extraordinary, or Z, reflection level." This is the L-X mode cut-off in Fig. 4.4. (Again we note that we have chosen to limit the Z-mode appellation to the slower-than-light branch.) But this method of generating the radiation through linear mode conversion is inherently inefficient.

A more sophisticated mode-conversion process was suggested by Palmadesso et al. (1976). They suggested that AKR could be generated through beat resonance with a virtual mode. If the AKR frequency is ω with wave number $\mathbf{k}$, and low frequency, short wavelength, ion turbulence is present, with frequency ω_I and wave number $\mathbf{k}_I$, then the beat wave is given by $\omega_0 = \omega - \omega_I$, $\mathbf{k}_0 = \mathbf{k} - \mathbf{k}_I$. Because the ion turbulence has a low frequency, and the AKR wavelength is long, $\omega_0 \approx \omega$ and $k_0 \approx -k_I$. Palmadesso et al. argue that if the resultant beat wave is a normal mode for the plasma, then the coupling is very efficient. In this case, the grey circle marks the beat wave, and the coupling is shown schematically by the black arrows in Fig. 4.4. Note that we have exaggerated the frequency of the ion wave so as to make the sketch more readable. The ion wave and the beat wave have their wave vectors in opposite directions so that their sum is small.

The major drawback with both the linear-mode conversion and virtual-mode coupling is that they both generate waves on the L-O branch. As noted in the introduction, AKR is primarily observed as R-X radiation. There are reports of L-O mode AKR (e.g., Mellott et al. 1984). It is possible that the mode conversion processes outlined above could generate the L-O mode AKR, but it is also possible the L-O mode is generated through mode conversion of the primary R-X mode emission. But this still leaves the question as to how the primary R-X mode radiation is generated.

Figure 4.4 shows that it is possible for electrons to be in gyro-resonance with the R-X mode, and the corresponding electron energy is smaller (i.e., the slope of the gyro-resonance line in Fig. 4.4 is less) for smaller ω_{pe}/Ω_e ratios. However, prior to the work of Melrose (1976) and Wu and Lee (1979), it was considered unlikely that direct gyro-resonance would be effective, for reasons outlined below.

For simplicity we will consider a transverse mode that is propagating parallel to the ambient magnetic field. In that case, the waves are either R- or L-mode waves, and are right-hand or left-hand circularly polarized. Letting $\mathbf{E}_r$ represent the right-hand circularly polarized electric field that is varying as $\exp\left[-i\left(\omega t - k_{\|}z\right)\right]$ (i.e., a harmonic perturbation), then the current density induced by this electric field is

$$\mathbf{j}_r = -\frac{ie^2}{2m_e}\mathbf{E}_r\int d\mathbf{v}v_\perp \frac{\left[\left(1-\frac{k_\| v_\|}{\omega}\right)\frac{\partial f_0}{\partial v_\perp}+\frac{k_\| v_\perp}{\omega}\frac{\partial f_0}{\partial v_\|}\right]}{\left(\omega - k_\| v_\| - \Omega_e\right)}, \tag{4.16}$$

assuming no relativistic effects are included. We have also only considered electrons since we are considering high frequency modes. In Eq. (4.16) f_0 is the unperturbed phase space density, and f_0 is a function of parallel $\left(v_{||}\right)$ and perpendicular velocity $(v_\perp)$ only, with no dependence on gyrophase.

The denominator within the integral in Eq. (4.16) gives the non-relativistic gyro-resonance condition, and following the Landau prescription, the growth or damping rate for waves is given by integrating along a contour defined by $\omega - k_{||}v_{||} - \Omega_e = 0$. An example of such a contour is shown in Fig. 4.3, indicated by the vertical dashed line. For very high phase velocity waves the $\partial f_0/\partial v_\perp$ term dominates in Eq. (4.16). The integral over $v_\perp$ is

$$\int_0^\infty v_\perp^2 \frac{\partial f_0}{\partial v_\perp} dv_\perp = -2\int_0^\infty v_\perp f_0 dv_\perp \tag{4.17}$$

and this is negative regardless of the form of the distribution function. Since a Maxwellian must be stable, then a distribution such as that shown in Fig. 4.3 must also be stable, even though the distribution has a positive slope in $\partial f_0/\partial v_\perp$ that could be expected to contribute to growth. The contribution to the integral from the higher velocity tail of the distribution, which has a negative slope, always exceeds the positive contribution to the integral.

This argument is incorrect, however, because relativistic effects must be included. In that case Eq. (4.16) becomes

$$\mathbf{j}_r = -\frac{ie^2}{2m_e}\mathbf{E}_r \int d\mathbf{p}\, p_\perp \frac{\left[\left(1 - \frac{k_{||}p_{||}}{\omega\gamma}\right)\frac{\partial f_0}{\partial p_\perp} + \frac{k_{||}p_\perp}{\omega\gamma}\frac{\partial f_0}{\partial p_{||}}\right]}{\left(\omega\gamma - k_{||}p_{||} - \Omega_e\right)} \tag{4.18}$$

In Eq. (4.18) m_e is again the electron rest mass, and, as before, $\Omega_e = eB/m_e$. In addition, $\mathbf{p}$ is the momentum per unit rest mass, $\mathbf{p} = \gamma\mathbf{v}$. The denominator in Eq. (4.18) corresponds to the resonance condition, Eq. (4.15), with $n = 1$. This is an ellipse in 2-D velocity space,

$$\begin{aligned} &\left(\frac{v_{||}}{c} - \frac{ck_{||}\omega}{\left(c^2k_{||}^2 + \Omega_e^2\right)}\right)^2 \left(c^2k_{||}^2 + \Omega_e^2\right) + \frac{v_\perp^2}{c^2}\Omega_e^2 \\ &= \frac{\Omega_e^2}{\left(c^2k_{||}^2 + \Omega_e^2\right)}\left(c^2k_{||}^2 + \Omega_e^2 - \omega^2\right), \end{aligned} \tag{4.19}$$

an example of which is shown in Fig. 4.3. Because the resonance condition now restricts the growth-rate integral to finite values of $v_\perp$, the tail of the distribution, which would contribute to damping, may not have to be included. The distribution could therefore be unstable to gyro-resonance with the R-X mode. This was explored by Melrose (1976), in the context of Jovian decametric radiation. He used a drifting bi-Maxwellian, and found that very high temperature anisotropies (i.e., $T_\perp >> T_{||}$) were required.

The breakthrough was the work of Wu and Lee (1979), who recognized that under appropriate conditions the resonance ellipse could lie entirely within the loss-cone of a distribution such as that shown in Fig. 4.3. In that case the distribution function only has a positive slope in $v_\perp$, allowing for maximum growth of the waves. The primary condition is that the right-hand side of Eq. (4.19) be greater than zero. Wu and Lee assumed the wave dispersion was given by the standard cold plasma dispersion, with the energetic electrons only contributing to the growth of the waves. For a cold plasma the R-X mode cut-off is

$$\omega_{RX} = \frac{\Omega_e}{2}\left(1 + \sqrt{1 + 4\omega_{pe}^2/\Omega_e^2}\right). \tag{4.20}$$

From Eq. (4.20) $\omega_{RX} > \Omega_e$, and $ck_{||}$ must therefore be large enough that the right-hand side of Eq. (4.19) is positive. However, the semi-major axis of the resonance ellipse is given by

$$\frac{v_{\perp m}}{c} \approx \frac{ck_{||}}{\sqrt{c^2k_{||}^2 + \Omega_e^2}}. \tag{4.21}$$

Since we require the resonance ellipse to lie within the loss-cone of the distribution, we expect $v_{\perp m} \approx O(0.1c)$, which in turn implies $ck_{||} \approx O(0.1\Omega_e)$. This in turn places a limit on ω_{RXe}, as $\omega^2 - \Omega_e^2$ cannot be too large. This requires $\omega_{pe}^2/\Omega_e^2 << 1$, in which case $\omega_{RX} \approx \Omega_e\left(1 + \omega_{pe}^2/\Omega_e^2\right)$.

Wu and Lee (1979) pointed out that the potential that accelerates the electrons also reduces the plasma density as secondary and ionospheric electrons are reflected by the electric field. Thus the potential that makes the horseshoe distribution as shown in Fig. 4.3 also establishes the necessary low-density condition for the loss-cone instability to operate. There is just one problem with this scenario—one of the consequences of wave generation is to remove the gradients in the phase space density that led to wave growth. Thus generating waves through the loss-cone has the effect of filling the loss-cone. But it is the presence of the loss-cone itself that allows the plasma to carry field-aligned current.

The way out of this quandary is to allow $\omega < \Omega_e$, and then it is possible for $ck_{||} = 0$ in Eq. (4.19), and the resonance ellipse becomes a resonance circle, as shown in Fig. 4.3. One immediate advantage in terms of the macro-physical requirement that the plasma carry field-aligned current is the symmetry of any diffusion induced by the wave. Both upwards and downwards travelling electrons contribute to growth, and the current carried by the particles should be unaltered.

In order to allow the R-X mode cut-off to be below the cold (non-relativistic) electron gyro-frequency it is necessary to include relativistic effects not only in the resonance condition but also in the wave dispersion. In the 1980s several authors (e.g., Wu et al. 1982; Winglee 1983; Pritchett 1984a, b; Le Quéau et al. 1984a, b; Strangeway 1985, 1986; Le Queau and Louarn 1989) included relativistic effects using a variety of distributions, such as perpendicular ring distributions, Dory-Guest-Harris distributions (Dory et al. 1965), and spherical shell distributions. In addition, Pritchett and Strangeway (1985) presented numerical simulations using distributions that had features similar to the horseshoe distribution shown in Fig. 4.3. The inclusion of relativistic effects in the dispersion did allow for wave growth for perpendicular or nearly perpendicular propagation, and this was generally more efficient.

It was noted, however, that if both energetic and cold electrons were present, then the most unstable mode was a new X-mode wave that was essentially trapped between the energetic electron species gyro-frequency and the cold electron species R-X mode cut-off (e.g., Winglee 1983; Pritchett 1984b; Strangeway 1986). At the time these papers were published it was not clear if only energetic electrons were present in the cavity. If the low energy electrons were significant, then R-X mode wave generation for nearly perpendicular propagation had the same issue as the previously discussed hypotheses, namely that the waves had to be mode-converted to the freely propagating R-X mode. We will discuss the resolution of this issue in the next section.

4.4 Evidence that AKR Is Driven by Accelerated and Mirroring Electrons

As noted earlier, the prime objective for the FAST mission was a detailed exploration of the auroral acceleration region (Carlson et al. 1998; Pfaff et al. 2001). An example of the data acquired by FAST in the auroral acceleration region is shown in Fig. 4.2, and an averaged phase space electron distribution in Fig. 4.3. Similar phase space distributions were shown by Delory et al. (1998), indicating that the characteristic feature of the auroral electrons is indeed the horseshoe-like distribution that arises out of the acceleration process invoked by Knight (1973), and elucidated by Chiu and Schulz (1978).

The other aspect of the AKR generation process confirmed by FAST can also be seen in Fig. 4.2. The AKR signal is shown in the top panel of Fig. 4.2, with the white line marking the cold electron gyro-frequency, as determined from the measured magnetic field. In the intervals were the spacecraft is within the acceleration region, as indicated by the upward ion beam (i.e., from 06:43:50 to 06:44:40, and 06:45:00 to 06:45:20), the AKR signals drop below the non-relativistic electron gyro-frequency (see also, Ergun et al. 1998). This strongly indicates that within the acceleration region the wave dispersion is dominated by the accelerated electrons.

Confirmation that hot electrons are the major contributor to the plasma density was presented by Strangeway et al. (1998). The second panel in Fig. 4.2 shows VLF electric field data, where the most striking feature is the cut-off around 4 kHz when the spacecraft is within the acceleration region, as indicated by the narrow ion beam, and reduced electron secondaries. The VLF data show spin modulation, and the phase of the modulation is dependent on frequency. The FAST spacecraft orientation is such that when the spacecraft is at high latitudes the ambient magnetic field lies close to the spin plane, and a wave will show spin-modulation with the spin-phase of the maxima and minima in the signals depending on whether the wave electric field is mainly parallel or perpendicular to the ambient magnetic field. Strangeway et al. (1998) showed that the change in spin-phase is consistent with whistler-mode waves propagating on the resonance cone. (The resonance cone corresponds to the large wave number whistler-mode waves in Fig. 4.4, where the dispersion curves are essentially flat.) The upper frequency cut-off is therefore given by the plasma frequency, corresponding to a plasma density of 0.2 cm^{-3}. Calvert (1981) reported that the density within the cavity can be <1 cm^{-3}. Strangeway et al. analyzed some 28 cavities, comparing the density deduced from the VLF wave data and the density calculated from the electron measurements. They argued that their results were consistent with low energy electrons being almost entirely excluded from the acceleration region.

As pointed out by Strangeway et al. (2001), the free energy for the generation of AKR is created by the parallel electric field and magnetic mirror. This allows for AKR to be generated throughout the acceleration region. While the waves have the effect of removing free energy, the re-acceleration of these particles by the combination of the electric field and mirror force puts free energy back into the distribution. If, on the other hand, the loss-cone is the free energy source for the waves, then that cannot be replenished. A similar point was made by Louarn et al. (1990), who presented a case study from Viking observations of the auroral acceleration region at higher altitudes than FAST.

4.5 Remaining Issues

While the horseshoe-driven instability appears to be the primary means for generating AKR, there are still some remaining issues. Most notable of these is the fine structure of the emissions. Figure 4.5 shows one example of AKR fine structure. The top panel shows spectra generated from the Plasma Wave Tracker (PWT) data acquired by FAST. The PWT has a 16 kHz bandwidth that is mixed down to baseband with a mixing frequency that in this case is adjusted so the band includes the local electron gyro-frequency, as indicated by the white line. The bottom panel shows the electron energy spectra, indicating that FAST is in the auroral acceleration region (cf. Fig. 4.2, where the acceleration region is characterized by reduced low energy fluxes).

Inspection of Fig. 4.5 shows at least three different types of fine structure. The first is the slowly varying, almost linear, upper cut-off to the emissions near 345 kHz (note that the 1-s period notch in this cut-off is an artifact of the PWT).

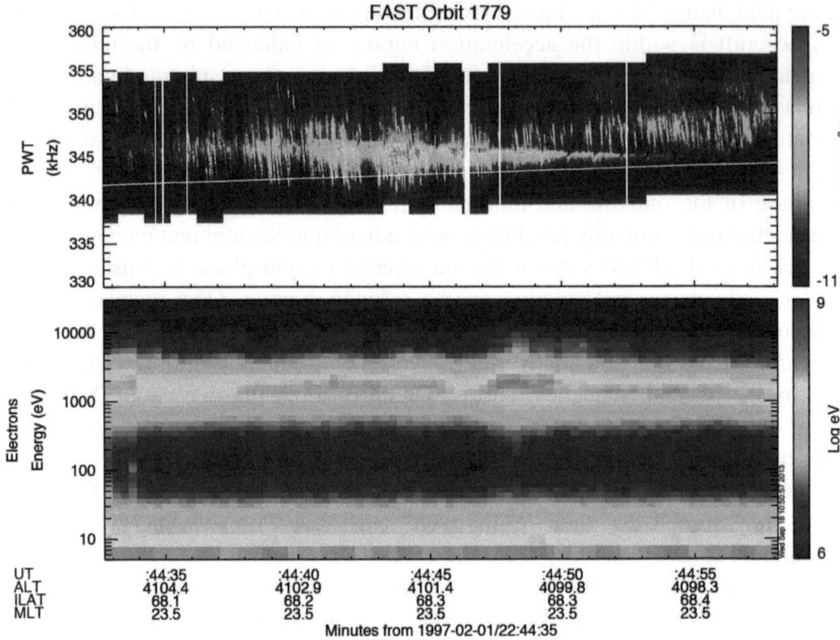

Fig. 4.5 AKR fine structure as observed by FAST in the auroral acceleration region. The *top panel* shows spectra generated by Fourier transforming the 32,768 samples per second Plasma Wave Tracker (PWT) data. The base frequency of the PWT is shifted so that the 16 kHz bandwidth of the PWT captures the local electron gyro-frequency, shown by the *white line*. The notches in the PWT spectra that occur at about 1-s intervals, most notably around 22:45:50, are an artifact. The *lower panel* shows the electron energy spectra

It is likely that this cut-off is a propagation effect. The second type of fine structure is the rapidly falling tones. The frequency drift rate for these tones is ~50 kHz/s. The final type of fine structure is simply the burstiness of the signal, as evidenced by the speckling within the spectra.

Not apparent in the figure is any evidence for very narrow tones. Through Fourier transforms of the autocorrelation function Baumback and Calvert (1987) argued that the bandwidth of AKR could be as narrow as 5 Hz, indicating that AKR requires a true lasing mechanism, where the waves are reflected between the edges of the auroral cavity. This conclusion was based on observations far from the source region, but detailed analysis of the AKR waveforms acquired within the source region shows that AKR is generated in packets, and these packets have intrinsic time scales of a few ms (Strangeway et al. 2001), corresponding to a bandwidth of a few 100 Hz. Simulations by Pritchett et al. (1999, 2002) indicate that the natural bandwidth of AKR should be slightly larger than this, so some additional mechanism is required to give narrow bandwidths, but the lasing mechanism that produces extremely narrow tones does not appear to be necessary.

Assuming the frequency of the falling tones corresponds roughly to the local gyro-frequency at the source region, then the sources are moving away from the

Earth with a velocity of the order 500 km/s. This would correspond to a proton energy of ~1.3 keV, comparable with the ion beam energy. Mutel et al. (2006) have invoked ion holes propagating along the field, the potential structure of which perturbs the electron distribution so as to create a burst of radiation. It has also been argued that ion acoustic waves can become non-linear, generating electrostatic shocks that perturb the electrons within the auroral acceleration region (Pottelette et al. 2003). In addition, interactions with electron holes have been invoked (Pottelette et al. 2001; Treumann et al. 2011) to explain drifting tones, especially tones that increase in frequency with time.

Understanding the processes responsible for AKR fine structure therefore tells us something about the micro-structure of the auroral acceleration region. In his work, Knight (1973) made no assumptions about the structure of the potential within the acceleration region, other than the requirement that $d^2\phi/dB^2 \leq 0$. If small scale structures, such as electrostatic shocks are indeed present within the auroral acceleration region, then the electron phase space structure may be more complicated than that implied by Fig. 4.1. But observations such as those shown in Fig. 4.3 suggest that the wave processes have smoothed out any fine scale structure, and the current density given by Eq. (4.13) may still be a reasonable estimate.

4.6 Concluding Remarks

As noted in the introduction I was a graduate student with Prof. Jim Dungey in the 1970s. My work was mainly concerned with the stability of narrow field-aligned electron beams. AKR was discovered around that time, and for several years the generation mechanism was a puzzle. The resolution of the problem came from a synthesis of the work of Jim Dungey's Masters student, Stephen Elliott, and the insight of C. S. Wu, who visited Imperial College while I was a graduate student. Knight (1973) showed how a parallel electric field on an auroral flux tube could enhance the ability of that flux tube to carry field-aligned current. Wu and Lee (1979) showed how features in the auroral distribution function, most notably the loss-cone, could generate AKR. It is the presence of the loss-cone together with the acceleration of precipitating electrons that allows the flux tube to carry upward current. A further refinement of the theory, were relativistic corrections affect the wave dispersion (e.g., Pritchett 1984a, b; Pritchett and Strangeway 1985; Strangeway 1985,1986) in addition to the gyro-resonance condition, has demonstrated that the accelerated and mirroring electrons are the primary free energy source for AKR (Strangeway et al. 2001).

In this paper I have emphasized the role of field-aligned currents in magnetosphere—ionosphere coupling. Understanding the generation of AKR has potential consequences for any object that emits radiation, if that object consists of a coupled system such as the magnetosphere—ionosphere system of the Earth. At Jupiter and Saturn, for example, the moons are a source of material that becomes ionized. The ionized material initially has the Keplerian velocity associated with orbital motion, but Jupiter and Saturn are rapid rotators, and the corotation speed where the material is ionized is much faster than the Keplerian velocity. This addition of material is known as mass loading, and the planetary rotation is communicated to the mass-loaded plasma through currents. In this case the source of the currents is the ionosphere, rather than the magnetosphere. But again, the ability of a flux tube to carry current away from the ionosphere may be restricted by the density of the available magnetospheric electrons to carry the current. For example, Ray et al. (2009) have investigated the current-voltage relation for the case of mass loading in the Jovian magnetosphere.

The ubiquity of the AKR-like emission process has been discussed for the solar system by Zarka (1998), and for astrophysical objects by Ergun et al. (2000) and Treumann (2006). Based on the knowledge gleaned from terrestrial observations, this type of radio emission is related to the presence of accelerating potentials that arise out of the need for electric current to be carried between two different plasma regimes. The need for that current in turn arises out of differential motion between the two plasmas, with one plasma driving flows in the other. At the Earth this is ultimately driven by reconnection between the IMF and the Earth's magnetic field, i.e., externally driven. At other objects this may be because of mass loading and differential rotation, which is an internally driven process. As a closing remark, we can use the detailed structure of these emissions as an indicator of microphysical processes occurring within the source region.

Acknowledgements This work was supported by the THEMIS project, Principal Investigator V. Angelopoulos, through NASA contract NAS5-02099.

References

Alfvén, H.: On the existence of electromagnetic-hydrodynamic waves. Arkiv Mat. Astron. Fys. **29B**(2), 1–7 (1943)

Baumback, M.M., Calvert, W.: The minimum bandwidth of auroral kilometric radiation. Geophys. Res. Lett. **14**, 119–122 (1987)

Bernstein, W., Leinbach, H., Cohen, H., Wilson, P.S., Davis, T.N., Hallinan, T., Baker, B., Martz, J., Zeimke, R., Huber, W.: Laboratory observations of RF emissions at ω_{pe} and $(n + ½)\omega_{ce}$ in electron beam-plasma and beam-beam interactions. J. Geophys. Res. **80**(31), 4375–4379 (1975). doi:10.1029/JA080i031p04375

Boyd, T.J.M., Sanderson, J.J.: Plasma Dynamics. Nelson, London (1969)

Calvert, W.: The auroral plasma cavity. Geophys. Res. Lett. **8**, 919–921 (1981)

Carlson, C.W., Pfaff, R.F., Watzin, J.G.: The Fast Auroral Snapshot (FAST) mission. Geophys. Res. Lett. **25**, 2013–2016 (1998)

Chiu, Y.T., Schulz, M.: Self-consistent particle and parallel electrostatic field distributions in the magnetospheric-ionospheric auroral region. J. Geophys. Res. **83**, 629–642 (1978)

Delory, G.T., Ergun, R.E., Carlson, C.W., Muschietti, L., Chaston, C.C., Peria, W., McFadden, J. P., Strangeway, R.: FAST observations of electron distributions within AKR source regions. Geophys. Res. Lett. **25**, 2069–2072 (1998)

Dory, R.A., Guest, G.E., Harris, E.G.: Unstable electrostatic plasma waves propagating perpendicular to a magnetic field. Phys. Rev. Lett. **14**(5), 131–133 (1965)

Dungey, J.W.: Interplanetary magnetic field and the auroral zones. Phys. Rev. Lett. **6**(2), 47–48 (1961)

Dungey, J.W., Strangeway, R.J.: Instability of a thin field-aligned electron beam in a plasma. Planet. Space Sci. **24**, 731–738 (1976)

Elliott, D.T.: The ducting of wave energy by field-aligned current sheets. Planet. Space Sci. **23**, 751–761 (1975)

Ellis, G.R.: The Z propagation hole in the ionosphere. J. Atmos. Terr. Phys. **8**, 43–54 (1956)

Ergun, R.E., Carlson, C.W., McFadden, J.P., Mozer, F.S., Delory, G.T., Peria, W., Chaston, C.C., Temerin, M., Elphic, R., Strangeway, R., Pfaff, R., Cattell, C.A., Klumpar, D., Shelley, E., Peterson, W., Moebius, E., Kistler, L.: FAST satellite wave observations in the AKR source region. Geophys. Res. Lett. **25**, 2061–2064 (1998)

Ergun, R.E., Carlson, C.W., McFadden, J.P., Delory, G.T., Strangeway, R.J., Pritchett, P.L.: Electron-cyclotron maser driven by charged-particle acceleration from magnetic field-aligned electric fields. Astrophys. J. **538**, 456–466 (2000)

Gurnett, D.A.: The Earth as a radio source: terrestrial kilometric radiation. J. Geophys. Res. **79**, 4227–4238 (1974)

Haerendel, G.: Six auroral generators: a review. J. Geophys. Res. **116**, A00K05 (2011). doi:10.1029/2010JA016425 [Printed 117(A1), 2012]

Hasegawa, A., Sato, T.: Generation of field aligned current during substorm. In: Akasofu, S.-I. (ed.) Dynamics of the Magnetosphere, pp. 529–542. D. Reidel, Dordrecht (1979)

Iijima, T., Potemra, T.A.: Large-scale characteristics of field-aligned currents associated with substorms. J. Geophys. Res. **83**, 599–615 (1978)

International Association of Geomagnetism and Aeronomy, Working Group V-MOD. Participating members, Finlay, C.C., Maus, S., Beggan, C.D., Bondar, T.N., Chambodut, A., Chernova, T.A., Chulliat, A., Golovkov, V.P., Hamilton, B., Hamoudi, M., Holme, R., Hulot, G., Kuang, W., Langlais, B., Lesur, V., Lowes, F.J., Lühr, H., Macmillan, S., Mandea, M., McLean, S., Manoj, C., Menvielle, M., Michaelis, I., Olsen, N., Rauberg, J., Rother, M., Sabaka, T.J., Tangborn, A., Tøffner-Clausen, L., Thébault, E., Thomson, A.W.P., Wardinski, I., Wei, Z., Zvereva, T.I.: International Geomagnetic Reference Field: the eleventh generation. Geophys. J. Int. **183**, 1216–1230 (2010). doi:10.1111/j.1365-246X.2010.04804.x

Jones, D.: Mode-coupling of Z-mode waves as a source of terrestrial kilometric and Jovian decametric radiations. Astron. Astrophys. **55**, 245–252 (1977)

Kaiser, M.L., Alexander, J.K., Riddle, A.C., Pearce, J.B., Warwick, J.W.: Direct measurements by Voyagers 1 and 2 of the polarization of terrestrial kilometric radiation. Geophys. Res. Lett. **5**, 857–860 (1978)

Kaiser, M.L., Desch, M.D., Warwick, J.W., Pearce, J.B.: Voyager detection of nonthermal radio emission from Saturn. Science **209**(12), 1238–1240 (1980)

Knight, S.: Parallel electric fields. Planet. Space Sci. **21**, 741–750 (1973)

Kurth, W.S., Baumback, M.K., Gurnett, D.A.: Direction-finding measurements of Auroral Kilometric Radiation. J. Geophys. Res. **80**, 2764–2770 (1975)

Le Queau, D., Louarn, P.: Analytical study of the relativistic dispersion: application to the generation of the auroral kilometric radiation. J. Geophys. Res. **94**(A3), 2605–2616 (1989). doi:10.1029/JA094iA03p02605

Le Quéau, D., Pellat, R., Roux, A.: Direct generation of the auroral kilometric radiation by the maser synchrotron instability—an analytical approach. Phys. Fluids **27**, 247–265 (1984a)

Le Quéau, D., Pellat, R., Roux, A.: Direct generation of the auroral kilometric radiation by the maser synchrotron instability: physical mechanism and parametric study. J. Geophys. Res. **89** (A5), 2831–2841 (1984b)

Louarn, P., Roux, A., de Féraudy, H., Le Quéau, D., André, M., Matson, L.: Trapped electrons as a free energy source for the auroral kilometric radiation. J. Geophys. Res. **95**(A5), 5983–5995 (1990)

Mellott, M.M., Calvert, W., Huff, R.L., Gurnett, D.A., Shawhan, S.D.: DE-1 observations of ordinary mode and extraordinary mode auroral kilometric radiation. Geophys. Res. Lett. **11**, 1188–1191 (1984). doi:10.1029/GL011i012p01188

Melrose, D.B.: An interpretation of Jupiter's decametric radiation and the terrestrial kilometric radiation as direct amplified gyroemission. Astrophys. J. **207**, 651–662 (1976)

Mutel, R.L., Menietti, J.D., Christopher, I.W., Gurnett, D.A., Cook, J.M.: Striated auroral kilometric radiation emission: a remote tracer of ion solitary structures. J. Geophys. Res. **111**, A10203 (2006). doi:10.1029/2006JA011660

Oya, H.: Origin of Jovian decameter wave emissions—conversion from the electron cyclotron plasma wave to the ordinary mode electromagnetic wave. Planet. Space Sci. **22**, 687–708 (1974)

Palmadesso, P., Coffey, T.P., Ossakow, S.L., Papadopoulos, K.: Generation of terrestrial kilometric radiation by a beam-driven electromagnetic instability. J. Geophys. Res. **81**, 1762–1770 (1976)

Parker, E.N.: The alternative paradigm for magnetospheric physics. J. Geophys. Res. **101**, 10587–10625 (1996). doi:10.1029/95JA02866

Pfaff, R., Carlson, C., Watzin, J., Everett, D., Gruner, T.: An overview of the Fast Auroral Snapshot (FAST) satellite. Space Sci. Rev. **98**, 1–32 (2001)

Pottelette, R., Treumann, R.A., Berthomier, M.: Auroral plasma turbulence and the cause of auroral kilometric radiation fine structure. J. Geophys. Res. **106**(A5), 8465–8476 (2001)

Pottelette, R., Treumann, R.A., Berthomier, M., Jasperse, J.: Electrostatic shock properties inferred from AKR fine structure. Nonlinear Process. Geophys. **10**, 87–92 (2003)

Pritchett, P.L.: Relativistic dispersion and the generation of auroral kilometric radiation. Geophys. Res. Lett. **11**, 143–146 (1984a)

Pritchett, P.L.: Relativistic dispersion, the cyclotron maser instability, and auroral kilometric radiation. J. Geophys. Res. **89**(A10), 8957–8970 (1984b)

Pritchett, P.L., Strangeway, R.J.: A simulation study of kilometric radiation generation along an auroral field line. J. Geophys. Res. **90**(A10), 9650–9662 (1985)

Pritchett, P.L., Strangeway, R.J., Carlson, C.W., Ergun, R.E., McFadden, J.P., Delory, G.T.: Free energy sources and frequency bandwidth for the auroral kilometric radiation. J. Geophys. Res. **104**(A5), 10317–10326 (1999). doi:10.1029/1998JA900179

Pritchett, P.L., Strangeway, R.J., Ergun, R.E., Carlson, C.W.: Generation and propagation of cyclotron maser emissions in the finite auroral kilometric radiation source cavity. J. Geophys. Res. **107**(A12), 1437 (2002). doi:10.1029/2002JA009403

Raeder, J., Larson, D., Li, W., Kepko, E.L., Fuller-Rowell, T.: OpenGGCM simulations for the THEMIS mission. Space Sci. Rev. **141**, 535–555 (2008)

Ray, L.C., Su, Y.-J., Ergun, R.E., Delamere, P.A., Bagenal, F.: Current-voltage relation of a centrifugally confined plasma. J. Geophys. Res. **114**, A04214 (2009). doi:10.1029/2008JA013969

Robinson, R.M., Vondrak, R.R., Miller, K., Dabbs, T., Hardy, D.: On calculating ionospheric conductances from the flux and energy of precipitating electrons. J. Geophys. Res. **92**, 2565–2569 (1987)

Strangeway, R.J.: Further results from a model for a thin field-aligned electron beam in a plasma. Planet. Space Sci. **25**, 795–797 (1977)

Strangeway, R.J.: Wave dispersion and ray propagation in a weakly relativistic electron plasma: implications for the generation of auroral kilometric radiation. J. Geophys. Res. **90**(A10), 9675–9687 (1985)

Strangeway, R.J.: On the applicability of relativistic dispersion to auroral zone electron distributions. J. Geophys. Res. **91**(A3), 3152–3166 (1986)

Strangeway, R.J.: The relationship between magnetospheric processes and auroral field-aligned current morphology. In: Keiling, A., Donovan, E., Bagenal, F., Karlsson, T. (eds.) Auroral Phenomenology and Magnetospheric Processes: Earth And Other Planets. American Geophysical Union, Washington, DC (2012). doi:10.1029/2012GM001211

Strangeway, R.J., Raeder, J.: On the transition from collisionless to collisional magnetohydrodynamics. J. Geophys. Res. **106**(A2), 1955–1960 (2001). doi:10.1029/2000JA900116

Strangeway, R.J., Kepko, L., Elphic, R.C., Carlson, C.W., Ergun, R.E., McFadden, J.P., Peria, W. J., Delory, G.T., Chaston, C.C., Temerin, M., Cattell, C.A., Möbius, E., Kistler, L.M., Klumpar, D.M., Peterson, W.K., Shelley, E.G., Pfaff, R.F.: FAST observations of VLF waves in the auroral zone: evidence of very low plasma densities. Geophys. Res. Lett. **25**, 2065–2068 (1998)

Strangeway, R.J., Ergun, R.E., Carlson, C.W., McFadden, J.P., Delory, G.T., Pritchett, P.L.: Accelerated electrons as the source of auroral kilometric radiation. Phys. Chem. Earth (C) **26**, 145–149 (2001)

Treumann, R.A.: The electron–cyclotron maser for astrophysical application. Astron. Astrophys. Rev. **13**, 229–315 (2006). doi:10.1007/s00159-006-0001-y

Treumann, R.A., Baumjohann, W., Pottelette, R.: Electron-cyclotron maser radiation from electron holes: upward current region. Ann. Geophys. **29**, 1885–1904 (2011). doi:10.5194/angeo-29-1885-2011

Vasyliūnas, V.M.: Electric field and plasma flow: what drives what? Geophys. Res. Lett. **28**(11), 2177–2180 (2001). doi:10.1029/2001GL013014

Warwick, J.W.: Radiophysics of Jupiter. Space Sci. Rev. **6**, 841–891 (1967)

Winglee, R.M.: Interrelation between azimuthal bunching and semirelativistic maser cyclotron instabilities. Plasma Phys. **25**, 217–235 (1983)

Wu, C.S., Lee, L.C.: A theory of the terrestrial kilometric radiation. Astrophys. J. **230**, 621–626 (1979)

Wu, C.S., Wong, H.K., Gorney, D.J., Lee, L.C.: Generation of the auroral kilometric radiation. J. Geophys. Res. **87**(A6), 4476–4499 (1982)

Zarka, P.: Auroral radio emissions at the outer planets: observations and theories. J. Geophys. Res. **103**(E9), 20159–20194 (1998). doi:10.1029/98JE01323

Chapter 5
A Simulation Study of the Relationship Between Tail Dynamics and the Aurora

Maha Ashour-Abdalla

Abstract The Earth's magnetotail goes through large scale changes during geomagnetically disturbed times that result in the earthward propagation of dipolarization events in the magnetotail containing highly energetic particles and a heightened electron precipitation into the ionosphere. During the 11 March 2008 substorm, earthward-propagating dipolarization fronts were observed by THEMIS spacecraft in the near-Earth magnetotail plasma sheet, and auroral brightening was observed by all-sky cameras at high northern latitudes around 70° in the pre-midnight sector. Using large-scale kinetic simulations along with spacecraft and ground-based observations, the properties (location, flux, energy, etc.) of precipitating particles were determined. We find that ~keV electrons in the region modeled by the simulation precipitate into the pre-midnight sector between about 68° and 72° due to two different physical mechanisms. Precipitation at higher latitudes is due to non-adiabatic pitch angle scattering that occurs at about 20–25 R_E down tail, eastward of the reconnection region. The lower latitude precipitation is due to Fermi acceleration, which causes adiabatic electrons to enter the loss cone closer to the Earth, 10–15 R_E downtail of the dipolarization front. The location, timing, and energy of electrons precipitating via these two mechanisms are in good agreement with all-sky camera images of auroral brightening observed at substorm onset.

5.1 Introduction

It has only been 80 years since Chapman and Ferraro (1931) first suggested that the Earth's magnetic field would be confined to a cavity in the interplanetary medium, the magnetosphere. Observations (Hoffmeister 1943; Biermann 1950) and Alfvén's

M. Ashour-Abdalla (✉)
Institute of Geophysics and Planetary Physics, UCLA, Los Angeles, CA, USA

Department of Physics and Astronomy, UCLA, Los Angeles, CA, USA
e-mail: abdalla@physics.ucla.edu

D. Southwood et al. (eds.), *Magnetospheric Plasma Physics: The Impact of Jim Dungey's Research*, Astrophysics and Space Science Proceedings 41,
DOI 10.1007/978-3-319-18359-6_5

hypothesis on the role of magnetic field draping in forming cometary tails (Alfven 1957) laid the ground work for Parker's theoretical explanation of a magnetized solar wind (Parker 1958). This was soon followed by observations in 1963 by Interplanetary Monitoring Platform (IMP)-1 confirming the existence of the magnetosphere and the magnetotail. Two hypotheses were put forth to explain the coupling of the magnetosphere with the solar wind: Axford and Hines (1961) suggested a viscous interaction driving magnetospheric circulation, and Dungey (1961), based on work done on solar flares (Giovanelli 1947, 1948; Hoyle 1949; Dungey 1958), postulated reconnection-driven convection as the engine of the magnetosphere. Both processes appear to be involved in the generation of geomagnetic activity (Baker et al. 1996) with the reconnection process being much more important for the substorm than the viscous interaction (Cowley 1982). Since then, there have been detailed measurements from numerous satellites of the different plasma regions of the magnetosphere and they show a very complex and dynamic system (Frank 1985).

Figure 5.1a shows Dungey's illustration of the interaction of the solar wind with the Earth's magnetic field (Dungey 1961). Dungey noted that when the interplanetary magnetic field (IMF) was southward, reconnection would occur at the subsolar point. Magnetic flux would then be convected across the polar cap and reconnection would occur once again behind the Earth, returning the flux back to the dayside. The advent of supercomputers in the 1980s enabled the use of numerical magnetohydrodynamic (MHD) models to simulate the dynamics of the solar wind interaction with the Earth's magnetosphere in three dimensions. In particular, our group at UCLA has carried out global MHD simulations to study magnetic reconnection at the dayside magnetopause, the structure of the distant geomagnetic tail, and the ionospheric convection pattern (Berchem et al. 1995; Raeder et al. 1995; El-Alaoui et al. 2010). The magnetic field lines from a simulation with constant southward IMF are shown in Fig. 5.1b. This simulation was run with a purely southward IMF of −5 nT, a solar wind velocity along the Sun-Earth line of 500 km/s, a density of 5 cm^{-3} and a thermal pressure of 4.9 nPa. Solar wind field lines are in red, field lines which are closed at both ends are blue, and interplanetary field lines that are connected to the Earth at one end are colored yellow. The striking similarities between these simulated field lines and the original Dungey schematic are clearly evident, as are the similarities between the Dungey convection patterns (Fig. 5.2a) and ionospheric equipotentials obtained from the MHD simulation (Fig. 5.2b). Both Fig. 5.2a, b show a two cell convection pattern (Heppner and Maynard 1987).

One vital consequence of the reconnecting magnetosphere is the convection of the footpoints of the field lines in the ionosphere. The progression of reconnected flux tubes following reconnection is represented by the numbered field lines in Fig. 5.3 (Hughes 1995). The inset shows the corresponding motion of the footpoints in the ionosphere. Auroral generation mechanisms were suggested as early as 1900 (Birkeland 1900), but it was after satellite observations in the magnetosphere were available that it was recognized that these footpoints would be where energetic ions and electrons would impact the ionosphere, resulting in the Aurora Borealis (e.g., Feldstein and Starkov 1967; Lui and Anger 1973; Lui et al. 1975). As we now

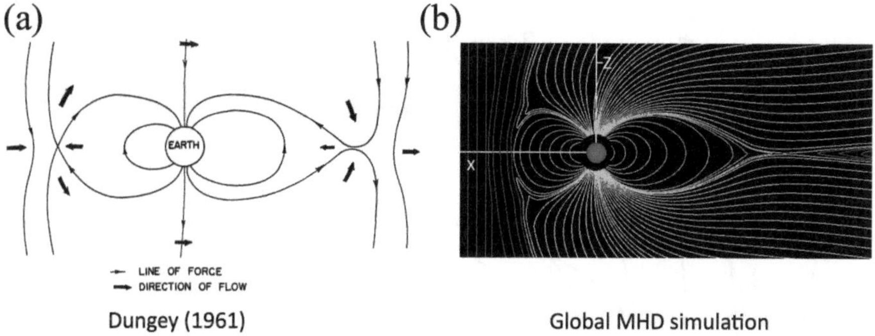

Fig. 5.1 Dungey's illustration of the interaction of the solar wind with Earth's magnetic field is shown in panel (**a**). The interplanetary magnetic field (IMF) is southward. The magnetic field lines from a simulation with constant southward IMF are shown in panel (**b**). Solar wind field lines are in *red*, field lines which are closed at both ends are *blue*, and interplanetary field lines that are connected to the Earth at one end are colored *yellow*

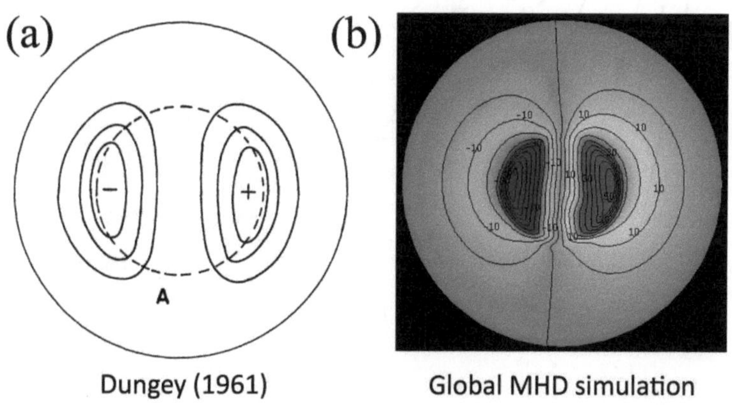

Fig. 5.2 Dungey's convection patterns for southward IMF are shown in panel (**a**). The ionospheric equipotentials obtained from the MHD simulation with constant southward IMF are shown in panel (**b**). Both (**a**, **b**) show a two cell convection pattern

know, the aurorae are most interesting and dynamic during substorms (e.g. Akasofu 1976). Even though the beautiful images we see are at the footpoints, these are really the signatures of processes happening in the magnetotail, processes that are driven by Dungey's reconnection and convection. In this paper, we will consider some of the ionospheric consequences of a reconnecting magnetosphere and tie particle acceleration in the magnetotail to their auroral signatures.

The visible aurora at high latitudes provide a window into the dynamic plasma processes that occur in the Earth's magnetosphere. Although most researchers agree that particle flow, field-aligned currents, and electromagnetic fields from the magnetotail drive auroral dynamics, a direct connection between processes in the magnetotail 10 to 25 R_E from the Earth and the low altitude auroral zone is still the

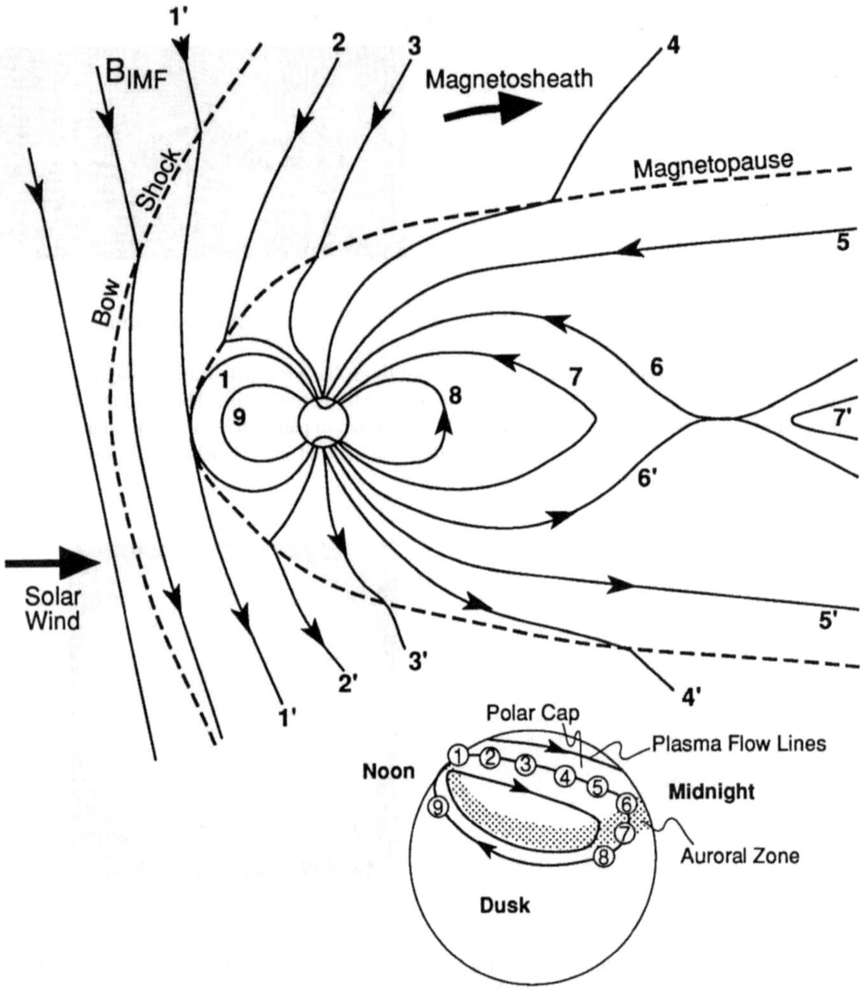

Fig. 5.3 Schematic showing the progression of reconnection-driven convection. A magnetospheric field line (1) reconnects with a solar wind field line (1′), creating an open field line (2–2′) that then convects across the polar cap (3–3′ through 5–5′). Oppositely directed open field lines reconnect in the magnetotail (6–6′), creating a closed field line (7) and an IMF field line (7′). The closed field line convects toward the dayside (8, 9), and returns the magnetic flux to the dayside. The ionospheric footpoints of the field lines shown in the schematic are shown in the *inset* (Hughes 1995)

subject of intense research. Theories suggested to explain auroral dynamics involve large-scale parallel electric fields, double layers, solitary structures, Alfvén waves, and various types of wave-particle interactions (Borovsky 1993; Schriver et al. 2003).

It has been known for some time that field-aligned precipitating electrons with keV energies cause discrete aurora (McIlwain 1960; Hoffman and Evans 1968;

Hultqvist 1971; Frank and Ackerson 1971; Rees and Luckey 1974; Evans 1974; Christensen et al. 1987) and it is a reasonable hypothesis that such electron precipitation ultimately results from processes occurring in the magnetotail (e.g., Frank 1985; Lyons et al. 1999). The basic idea is that earthward-propagating currents, particles, and waves that originate in the magnetotail transfer energy into field-aligned, accelerated electrons in the auroral zone at relatively low altitudes (<10,000 km). It is well established that quasi-static (dc) parallel electric fields (inverted V structures and parallel potential drops) occur in regions of field-aligned current (Mozer et al. 1977; Elphic et al. 1998; Ergun et al. 2000; Nakamura et al. 2001; Kepko et al. 2009). While small scale auroral structure is probably related to low altitude processes, the broader scale distribution of auroral precipitation is governed by field-aligned current systems that map into the magnetotail (e.g., Haerendel 1989). The existence of a magnetotail driver for kinetic Alfvén wave auroral acceleration is supported by observations of intense Poynting flux detected in a region above, but magnetically connected to the discrete auroral precipitation region (Wygant et al. 2000; Keiling et al. 2001; Shay et al. 2011).

In this paper we present a scenario for auroral precipitation during disturbed times in which precipitating electrons of ~few keV energy are scattered into the loss cone in the magnetotail current sheet, relatively far from Earth between $x_{GSM} \sim -10$ and -25 R_E. This paper follows on the work of Ashour-Abdalla et al. (2011a, 2013) where we studied a different event on 15 February 2008. Results are discussed from a substorm event observed on 11 March 2008 in which we model the relationship between enhanced auroral brightening and dipolarization fronts that propagate earthward from magnetotail reconnection regions. Using large-scale kinetic electron particle tracing within a global magnetohydrodynamic (MHD) model of this event, it is shown that ~keV electrons, precipitating between about 68° and 72° latitude in the pre-midnight sector, map directly to a relatively wide region of the magnetotail current sheet mid-plane between about 10 and 20 R_E from the Earth. These electrons enter the loss cone due to two distinctly different physical processes occurring at different magnetotail locations. Earthward of the reconnection region at about $x \sim -20$ to -25 R_E, we find that some electrons that started with relatively large pitch angles (well outside the loss cone) experience non-adiabatic motion when mirror bouncing through the current sheet in a region where the magnetic field bends sharply. Such electrons are stochastically pitch angle scattered into the loss cone (Zelenyi et al. 1990), and consequently precipitate into the ionosphere at latitudes >70° in the pre-midnight sector. A very different physical mechanism that leads to the direct precipitation of electrons from the magnetotail at latitudes <70° involves Fermi acceleration. These precipitating electrons start at about 20 R_E with relatively small pitch angles (~10°), but outside of the loss cone, and thus, they have fairly long bounce paths. Convection transports these electrons earthward with each subsequent bounce through the current sheet. As they move onto shorter geomagnetic field lines, conservation of the second adiabatic invariant ($I = \int p_{||} ds$, where the parallel momentum $p_{||}$ is integrated along the path length s over a mirror bounce) leads to a gradual increase in parallel energy. Therefore, the pitch angle gradually decreases until the electron enters the loss cone earthward of about 15 R_E in the magnetotail and precipitates into the ionosphere. Electrons from these two processes precipitate at locations that are in general agreement with ground-based all-sky camera observations of auroral brightening during this event.

In Sect. 5.2 of this paper, the substorm event on 11 March 2008 and the simulation model are described. In Sect. 5.3, simulation results are presented and compared to all-sky camera images. Summary and conclusions are found in Sect. 5.4.

5.2 Substorm Characteristics and Simulation Model

During the substorm on 11 March the THEMIS satellites in the near-Earth tail observed several earthward-propagating dipolarization fronts. Three of the THEMIS satellites were grouped between $x_{\mathrm{GSM}} = -15\ \mathrm{R_E}$ and $x_{GSM} = -10\ \mathrm{R_E}$ in Geocentric Solar Magnetospheric (GSM) coordinates. They were near the equator with $y_{GSM} \sim 5\ \mathrm{R_E}$ duskward of midnight. During the substorm expansion phase, THEMIS P4 (Fig. 5.4) detected two dipolarization fronts, one at 6:23:40 UT and the second one at 6:24:40 UT. The magnetic field is shown in panel (a) and the red line

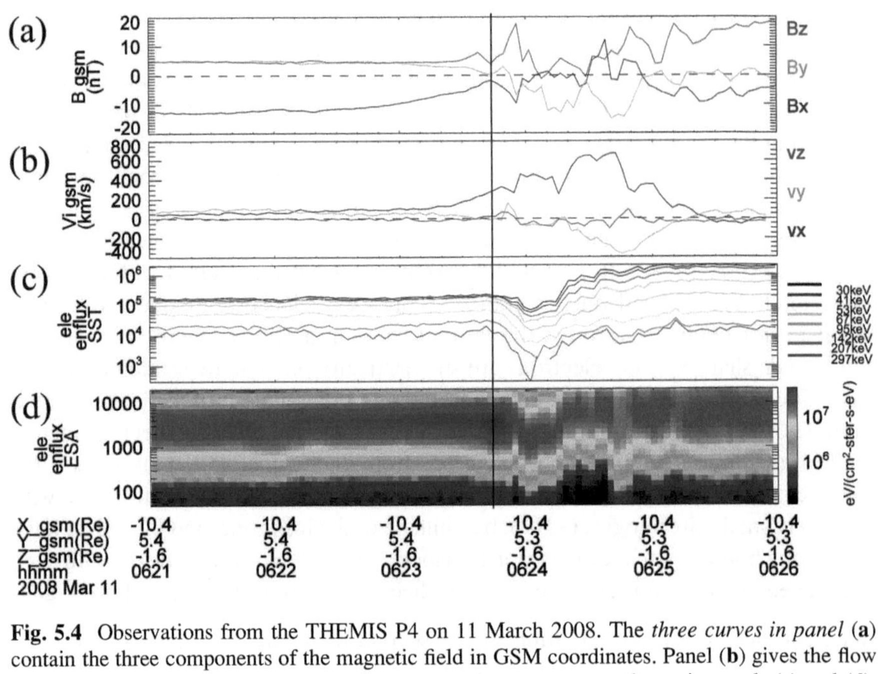

Fig. 5.4 Observations from the THEMIS P4 on 11 March 2008. The *three curves in panel* (**a**) contain the three components of the magnetic field in GSM coordinates. Panel (**b**) gives the flow velocity. The energy fluxes from the SST and ESA instruments are shown in panels (**c**) and (**d**), respectively. The first dipolarization front is highlighted by a *black vertical line*

(B_z in GSM coordinates) shows a series of dipolarization fronts. The first dipolarization front is highlighted by a black vertical line. Dipolarization fronts are characterized by rapid increases in B_z associated with strong earthward flows that in this case exceeded 600 km/s. In panels (c) and (d), we plot the energy flux from the Solid State Telescope (SST) which measures energetic particles and a spectrogram of the thermal energy particles from the Electrostatic Analyzer (ESA). Following the passage of the dipolarization front (black line) the energy flux of both the thermal electrons and the energetic electrons decrease. About 20 s later (around 0624:00 UT), the energy flux for electrons with energies >10 keV increases dramatically. All-sky images from several different stations at high northern latitudes show enhanced auroral brightening in the pre-midnight sector at latitudes within a few degrees of 70°, starting at about 06:24:00 UT.

We modeled the solar-wind magnetosphere interaction for this event using a global magnetohydrodynamic (MHD) simulation driven by observed upstream solar wind conditions (Raeder et al. 1995; El-Alaoui 2001). For this event, we used Geotail solar wind observations from just outside the bow shock to drive the global MHD simulation. The earthward-propagating dipolarization fronts observed by the THEMIS satellites were seen in the MHD simulations and are shown in Fig. 5.5. This figure shows results from the MHD simulation at four different times when dipolarization was occurring, from 06:22:20 to 06:24:00 UT. The color coding in Fig. 5.5 shows the north-south field component (B_z) in the maximum pressure surface in the magnetotail. Flow vectors are superimposed in black. The locations of the three THEMIS spacecraft (P2, P3, and P4) are also shown. The white contours give an indication of the location of the neutral line. The multiple dipolarization fronts (DFs) formed in front of the earthward flow [red regions propagating Earthward, e.g., at 06:23:00, panel (b)]. These DFs can be seen moving inwards with time and reaching P3 and P4, the last of which reaches P4 at 06:24:00.

Since MHD simulations approximate the plasma as a single ion-dominated fluid, electron kinetic effects are not included. Thus, to study electron particle dynamics in the magnetotail we used the large scale kinetic (LSK) technique. In a LSK simulation, a large number of charged particle trajectories are followed in the time-dependent electric and magnetic fields obtained from the MHD simulations (Ashour-Abdalla et al. 1997; Birn et al. 1998; El-Alaoui et al. 1998).

The electron trajectories were followed by using a method that switches between full particle and guiding center equations of motion where the local conditions determine which set of motion equations to use. This method has been discussed in detail in previous studies (e.g., Ashour-Abdalla et al. 1993, 2011a; Birn et al. 2004; Schriver et al. 1998, 2011). The adiabaticity parameter κ, defined as the square root of local magnetic field radius of curvature divided by the local gyroradius of the electron (Büchner and Zelenyi 1986), is used to determine when and where to switch between guiding center and full particle orbit calculations. In the simulations here, when $\kappa < 10$, the full particle orbit calculation is used; otherwise, the guiding center equations are used. In guiding-center motion there is no gyrophase information, but the phase angle is needed when initializing full particle motion. This need is handled by assuming a random phase angle when making the switch from

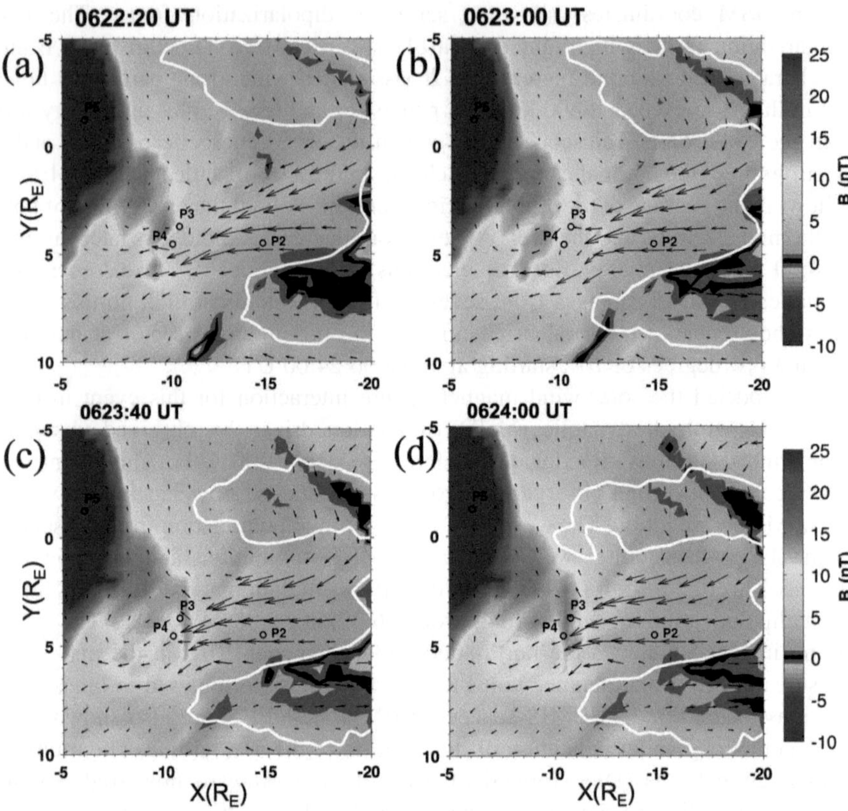

Fig. 5.5 MHD simulation for 11 March 2008. Plots of B_z on the maximum pressure surface with superimposed flow vectors. Locations of the four THEMIS satellites are given by the *circles*. *White contours* show the flow reversal region. Dipolarization fronts of interest cross P3 at 0623:40 [panel (**c**)] and P4 at 0624:00 [panel (**d**)]

guiding-center to full particle motion. For this study, we used an electron launch scheme similar to that in Ashour-Abdalla et al. (2011a). Specifically, we launched 3.75×10^4 particles with a 1 keV energy Maxwellian distribution every 20 s from 06:13:00 UT to 06:28:40 UT (1.80 million particles in total) in the plasma sheet earthward of the neutral line. The launch locations were spread evenly between $X = -16$ to -18 R_E, $Y = -1.0$ to -5.0 R_E, $Z = -2.5$ R_E. These launch locations are in the center of the current sheet, and earthward of the reconnection site ($x \sim -20$ to -25 R_E).

Reconnection occurs continuously until the end of the run (06:28:40 UT), though the rate and location varies. However, the location is always tailward of the launch region. Our previous studies examined the high energy enhancement in the near-Earth plasma sheet associated with the dipolarization front during this event (Ashour-Abdalla et al. 2011b; Pan et al. 2012). In this study we concentrate on the electrons that are lost from the magnetotail and precipitate into the ionosphere.

The inner boundary of the simulation is a spherical shell at a radial distance of 3.5 R_E. Particles that would have mirrored earthward of 100 km in a dipole magnetic field are considered to have precipitated. The distribution of those particles mapped along the dipole to 100 km are plotted in the precipitation plots.

5.3 Auroral Precipitation: Simulation and Data Results

Results from the LSK simulations for the precipitating electrons at 06:24:00 UT are shown in Fig. 5.6. Panel (a) shows a region of electron precipitation in the pre-midnight sector between about 68° and 72° geomagnetic latitude with a peak energy flux of 3×10^6 eV/cm^2-s-eV-sr. The average particle energy of these electrons is 5–10 keV. In panel (b) of Fig. 5.6, we show the locations where the corresponding particles last crossed the maximum pressure surface before precipitating. To determine where the precipitating electrons came from in the magnetotail, they were mapped back to their last crossing of the maximum pressure surface (e.g., Ashour-Abdalla et al. 2002) and collected in 0.5 R_E by 0.5 R_E planar bins, with the corresponding energy flux of each bin shown by using a gray scale, as shown in panel (b). The electron crossing locations are superimposed on color contours of the MHD magnetic field B_z component, with arrows that show the MHD bulk plasma flow direction at 06:24:00 UT on 11 March 2008. The precipitating electrons were found to originate from two distinct regions in the equatorial magnetosphere, as indicated by the red and blue circled regions in panel (b). These regions lie between about $x \sim -20$ R_E and $x \sim -10$ R_E. We will show that the electrons from these regions precipitate due to two different physical mechanisms. The electrons enclosed by the blue circle precipitate from the deeper magnetotail at ~−20 to −15 R_E due to non-adiabatic stochastic pitch angle scattering in the region earthward of the neutral line (Zelenyi et al. 1990). A typical electron of this type is shown in Fig. 5.7. Panel (a) shows the trajectory projected onto the *XZ* plane, while panel (c) shows the three-dimensional trajectory. Both plots are color-coded in energy. The time history of the electron has been plotted in panel (c). The particle pitch angle is plotted in blue with the scale on the left, and the particle κ is plotted in red with the scale on the right. Recall that the kappa parameter, κ, is defined as the square root of the radius of magnetic field curvature R_c divided by the gyroradius ρ, i.e., $\kappa = (R/\rho)^{1/2}$ at the local position of the particle (Büchner and Zelenyi 1986). The kappa parameter is a measure of a particle's adiabaticity, with smaller values, i.e., $\kappa < 5$ indicating that its motion is quasi- or non-adiabatic, without a well-defined gyro-center (Büchner and Zelenyi 1986, 1987; Chen and Palmadesso 1986; Ashour-Abdalla et al. 1990; Delcourt et al. 1995). If $\kappa > 10$, then generally the particle motion is highly adiabatic and follows guiding center equations of motion (Northrop 1963; Baños 1967). Both traces give the values as the particle crosses the equatorial current sheet and both show a decreasing trend. In panel (a) of Fig. 5.7 the electron trajectory is sharply kinked at the current sheet crossings, indicating the smaller magnetic field

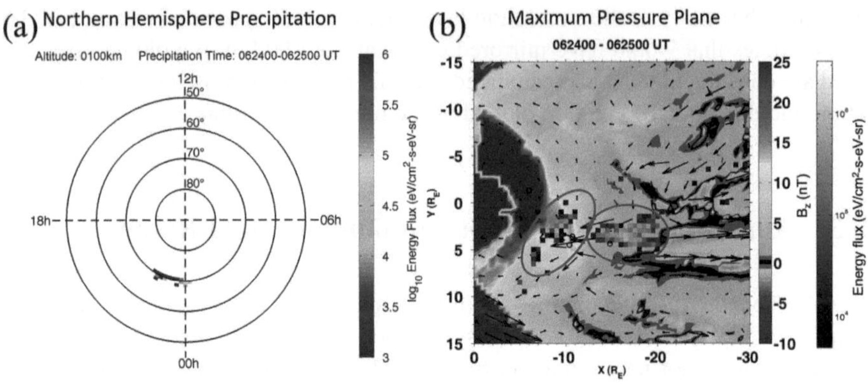

Fig. 5.6 Results from the LSK simulations for the precipitating electrons at 06:24:00 UT. Panel (**a**) shows the location of the precipitating electrons on a polar plot that are color-coded in energy flux. The locations where the corresponding particles last crossed the maximum pressure surface before precipitating are shown in panel (**b**). The *red* and *blue ovals* differentiate electrons that precipitate due to two different physical mechanisms

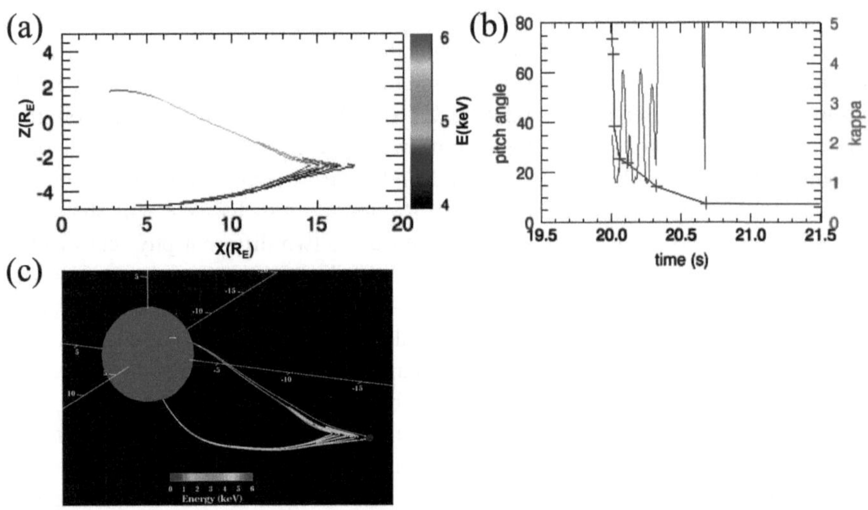

Fig. 5.7 The trajectory of an electron experiencing non-adiabatic stochastic pitch angle scattering is shown in panels (**a**) and (**c**). Panel (**b**) displays the lower envelopes of time histories of the pitch angle in *blue* (*scale on left*) and kappa parameter $\kappa = (R/\rho)^{1/2}$ in *red* (*scale on right*), where R is the magnetic field curvature and ρ is the gyroradius

curvature radius at those positions. This electron initially has a relatively large pitch angle (>45°), and thus, it has a relatively short bounce path between magnetic mirror points. The electron κ is about 10 at this time, but as the electron position in the current sheet drifts in y to a region of weaker magnetic field, κ decreases, and on subsequent current sheet crossings, stochastic pitch angle scattering occurs, which further decreases the pitch angle. By the time κ decreases to <5, the pitch angle has been reduced to <1°, at which time the electron is within the loss cone and precipitates. This occurs at about 15 R_E downtail, and the particle precipitates into the ionosphere at ~72° latitude and is lost from the system. In general, non-adiabatic pitch angle scattering can either increase or decrease the pitch angle, and only a few percent of the launched electrons experience a decrease in pitch angle such that they precipitate. Non-adiabatic electrons whose overall pitch angles increased remained in the system and contributed to the heated central plasma sheet (Ashour-Abdalla et al. 2011a).

Precipitating electrons that enter the loss cone closer to the Earth, i.e., $x > -15$ R_E and are circled in red in Fig. 5.6b, behave very differently. One such electron is illustrated in Fig. 5.8. The formats of panels (a) and (c) are the same as in Fig. 5.7. The time history of the second adiabatic invariant I, path length s, and pitch angle is in panel (b), and the time history of the particle energy (total, parallel and perpendicular) is in panel (d). This particle initially started near the neutral line earthward of -20 R_E. It has a relatively small pitch angle to begin with (~8°) and has a long bounce path between mirror points that are relatively close to the Earth [panels (a) and (c)]. Even near the neutral line, it did not pass through a region where κ became small. Particle motion is highly adiabatic with $\kappa > 50$ (not shown). In panel (b), the second invariant of motion I, defined as the parallel momentum integrated over a bounce path length, i.e., $I = \int p_{||} ds$, is well conserved. As the electron convects earthward, it moves onto shorter field lines and its path length decreases, causing the parallel momentum to increase. This increase can be seen in panel (d) of Fig. 5.8 where the parallel energy increases gradually with time due to Fermi acceleration. When coupled with a relatively small increase in the perpendicular energy due to betatron acceleration, this results in an ever decreasing pitch angle as the particle mirror bounces and moves closer to the Earth. Eventually the electron enters the loss cone earthward of -12 R_E causing it to precipitate at a relatively lower latitude (<70°) than the non-adiabatically scattered particle discussed in Fig. 5.7. Note that the loss cone gets wider with decreasing distance to the Earth, thus allowing particles with larger pitch angles to precipitate.

The locations determined by the simulations for all of the electrons that precipitate show good correspondence in latitude and local time with the auroral brightening during the substorm event. Figures 5.9 and 5.10 show images from THEMIS-ASI stations (see, e.g., Donovan et al. (2006) and Mende et al. (2008)) displayed in magnetic latitude versus magnetic longitude format. In Fig. 5.9, we show the images from 06:22:00 UT. We have superimposed the precipitating electron locations and energy flux from our simulations on the images. This snapshot was taken just prior to an auroral brightening. For comparison, we show the equatorial location from which the precipitating electrons originated [panel (b)], using the same format as Fig. 5.6b. For the auroral brightening at 06:24:00 UT, shown in Fig. 5.10a, we have plotted the precipitating energy flux on the images. Note that the simulated electron energy flux increases by over an order of magnitude between 06:22:00 UT and 06:24:00 UT. Those electrons precipitating at lower latitudes tend

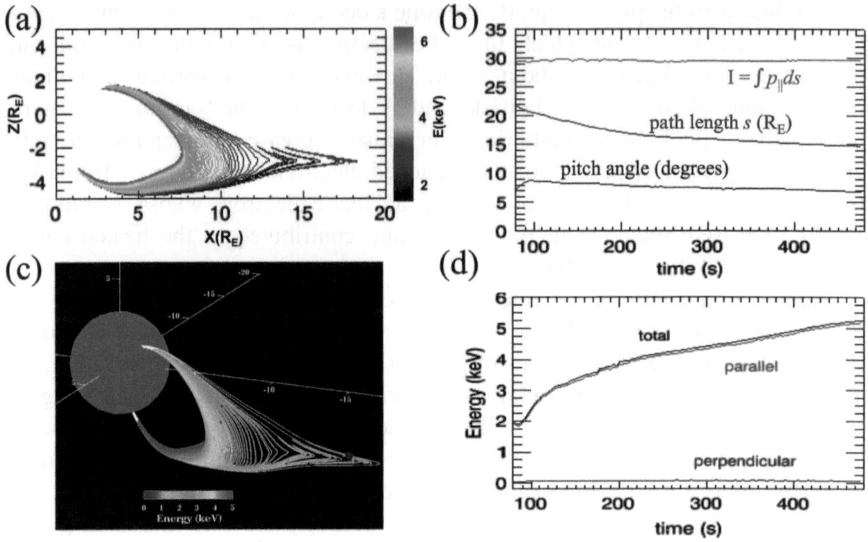

Fig. 5.8 The trajectory of a Fermi accelerated particle is shown in panels (**a**) and (**c**). Panel (**b**) shows the time history of the second adiabatic invariant ($I = \int p_{||} ds$ shown by the *red trace*), the bounce path length in *blue* (*scale in $R_E$on the left*), and pitch angle in *black* (*scale in degrees on the left*). Panel (**d**) shows the time history of the components of the particle energy (total in *black*, parallel in *red*, perpendicular in *blue*)

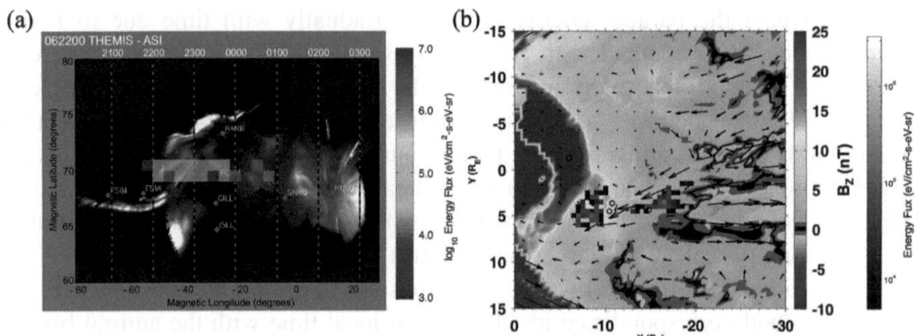

Fig. 5.9 Precipitation mapping for 06:22:00 UT. Images from the Kuujjuaq, Sanikiluaq, Rankin Inlet, Gillam, Fort Smith, and Fort Simpson THEMIS-ASI sites for 06:22:00 UT on 11 March 2008 are shown in Panel (**a**). The precipitating electron energy fluxes (color-coded in units of eV/cm^2-s-eV-sr) at 100 km altitude from the simulations are superimposed on the images. Panel (**b**) shows the locations of the last crossings of the precipitating electrons in the magnetotail maximum pressure plane binned in 0.5 R_E by 0.5 R_E squares, with corresponding energy fluxes indicated by the *white-gray scale*

to be Fermi accelerated, and those at the higher latitudes are primarily stochastically pitch angle scattered. There is a good correlation between the location of the precipitating electrons and an auroral brightening seen at RANK (Rankin Inlet) station, as well as at Fort Smith (FSMI) station, just before midnight. Between

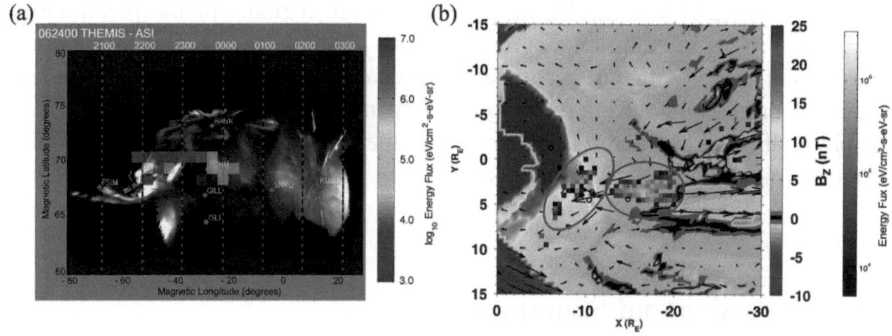

Fig. 5.10 Precipitation mapping for 06:24:00 UT. Images from the Kuujjuaq, Sanikiluaq, Rankin Inlet, Gillam, Fort Smith, and Fort Simpson THEMIS-ASI sites for 06:24:00 UT on 11 March 2008 are shown in Panel (**a**). The color-coded precipitating electron energy fluxes (in units of eV/cm^2-s-eV-sr) at 100 km altitude from the simulations are superimposed on the images. In Panel (**b**), the locations of the last crossings of the precipitating electrons in the magnetotail maximum pressure plane are binned in 0.5 R_E by 0.5 R_E squares, with energy fluxes indicated by the *white-gray scale* (*white* represents a peak energy flux of 10^7 eV/cm^2-s-eV-sr). The *red* and *blue* ovals differentiate electrons precipitating due to two different physical mechanisms

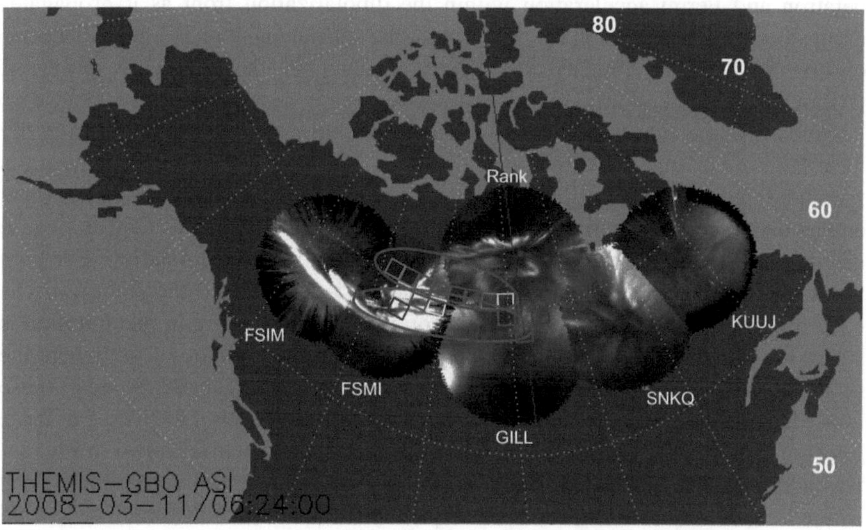

Fig. 5.11 Simulated precipitating electron locations at 100 km altitude are shown as *boxes* superimposed on all-sky camera images. The *boxes* are color-coded for energy flux, the same as in Fig. 5.10a. The *red* and *blue* ovals correspond to the two electron classes that are also shown in Fig. 5.10b

06:22:00 UT and 06:24:00 UT the aurora observed by the ASI brightened markedly. At the same time the precipitating electron energy flux from the LSK calculation increased by more than an order of magnitude. The largest increase was from the outer region nearest the neutral line. The auroral images and simulated energy fluxes (color-coded boxes) are shown in geographic coordinates in Fig. 5.11. As before, the red and blue circles indicate the source regions for the precipitating electrons. This is in qualitative agreement with the observed auroral brightening. These results suggest that the precipitation mechanisms, as described here, that lead to auroral brightening during the substorm expansion phase may be different from the physical processes that lead to the quiet time auroral arc electron precipitation occurring prior to the expansion phase.

5.4 Discussion and Summary

During a substorm on 11 March 2008, earthward-propagating dipolarization fronts were observed by THEMIS spacecraft in the near-Earth magnetotail plasma sheet, and auroral brightening was observed by all-sky cameras at high northern latitudes around 70° in the pre-midnight sector. Previous large-scale kinetic simulations of this event have shown that energized electrons can be accelerated to ~100 keV within the dipolarization front in the plasma sheet by a combination of non-adiabatic acceleration near the magnetotail reconnection region and adiabatic betatron and Fermi acceleration within the dipolarization front as it propagates earthward. This acceleration can account for the enhancement in energetic electrons observed when the dipolarization front passes over the THEMIS spacecraft (Ashour-Abdalla et al. 2011b; Pan et al. 2012).

Some of the electrons are strongly energized during this event, these primarily have large pitch angles (Ashour-Abdalla et al. 2011b; Pan et al. 2012), while it is seen here that electrons with more modest energies (~keV) can have smaller pitch angles and precipitate into the ionosphere. Two acceleration mechanisms cause electrons to enter the loss cone from the magnetotail in the simulation. For both types, the average precipitating energy is about 5–10 keV and the energy flux is ~10^6 eV/cm^2-s-eV-sr. The first acceleration mechanism leading to precipitation is stochastic pitch angle scattering that occurs at about 20–25 R_E down tail where the magnetic field is relatively weak and particles can be non-adiabatic. Some of these non-adiabatic electrons are scattered into the loss cone, reducing the fluxes of these lower energy (~keV) electrons in the plasma sheet. The electron shown in Fig. 5.7 was adiabatic until the last three bounces. It then became non-adiabatic when it entered the current sheet and was subsequently transported earthward. The second acceleration mechanism is Fermi acceleration, which causes adiabatic electrons to enter the loss cone closer to the Earth, between about -10 R_E and -15 R_E. THEMIS observations near -10 R_E in the magnetotail show a dropout in lower energy (5–10 keV) electrons after the dipolarization front passes the satellite and the electrons in this energy range are observed to have low pitch angles (more field-aligned). This is consistent with our results which show that the low pitch angle electrons tend to precipitate tailward of -10 R_E due to the combined non-adiabatic and Fermi acceleration processes, leaving a relatively few of these particles in the plasma sheet. In contrast, betatron acceleration leads to an enhancement of flux at larger pitch angles and at energies up to 100 keV, as shown by Ashour-Abdalla et al. (2011a).

These two types of motion combine to form a precipitation pattern in the ionosphere in the pre-midnight sector between about 68° and 72°. To the precision of our simulation, the location, timing, and energy of these precipitating electrons are in good agreement with all sky camera images of auroral brightening that occur at substorm onset. The results here imply that auroral emissions can be caused by a number of mechanisms. Quiet time auroral arcs are likely to be caused by a combination of low altitude (<10,000 km) field-aligned electron acceleration processes that result in electron precipitation, including double layers, inverted V structures, inertial Alfvén waves and wave-particle interactions. At lower latitudes that map to the Earth's radiation belt region <9 R_E, wave-particle interactions due to electron cyclotron harmonic waves and whistler-chorus emissions can result in diffuse auroral precipitation (Kennel 1969) and low altitude electron microbursts (Lorentzen et al. 2001). The results here show that mechanisms that operate in the magnetotail current sheet between −10 and −25 R_E also can lead to electron precipitation and auroral brightening during substorms. Ultimately, all of these different processes result in the visually beautiful aurora.

Acknowledgements This paper is in celebration of Jim Dungey's 90th birthday. He was my graduate thesis advisor and an inspiration to all of us. His incredible ability to gain fundamental physical insight based on limited information and pure intellectual fire power set a standard that cannot be matched. This research was performed in collaboration with R. Walker, D. Schriver, M. El-Alaoui, and R. Richard. I gratefully acknowledge their contributions. I thank J. Berchem for supplying the global MHD results for a southward IMF, and E. Donovan for supplying the auroral images from the All-Sky Imager array. Research at UCLA was supported by the NASA Magnetospheric Multiscale (MMS) Interdisciplinary Science Program grant NNX08AO48G and NASA Geospace grant NNX12AD13G. I acknowledge the use of data from the THEMIS Mission. Computing was carried out on NASA Advanced Supercomputing (NAS) including Columbia and Pleiades machines and XSEDE machines at San Diego.

References

Akasofu, S.-I.: Physics of Magnetospheric Substorms. D. Reidel, Dordrecht (1976)

Alfven, H.: On the theory of comet tails. Tellus **9**(1), 92–96 (1957). doi:10.1111/j.2153-3490.1957.tb01855.x

Ashour-Abdalla, M., Berchem, J., Büchner, J., Zelenyi, L.M.: Chaotic scattering and acceleration of ions in the Earth's magnetotail. Geophys. Res. Lett. **17**(13), 2317–2320 (1990). doi:10.1029/GL017i013p02317

Ashour-Abdalla, M., Zelenyi, L.M., Peroomian, V., Richard, R.L.,: On the structure of the magnetotail current sheet, Geophys. Res. Lett., 20(19), 2019–2022 (1993). doi:10.1029/93GL01695

Ashour-Abdalla, M., El-Alaoui, M., Peroomian, V., Raeder, J., Walker, R.J., Richard, R.L., Zelenyi, L.M., Frank, L.A., Paterson, W.R., Bosqued, J.M., Lepping, R.P., Ogilvie, K., Kokubun, S., Yamamoto, T.: Ion sources and acceleration mechanisms inferred from local distribution functions. Geophys. Res. Lett. **24**(8), 955–958 (1997). doi:10.1029/97GL00060

Ashour-Abdalla, M., El-Alaoui, M., Coroniti, F.V., Walker, R.J., Peroomian, V.: A new convection state at substorm onset: results from an MHD study. Geophys. Res. Lett. **29**(20), 1965 (2002). doi:10.1029/2002GL015787

Ashour-Abdalla, M., El-Alaoui, M., Goldstein, M.L., Zhou, M., Schriver, D., Richard, R., Walker, R. J., Kivelson, M.G., Hwang, K.-J.: Observations and simulations of non-local acceleration of electrons in magnetotail magnetic reconnection events. Nature **7**, 360–365 (2011a). doi:10.1038/nphys1903

Ashour-Abdalla, M., El-Alaoui, M., Schriver, D., Pan, Q., Richard, R.L., Zhou, M., Walker, R.J.: Electron acceleration associated with earthward propagating dipolarization fronts, Abstract SM13C-2098, presented at 2011 Fall Meeting, AGU, San Francisco, CA, 4–9 Dec (2011b)

Ashour-Abdalla, M., Schriver, D., El-Alaoui, M., Richard, R., Walker, R., Goldstein, M.L., Donovan, E., Zhou, M.: Direct auroral precipitation from the magnetotail during substorms. Geophys. Res. Lett. **40**(15), 3787–3792 (2013). doi:10.1002/grl.50635

Axford, W.I., Hines, C.O.: A unifying theory of high-latitude geophysical phenomena and geomagnetic storms. Can. J. Phys. **39**(10), 1433–1464 (1961). doi:10.1139/p61-172

Baker, D.N., Pulkkinen, T.I., Angelopoulos, V., Baumjohann, W., McPherron, R.L.: Neutral line model of substorms: past results and present view. J. Geophys. Res. **101**(A6), 12975–13010 (1996). doi:10.1029/95JA03753

Baños Jr., A.: The guiding centre approximation in lowest order. J. Plasma Phys. **1**(3), 305–316 (1967). doi:10.1017/S0022377800003317

Biermann, L.: On the origin of magnetic fields in stars and interstellar space. Z. Naturforsch. A **5**, 65–71 (1950)

Berchem, J., Raeder, J., Ashour-Abdalla, M.: Reconnection at the magnetospheric boundary: results from global magnetohydrodynamic simulations. In: Song, P., Sonnerup, B.U.Ö., Thomsen, M.F. (eds.) Physics of the Magnetopause. American Geophysical Union, Washington, DC (1995). doi:10.1029/GM090p0205

Birkeland, K.: Physical constitution of the sun. Int. Phys. Congr. Paris **3**, 471–487 (1900)

Birn, J., Thomsen, M.F., Borovsky, J.E., Reeves, G.D., McComas, D.J., Belian, R.D., Hesse, M.: Substorm electron injections: geosynchronous observations and test particle simulations. J. Geophys. Res. **103**(A5), 9235–9248 (1998). doi:10.1029/97JA02635

Birn, J., Thomsen, M.F., Hesse, M.: Electron acceleration in the dynamic magnetotail: test particle orbits in three-dimensional magnetohydrodynamic simulation fields. Phys. Plasmas **11**, 1825–1833 (2004). doi:10.1063/1.1704641

Borovsky, J.E.: Auroral arc thicknesses as predicted by various theories. J. Geophys. Res. **98**(A4), 6101–6138 (1993). doi:10.1029/92JA02242

Büchner, J., Zelenyi, L.M.: Deterministic chaos in the dynamics of charged-particles near a magnetic-field reversal. Phys. Lett. A **118**, 395–399 (1986). doi:10.1016/0375-9601(86)90268-9

Büchner, J., Zelenyi, L.M.: Chaotization of the electron motion as the cause of an internal magnetotail instability and substorm onset. J. Geophys. Res. **92**(A12), 13456–13466 (1987). doi:10.1029/JA092iA12p13456

Chapman, S., Ferraro, V.C.A.: A new theory of magnetic storms. Terr. Magn. Atmos. Electr. **36**, 77–97 (1931). doi:10.1029/TE036i002p00077

Chen, J., Palmadesso, P.J.: Chaos and nonlinear dynamics of single-particle orbits in a magnetotail-like magnetic field. J. Geophys. Res. **91**(A2), 1499–1508 (1986). doi:10.1029/JA091iA02p01499

Christensen, A.B., Lyons, L.R., Hecht, J.H., Sivjee, G.G., Meier, R.R., Strickland, D.G.: Magnetic field-aligned electric field acceleration and the characteristics of the optical aurora. J. Geophys. Res. **92**(A6), 6163–6167 (1987). doi:10.1029/JA092iA06p06163

Cowley, S.W.H.: The causes of convection in the earth's magnetosphere: a review of developments during the IMS. Rev. Geophys. **20**(3), 531–565 (1982). doi:10.1029/RG020i003p00531

Delcourt, D.C., Sauvaud, J.A., Martin Jr., R.F., Moore, T.E.: Gyrophase effects in the centrifugal impulse model of particle motion in the magnetotail. J. Geophys. Res. **100**(A9), 17211–17220 (1995). doi:10.1029/95JA00657

Donovan, E., Mende, S., Jackel, B., Frey, H., Syrjäsuo, M., Voronkov, I., Trondsen, T., Peticolas, L., Angelopoulos, V., Harris, S., Greffen, M., Connors, M.: The THEMIS all-sky imaging array: system design and initial results from the prototype imager. J. Atmos. Sol. Terr. Phys. **68**(13), 1472–1487 (2006). doi:10.1016/j.jastp.2005.03.027

Dungey, J.W.: Cosmic Electrodynamics. Cambridge University Press, London (1958). doi:10.1002/qj.49708536628

Dungey, J.W.: Interplanetary magnetic field and the auroral zones. Phys. Rev. Lett. **6**, 47–48 (1961). doi:10.1103/PhysRevLett.6.47

El-Alaoui, M., Ashour-Abdalla, M., Raeder, J., Peroomian, V., Frank, L.A., Paterson, W.R., Bosqued, J.M.: Modeling magnetotail ion distributions with global magnetohydrodynamic and ion trajectory calculations. In: Horwitz, J.L., Gallagher, D.L., Peterson, W.K. (eds.) Geospace Mass and Energy Flow. American Geophysical Union, Washington, DC (1998). doi:10.1029/GM104p0291

El-Alaoui, M.: Current disruption during November 24, 1996, substorm. J. Geophys. Res. **106** (A4), 6229–6245 (2001). doi:10.1029/1999JA000260

El-Alaoui, M., Ashour-Abdalla, M., Richard, R.L., Goldstein, M.L., Weygand, J.M., Walker, R.J.: Global magnetohydrodynamic simulation of reconnection and turbulence in the plasma sheet. J. Geophys. Res. **115**, A12236 (2010). doi:10.1029/2010JA015653

Elphic, R. C., Bonnell, J.W., Strangeway, R.J., Kepko, L., Ergun, R.E., McFadden, J.P., Carlson, C.W., Peria, W., Cattell, C.A., Klumpar, D., Shelley, E., Peterson, W., Moebius, E., Kistler, L., Pfaff, R.,: The auroral current circuit and field-aligned currents observed by FAST, Geophys. Res. Lett., 25(12), 2033–2036 (1998). doi:10.1029/98GL01158

Ergun, R.E., Carlson, C.W., McFadden, J.P., Mozer, F.S., Strangeway, R.J.: Parallel electric fields in discrete arcs. Geophys. Res. Lett. **27**, 4053–4056 (2000). doi:10.1029/2000GL003819

Evans, D.S.: Precipitating electron fluxes formed by a magnetic field aligned potential difference. J. Geophys. Res. **79**(19), 2853–2858 (1974). doi:10.1029/JA079i019p02853

Feldstein, Y.I., Starkov, G.V.: Dynamics of auroral belt and polar geomagnetic disturbances. Planet. Space Sci. **15**, 209–229 (1967). doi:10.1016/0032-0633(67)90190-0

Frank, L.A., Ackerson, K.L.: Observations of charged particle precipitation into the auroral zone. J. Geophys. Res. **76**(16), 3612–3643 (1971). doi:10.1029/JA076i016p03612

Frank, L.A.: Plasmas in the earth's magnetotail. Space Sci. Rev. **42**, 211–240 (1985). doi:10.1007/BF00218233

Giovanelli, R.G.: Magnetic and electric phenomena in the sun's atmosphere associated with sunspots. Mon. Not. R. Astron. Soc. **107**, 338–355 (1947)

Giovanelli, R.G.: Chromospheric flares. Mon. Not. R. Astron. Soc. **108**, 163–175 (1948)

Haerendel, G.: Cosmic linear accelerators. In: Guyenne, T.D., Hunt, J.J. (eds.) Proceedings of an International School and Workshop on Plasma Astrophysics. Eur. Space Agency Spec. Publ., *ESA SP-285*, pp. 37–44 (1989).

Heppner, J.P., Maynard, N.C.: Empirical high-latitude electric field models. J. Geophys. Res. **92** (A5), 4467–4489 (1987). doi:10.1029/JA092iA05p04467

Hoffman, R.A., Evans, D.S.: Field-aligned electron bursts at high latitudes observed by OGO 4. J. Geophys. Res. **73**(19), 6201–6214 (1968). doi:10.1029/JA073i019p06201

Hoffmeister, C.: Physikalisch Untersuchungen auf Kometen. I. Die Beziehungen des primaren Schweifstrahl zum Radiusvektor. Z. Astrophys. **22**, 265–285 (1943)

Hoyle, F.: External sources of climatic variation. Q. J. Roy. Meteorol. Soc. **75**, 161–163 (1949). doi:10.1002/qj.49707532407

Hughes, W.J.: The magnetopause, magnetotail, and magnetic reconnection. In: Kivelson, M.G., Russell, C.T. (eds.) Introduction to Space Physics, pp. 227–287. Cambridge University Press, New York, NY (1995)

Hultqvist, B.: On the production of a magnetic-field-aligned electric field by the interaction between the hot magnetospheric plasma and the cold ionosphere. Planet. Space Sci. **19**, 749–759 (1971). doi:10.1016/0032-0633(71)90033-X

Keiling, A., Wygant, J.R., Cattell, C., Kim, K.-H., Russell, C.T., Milling, D.K., Temerin, M., Mozer, F.S., Kletzing, C.A.: Pi2 pulsations observed with the Polar satellite and ground stations: coupling of trapped and propagating fast mode waves to a midlatitude field line resonance. J. Geophys. Res. **106**(A11), 25891–25904 (2001). doi:10.1029/2001JA900082

Kennel, C.F.: Consequences of a magnetospheric plasma. Rev. Geophys. **7**, 379–419 (1969). doi:10.1029/RG007i001p00379

Kepko, L., Spanswick, E., Angelopoulos, V., Donovan, E., McFadden, J., Glassmeier, K.-H., Raeder, J., Singer, H.J.: Equatorward moving auroral signatures of a flow burst observed prior to auroral onset. Geophys. Res. Lett. **36**, L24104 (2009). doi:10.1029/2009GL041476

Lorentzen, K.R., Blake, J.B., Inan, U.S., Bortnik, J.: Observations of relativistic electron microbursts in association with VLF chorus. J. Geophys. Res. **106**(A4), 6017–6027 (2001). doi:10.1029/2000JA003018

Lui, A.T.Y., Anger, C.D.: A uniform belt of diffuse auroral emission seen by the ISIS-2 scanning photometer. Planet. Space Sci. **21**, 799–809 (1973). doi:10.1016/0032-0633(73)90097-4

Lui, A.T.Y., Anger, C.D., Venkatesan, D., Sawchuk, W., Akasofu, S.-I.: The topology of the auroral oval as seen by the Isis 2 scanning auroral photometer. J. Geophys. Res. **80**(13), 1795–1804 (1975). doi:10.1029/JA080i013p01795

Lyons, L.R., Koskinen, H.E.J., Blake, J.B., Egeland, A., Hirahara, M., Oieroset, M., Sandholt, P.E., Shiokawa, K.: Processes leading to plasma losses into the high-latitude atmosphere. Space Sci. Rev. **88**, 85–135 (1999). doi:10.1023/A:1005251700516

McIlwain, C.E.: Direct measurements of particles producing visible auroras. J. Geophys. Res. **65**, 2727–2747 (1960). doi:10.1029/JZ065i009p02727

Mende, S.B., Harris, S.E., Frey, H.U., Angelopoulos, V., Russell, C.T., Donovan, E., Jackel, B., Greffen, M., Peticolas, L.M.: The THEMIS array of ground-based observatories for the study of auroral substorms. Space Sci. Rev. **141**, 357–387 (2008). doi:10.1007/s11214-008-9380-x

Mozer, F.S., Carlson, C.W., Hudson, M.K., Torbert, R.B., Parady, B., Yatteau, J., Kelley, M.C.: Observations of paired electrostatic shocks in polar magnetosphere. Phys. Rev. Lett. **38**, 292–295 (1977). doi:10.1103/PhysRevLett.38.292

Nakamura, R., Baumjohann, W., Schodel, R., Brittnacher, M., Sergeev, V.A., Kubyshkina, M., Mukai, T., Liou, K.: Earthward flow bursts, auroral streamers, and small expansions. J. Geophys. Res. **106**(A6), 10791–10802 (2001). doi:10.1029/2000JA000306

Northrop, T.C.: The Adiabatic Motion of Charged Particles. Wiley, New York (1963)

Pan, Q., Ashour-Abdalla, M., El-Alaoui, M., Walker, R.J., Goldstein, M.L.: Adiabatic acceleration of suprathermal electrons associated with dipolarization fronts. J. Geophys. Res. **117**, A12224 (2012). doi:10.1029/2012JA018156

Parker, E.N.: Dynamics of the interplanetary gas and magnetic fields. Astrophys. J. **128**, 664–676 (1958). doi:10.1086/146579

Raeder, J., Walker, R.J., Ashour-Abdalla, M.: The structure of the distant geomagnetic tail during long periods of northward IMF. Geophys. Res. Lett. **22**, 349–352 (1995). doi:10.1029/94GL03380

Rees, M.H., Luckey, D.: J. Geophys. Res. **79**, 5181–5186 (1974). doi:10.1029/JA079i034p05181

Schriver, D., Ashour-Abdalla, M., Richard, R.L.: On the origin of the ion-electron temperature difference in the plasma sheet. J. Geophys. Res. **103**(A7), 14879–14895 (1998). doi:10.1029/98JA00017

Schriver, D., Ashour-Abdalla, M., Strangeway, R.J., Richard, R.L., Klezting, C., Dotan, Y., Wygant, J.: FAST/Polar conjunction study of field-aligned auroral acceleration and corresponding magnetotail drivers. J. Geophys. Res. **108**(A9), 8020 (2003). doi:10.1029/2002JA009426

Schriver, D., Trávníček, P., Ashour-Abdalla, M., Richard, R.L., Hellinger, P., Slavin, J.A., Anderson, B.J., Baker, D.N., Benna, M., Boardsen, S.A., Gold, R.E., Ho, G.C., Korth, H.,

Krimigis, S.M., McClintock, W.E., McLain, J.L., Orlando, T.M., Sarantos, M., Sprague, A.L., Starr, R.D.: Electron transport and precipitation at Mercury during the MESSENGER flybys: implications for electron-stimulated desorption. Planet. Space Sci. **59**, 2026–2036 (2011). doi:10.1016/j.104 pss.2011.03.008

Shay, M.A., Drake, J.F., Eastwood, J.P., Phan, T.D.: Super-alfvénic propagation of substorm reconnection signatures and poynting flux. Phys. Rev. Lett. **107**, 065001 (2011). doi:10.1103/PhysRevLett.107.065001

Wygant, J.R., Keiling, A., Cattell, C.A., Johnson, M., Lysak, R.L., Temerin, M., Mozer, F.S., Kletzing, C.A., Scudder, J.D., Peterson, W., Russell, C.T., Parks, G., Brittnacher, M., Germany, G., Spann, J.: Polar spacecraft based comparisons of intense electric fields and Poynting flux near and within the plasma sheet-tail lobe boundary to UVI images: an energy source for the aurora. J. Geophys. Res. **105**(A8), 18675–18692 (2000). doi:10.1029/1999JA900500

Zelenyi, L.M., Lominadze, J.G., Taktakishvili, A.L.: Generation of the energetic proton and electron bursts in planetary magnetotails. J. Geophys. Res. **95**(A4), 3883–3891 (1990). doi:10.1029/JA095iA04p03883

Chapter 6
Many-Body Calculations

James Eastwood

Abstract The computational cost of evaluating action-at-a-distance force or field sums and integrals has always been a major concern in many-body calculations. This paper outlines the evolution of FFT-based methods that have reduced the complexity of these N-body calculations from $O(N^2)$ to $O(N \log N)$ for a wide range of physical phenomena. These developments have made possible simulation of larger and more complex situations: collisionless plasmas with periodic, neumann and dirichlet boundary conditions; collisionless galaxy models with isolated boundary conditions; dense condensed matter; galaxy clustering in a Friedman cosmology; electromagnetic scattering from surface source distributions.

6.1 Introduction

This paper addresses complexity in many-body calculations. Investigation by computer simulation was in its infancy when I was one of Jim Dungey's students, but is now used in almost all areas of science and engineering. Jim initially pointed me to computational studies looking at non-adiabatic ion trajectories and their effects on currents in the current sheet of the geomagnetic tail (Eastwood 1972, 1974); this is where I first became involved in many-body calculations, using particle methods to model the interactions of plasma with electric and magnetic fields.

Simulation studies face a compromise between the detail of the model and computational resources. In principle classical and pseudo-classical systems may be described in terms of positions, velocities and force laws of the particles of which it is composed. Unfortunately, the vast number of particles involved in quite simple situations generally make such a detailed description computationally impractical, despite the enormous growth in computer power over the past four decades. I learnt from Jim the importance of assessing length- and time-scales and

J. Eastwood (✉)
Culham Electromagnetics Ltd, Culham Science Centre, Abingdon, Oxfordshire, OX14 3DB, UK
e-mail: James.Eastwood@CulhamEM.co.uk

D. Southwood et al. (eds.), *Magnetospheric Plasma Physics: The Impact of Jim Dungey's Research*, Astrophysics and Space Science Proceedings 41,
DOI 10.1007/978-3-319-18359-6_6

using them to reduce the mathematical models (e.g., as in hydromagnetics and kinetic plasma models). Applying this principle to many-body calculations can lead to models that are sufficiently detailed to reproduce important physical effects but not too detailed to make calculations impracticable.

Even with reduced models, a large part of the computational work in many-body calculations is the evaluation of interparticle forces. For example, given N charged particles, the Coulombic field at one particle has N-1 contributions from the other particles, so to compute fields at all N particles has complexity $O(N^2)$. This paper reviews methods of reducing this complexity to O(N log N) in different situations. Reduced to its basics, the problem is to evaluate in O(N log N) operations the convolution sum or integral:

$$\begin{aligned} \varphi_i &= \sum_{j=1}^{N} q_i G(\mathbf{x}_i - \mathbf{x}_j) \; ; \quad i \in [1, N] \\ \varphi(\mathbf{x}) &= \int d\mathbf{x}' \rho(\mathbf{x}') G(\mathbf{x} - \mathbf{x}') \end{aligned} \tag{6.1}$$

Central to this complexity reduction is the fast fourier transform. Simulation models relying on this include plasmas, vacuum electronic devices, galaxy evolution, structure of ionic liquids and biological molecules, clustering of galaxies and installed antenna performance.

6.2 The Particle-Mesh Method

The appropriate physical model for collisionless plasmas and galaxy evolution is Vlasov's equation (Clemmow and Dougherty 1969), where respectively charged particles and stars are represented by continuous phase fluids where all graininess is smoothed out. Provided that the physical phenomena have wavelength large compared to interparticle spacing and times short compared to collision time, then this description gives an accurate representation.

In the particle-mesh (PM) method (Hockney and Eastwood 1981; Birdsall and Langdon 1991), the one particle distribution function of Vlasov's equation is represented by a set of sample points—these samples ("superparticles") may be interpreted as finite sized clouds of charged particles or stars. The distribution function is constant along particle trajectories, so moving the superparticles according to the equations of motion is used to map the distribution function forward in time. The first moment of the distribution function is then used to give source terms for computing fields at the new time.

The PM method avoids the $O(N^2)$ force sum by using a mesh to compute interparticle forces. The smoothing out of short range interactions caused by the mesh has a beneficial effect of reducing collisional effects. In the plasma context, using a mesh that is sufficiently fine to resolve Debye length variations and is coarse

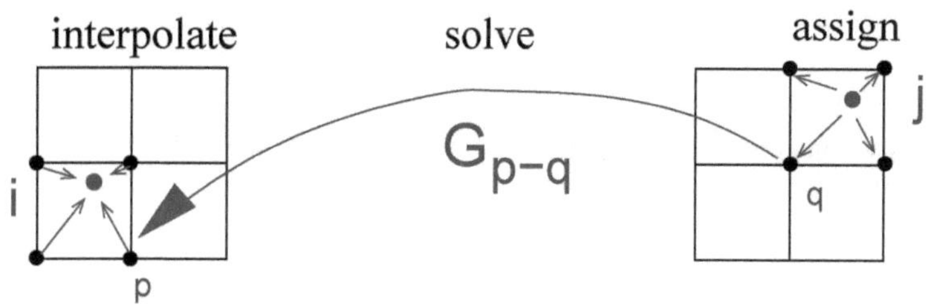

Fig. 6.1 The force calculation in the particle-mesh method assigns charge to the mesh from the particles, solves for the fields on the mesh and interpolates forces back to particle positions

enough to have tens of superparticles per cell (compared to billions of real charged particles) captures the physics of the plasma waves without being corrupted by collisional effects resulting from using only a few superparticles per Debye cube.

The sequence of operations in computing the interparticle forces are summarised in Fig. 6.1. Charge is assigned from the superparticles to the mesh, the fields are computed on the mesh and finally the fields are interpolated to superparticle positions to compute forces. The number of operations in the assign and interpolate step for each particle depends on the number of local mesh points involved, so the assign and interpolate step operation counts for all N superparticles is O(N).

There are a number of interpretations of the assign and interpolate operations, but we note here the dual finite element interpretation as it is important to cases considered later in this paper. The continuum charge density $\rho(\mathbf{x})$ is approximated in the first case using finite-element nodal amplitudes ρ_q and trial functions $W_q(\mathbf{x})$ and in the second by a monte-carlo samples at positions $\mathbf{x}_i$ of the first moment of the distribution function f with charge weight q_i:

$$\rho(\mathbf{x}) \simeq \sum_{\mathbf{q}} W_{\mathbf{q}}(\mathbf{x})\rho_{\mathbf{q}} \simeq \sum_{i} q_i \delta(\mathbf{x} - \mathbf{x}_i) \tag{6.2}$$

Testing with basis function $W_{\mathbf{p}}(\mathbf{x})$ relates the two approximations

$$\int d\mathbf{x} W_{\mathbf{p}}(\mathbf{x}) \sum_{\mathbf{q}} W_{\mathbf{q}}(\mathbf{x})\rho_{\mathbf{q}} = \sum_{i} q_i W_{\mathbf{p}}(\mathbf{x}_i) \tag{6.3}$$

and this is the prescription for charge assignment. It provides mapping from the particle positions to the mesh. Usually trial functions W_q are chosen to be low order spline functions. One generalisation of this is to use finite-sized particle profile rather than the delta function for the distribution function sampling (Langdon and Birdsall 1970).

Similarly, the interpolate step for momentum conserving schemes evaluates the mesh electric field at the superparticle position

$$\mathbf{E}(\mathbf{x}_i) = \sum_{\mathbf{q}} W_{\mathbf{q}}(\mathbf{x}_i)\mathbf{E}_{\mathbf{q}} \tag{6.4}$$

and energy conserving schemes (Lewis 1970) compute the field from the gradient of the finite element potential:

$$\mathbf{E}(\mathbf{x}_i) = \sum_{\mathbf{q}} \varphi_{\mathbf{q}} \nabla W_{\mathbf{q}}(\mathbf{x})\big|_{\mathbf{x}=\mathbf{x}_i} \tag{6.5}$$

Many PM implementations have used discrete approximations to Poisson's equation rather than that of the integral in Eq. (6.1), and a number of O(M log M) complexity sparse solvers that will solve the resulting discrete equations are known, e.g., DCR, FACR, nested dissection and MFT (Hockney 1965, Swarztrauber 1977, Hockney and Eastwood 1981). Since N/M is fixed for scaling to a larger system, it gives an overall complexity of O(N log N) for the PM method when the fast O (M log M) solver methods can be used.

We focus here on the multiple fourier transform (MFT) method. Optimised software (Frigo and Johnson 1998) implementing the fast fourier transform algorithm (Cooley et al. 1967; Bracewell 1965; Brigham 1988) is freely available, and will compute the discrete fourier transform of a periodic sequence of M numbers in O(M log M) operations.

If the boundary conditions are periodic, the discrete field equation $\varphi_{\mathbf{p}} = \sum_{\mathbf{q}} G_{\mathbf{p}-\mathbf{q}} \rho_{\mathbf{q}}$ transforms to a product for each harmonic $\hat{\varphi}_{\mathbf{k}} = \hat{G}_{\mathbf{k}} \hat{\rho}_{\mathbf{k}}$, where the influence function $\hat{G}_{\mathbf{k}}$ is the (pre-computed) FFT of the periodic Greens function $G_{\mathbf{p}}$. For a 3D MFT solver on a $M = P_1 \times P_2 \times P_3$ mesh the algorithm is

1. FFT source in all three directions O(M log M): $\{\rho_{\mathbf{p}}\} \supset \{\hat{\rho}_{\mathbf{k}}\}$
2. Multiply by the influence function O(M): $\hat{\varphi}_{\mathbf{k}} = \hat{G}_{\mathbf{k}} \hat{\rho}_{\mathbf{k}}$
3. Inverse FFT to get potential O(M log M): $\{\hat{\varphi}_{\mathbf{k}}\} \supset \{\varphi_{\mathbf{p}}\}$

Doubling the period length can also be used to relax the cyclic constraint on the discrete fourier transform convolution. This is used in galaxy simulation and in the Adaptive-P^3M and Block-P^3M method discussed later.

Figure 6.2 illustrates how this is done in galaxy simulations. The doubling of the period length allows a periodic Greens function that gives a 1/r interaction between particles that lie in the original quarter region. The FFT solver gives correct potentials for the isolated problem in the original quarter, and incorrect values that are discarded in the extended regions. Similar effects can be realised using interlaced sampling (Eastwood and Brownrigg 1979).

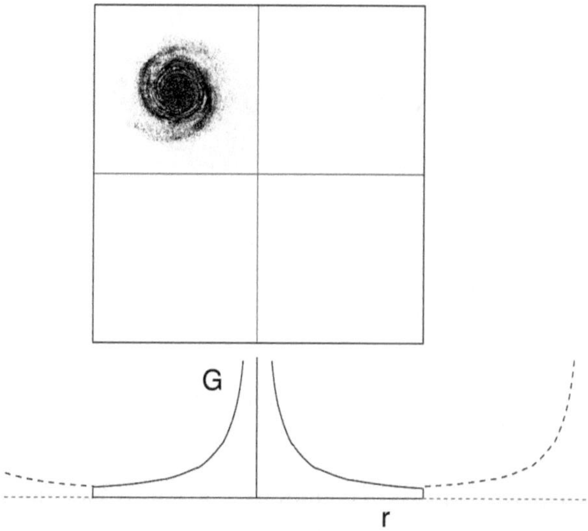

Fig. 6.2 Isolated boundary conditions are implemented by doubling the region widths, zero padding the sources and using a Greens function that is 1/r truncated to zero at ± region width and then periodically repeated

6.3 Electromagnetic Particle-Mesh Calculations

The particle-mesh method for potential fields has been extended to electromagnetic fields by replacing the Poisson equation solver by one which solve the Maxwell equations, and assigning currents to the mesh to provide the source terms. Gauss' law and magnetic flux conservation provide initial conditions for the electric and magnetic fields that are integrated forward in time using the time dependent Maxwell equations. Early versions of these electromagnetic PIC codes suffered from the problem that the continuity equation was not strictly satisfied, with the consequence that errors in the longitudinal part of the electric field grew. To correct for this, an equation of the form given by Eq. (6.1) is repeatedly solved for a correction to the electric field (Boris 1970), or the error is iteratively reduced (Marder 1987). Jim Dungey had suggested that I used the length of the trajectory in cells rather than the instantaneous particle velocity to compute the currents when studying the current sheet (Eastwood 1974). Using this idea (nearly two decades later!) in the context of electromagnetic particle-mesh algorithms allowed me to derive "charge conserving" current assignment schemes that strictly satisfied the continuity equation, and so ensured that if Gauss' Law was satisfied initially in a simulation, then it remained satisfied for later times, just as in the differential equations (Eastwood 1991, Eastwood et al. 1995).

In simple geometries where fast elliptic solvers can be used, the advantages of charge conserving assignment schemes are small, but in complex geometries using multi-block body conforming meshes on parallel computers (Eastwood et al. 1995) the advantages of avoiding repeated elliptic equation solutions is large. This has allowed detailed design studies of vacuum electronic devices, such as the

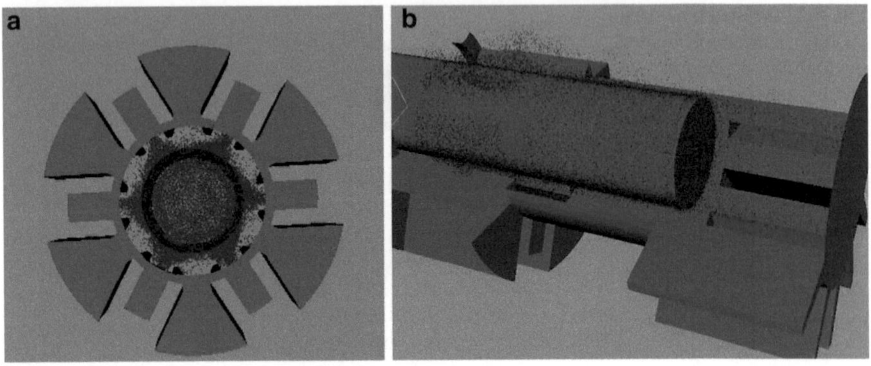

Fig. 6.3 Cutaway pictures of electron flow from 3D electromagnetic particle-mesh design simulations of a 250 MW tuneable relativistic magnetron

relativistic magnetron illustrated in Fig. 6.3 or the magnetically insulated line oscillator (Eastwood et al. 1998).

6.4 P^{3}M for Correlated Systems

A further class of N-body calculation for which the FFT-based convolution solver has been vital are correlated systems, for example, molecular dynamics of molten ionic salts, biological molecule modelling and galaxy clustering. The particle-mesh algorithm works well for collisionless systems, where the smoothing of interparticle forces at short range has a beneficial effect of reducing unwanted correlations. However, for dense systems, such as molten ionic salts, accurate treatment of short range forces is important in modelling correlations. If a particle-mesh method were to be used, then the mesh would have to be so fine as to make the mesh calculation more costly than the direct pairwise summation of forces.

The way to reduce complexity in this case is to split the Greens function into the sum of a short range and a smoothly varying part (Eastwood 1976; Eastwood et al. 1980; Hockney and Eastwood 1981). This is the particle-particle/particle-mesh (P^{3}M) method, where the short range part is computed by direct particle-particle summation in $O(N_nN)$ operations and the smoothly varying part is computed using the FFT-based particle-mesh method. N_n is the number of neighbours in the range of the short range force and is made as small as possible. Variational methods are used to perform an optimisation of the root mean squared errors in the force calculation versus computational costs, and the resulting algorithm is $O\left(N\sqrt{\log N}\right)$ for uniform density cases such as molten ionic salts. The use of FFT has the added advantage that corrections for interpolation and assignment errors may be built into the pre-computed mesh Greens function and so are applied at no extra computational cost. More recent implementations of P^{3}M using higher

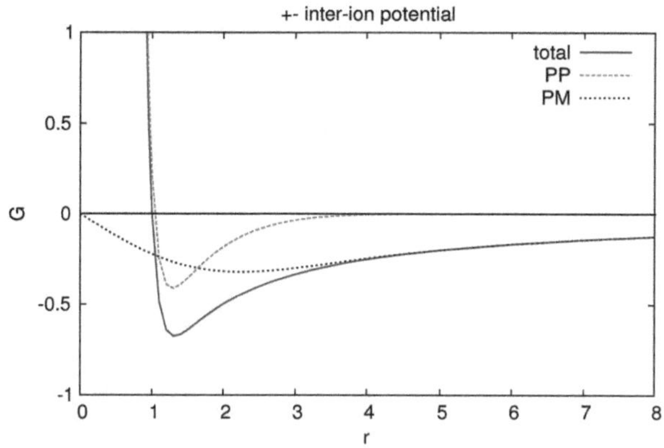

Fig. 6.4 P^3M splits the Greens function into a short range part (PP) and a smoothly varying part (PM)

order splines and point collocation have been renamed as the PME (Particle-Mesh Ewald) method (Darden et al. 1993).

Figure 6.4 illustrates how P^3M splits the potential and Green's function into sums of short-range (PP) parts and a smoothly varying (PM) parts

$$\varphi_i = \varphi_i^{sr} + \varphi_i^{mesh}; \quad G = G^{sr} + G^{mesh} \tag{6.6}$$

φ_i^{sr} is evaluated using direct pairwise sums over the set S_i of near neighbours of i

$$\varphi_i^{sr} = \sum_{j \in S_i} G^{sr}(\mathbf{x}_i - \mathbf{x}_j)\sigma_j; \quad i \in [1, N] \tag{6.7}$$

φ^{mesh} is computed using FFT on a mesh approximation. G^{mesh} is approximated by

$$G^{mesh}(\mathbf{x}_i - \mathbf{x}_j) \simeq W_p(\mathbf{x}_i) G^{fe}\left(\mathbf{x}_p - \mathbf{x}_{p'}\right) W_{p'}(\mathbf{x}_j) \tag{6.8}$$

Nodal amplitudes G^{fe} are precomputed using a Galerkin (or point collocation) method and their FTTs G^k are stored. The use of FFT on a regular lattice favours the use of spline functions for W_p rather than the more common finite element choices of Lagrange or Hermite polynomials. The mesh calculation reduces to the PM cycle

$$\text{assign}: \quad \sigma_{p'} = \sum_j W_{p'}(\mathbf{x}_j)\sigma_j \tag{6.9}$$

$$\text{solve}: \quad \varphi_p = \sum_{p'=0}^{P-1} G^{fe}_{p-p'}\sigma_{p'} \tag{6.10}$$

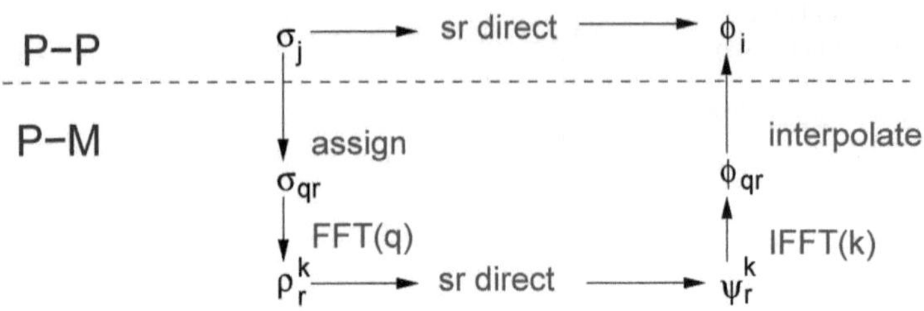

Fig. 6.5 A flow diagram for the P³M algorithm

$$\text{interpolate}: \ \varphi_i^{mesh} = \sum_p W_p(\mathbf{x}_i)\varphi_p \qquad (6.11)$$

where p and p' label mesh nodes. The mesh convolution sum [Eq. (6.10)] is evaluated using FFTs: the FFT of source values σ_p gives harmonics σ^k. Multiplication by Green's function harmonics G_k gives potential harmonics $\varphi^k = G^k\sigma^k$ and the inverse FFT converts these to values φ_p. If periodic boundary conditions are used, Eq. (6.10) can be transformed using length P FFT and short range contributions from periodic images need adding to the fields. If isolated boundary conditions are required, then zero padding sources and length $2P$ or greater FFT (Eastwood and Brownrigg 1979) are needed to diagonalise Eq. (6.10). The splitting of the Green's function is linear and additive, and so the total field is given by adding the contributions:

$$\varphi_i = \varphi_i^{sr} + \varphi_i^{mesh} \qquad (6.12)$$

Note the fast multipole method (Rokhlin 1985) splits short and long-range into mutually exclusive spatial regions rather than using this additive split, and so needs to take care to avoid double counting. The algorithm in two and three dimensions is given by interpreting $\mathbf{x}_p$ and p as 2D or 3D vectors and W_p as products of 1D polynomial functions. Figure 6.5 summarises the algorithm.

For spatially uniform distributions of sources, the operations count scales as $O(N\sqrt{\log N})$ when parameters are chosen so that the computational work in the short-range and long-range parts of the calculation are made equal. In molecular dynamics applications, where ion spacing is determined by the repulsive core forces, density remains almost constant and so a calculation that is optimally set up initially remains so throughout a simulation. However, in situations where the density of particles becomes strongly non-uniform, then the work in the short and long range parts becomes unbalanced and computational complexity rises. In the limit of a surface distribution of sources in a three dimensional volume the scaling of computational costs increases to $O(N^{3/2} \log N)$. In the following sections further refinements of the P³M that overcome this problem are discussed.

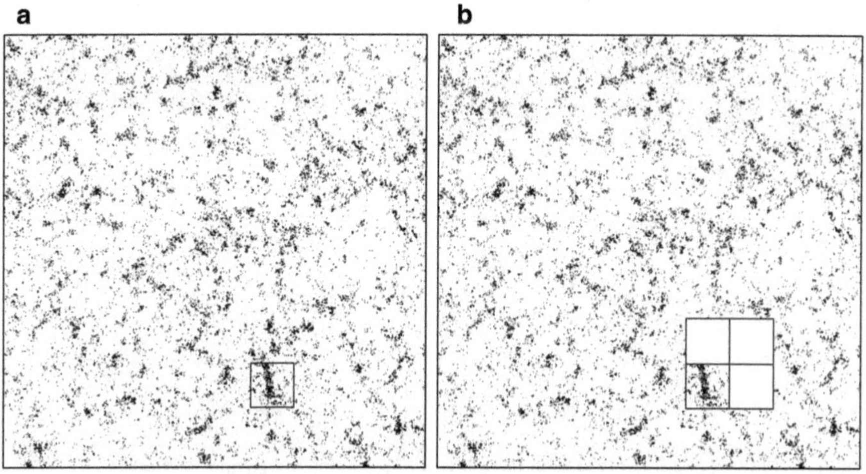

Fig. 6.6 The Adaptive-P^{3}M algorithm repeats the splitting into direct sum and particle-mesh part in regions of high density. A fine mesh is placed over a region of high density (*left*) and zero padding (*right*) is used to eliminate periodicity

6.5 Adaptive-P3M

In P^{3}M simulations of the clustering of galaxies in a Friedman cosmology (Efstathiou and Eastwood 1981), the initially uniform distribution of galaxies become highly non-uniform in later epochs and the computational time for the short range part of the calculation dominates. The Adaptive-P^{3}M algorithm (Couchman 1991; Thacker and Couchman 2006) overcomes this slowing down of computations by exploiting the scale-free nature of the Laplace Greens function kernel. The Greens function for the short-range part of the calculation is again split into short range and smoothly varying part in selected subregions where the density of galaxies is large. Couchman (1991) showed that this gave an order of magnitude speedup over conventional P^{3}M for typical galaxy clustering simulation parameters, and was more efficient than tree codes.

Figure 6.6 shows a snapshot of the distribution of galaxies in a P^{3}M simulation. In it is shown a selected region of higher density over which a finer mesh with zero-padding is laid. Zero padding is used (c.f., Fig. 6.2) to remove the periodic images in the refined mesh part of the calculation. For larger density contrast, the remaining short range part may again be split and the process repeated on a finer scale.

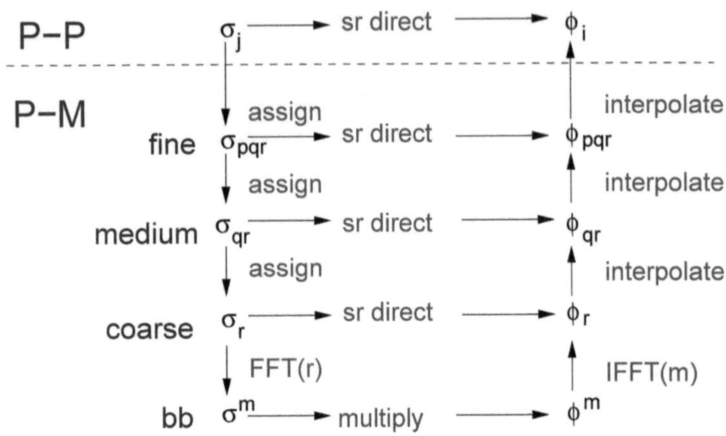

Fig. 6.7 Flow diagram for P^3M with three levels of recursive assign and interpolate

6.6 Recursive Fast Solvers

All fast solvers for the many-body problem rely on decomposing the Greens function into a sum of short range and long range parts. The convolution sums for short-range parts are evaluated explicitly, and the long-range parts are represented on meshes. P^3M variants and multigrid methods use interpolation, whereas multipole methods use extrapolation.

Brandt (1977), Rokhlin (1985) and Greengard and Rokhlin (1987) introduced the fast multipole method (FMM) for the potential many body problem in which the long-range Laplace Greens function is extrapolated from lattice points using spherical wave expansions. Recursive extrapolation on increasing scale lengths led to the tree-based multi-level fast multipole algorithm (MLFMA). More recently, Dehnen (2002) has shown that replacing the spherical wave expansion by a Cartesian Taylor expansion offer further speedup gains.

The Adaptive-P^3M algorithm brings together the nesting ideas of the tree algorithms (Appel 1985, Barnes and Hut 1986) and the FFT method. Another way of combining nesting with a FFT convolution sum evaluation is to use recursive interpolation. Inspection of Eqs. (6.6)–(6.11) shows that Eq. (6.10) has the same form as Eq. (6.7), but now for sources on the mesh. It can be decomposed in the same way as the original problem but on a coarser mesh. For the Laplace kernel as encountered in gravitational problems, this can be repeated to the size of the system and the FFT step could be eliminated since the Greens function is scale free. Figure 6.7 illustrates this for three levels of assign/interpolate. Indices p, q and r respectively label the fine, medium and coarse meshes and m is the index of the mesh harmonics.

If the method shown in Fig. 6.7 is applied to a system with a Helmholtz kernel, as arises for example in acoustic and electromagnetic scattering and quantum mechanical calculations, then the recursive interpolation is limited to meshes that are fine compared to the wavelength. The next two sections describes a further development of P^3M that overcomes this limitation.

6.7 MoM and Electromagnetic Scattering

The Block-P^3M algorithm (Eastwood 2008a, 2008b) described in the next section was motivated by the needs of electromagnetic scattering calculations using boundary integral methods, but is also applicable to many-body applications in astrophysics, molecular dynamics and biological molecule structure. It has advantages where the distribution of sources is primarily on surfaces rather than volume filling.

The convolution integrals in electromagnetic scattering are those arising in the Stratton-Chu equations (Stratton 1941). These are used in a boundary element method known as 'Method of Moments' or MoM (Harrington 1968; Gibson 2007) for antenna design, installed antenna performance and radar scattering. An attraction of this approach is that complex geometries can be modelled using only surface meshes. There are several variants of MoM using electric field, magnetic field and combined field formulations, but in its simplest electric field integral equation (EFIE) formulation the problem to be solved is to find surface currents on conductors using the condition that the tangential field is zero $\mathbf{n} \times \left(\mathbf{E}^i + \mathbf{E}^s\right) = 0$, where $\mathbf{n}$ is the surface normal, and superscripts i and s respectively refer to the (known) incident and (unknown) scattered fields.

The quadrature points for the integrals of MoM play the same role as 'particles' in particle applications of P^3M. For example, the EFIE scattered field $\mathbf{E}^s = -\nabla\varphi - i\omega\mathbf{A}$ is evaluated using the integral equations for φ and $\mathbf{A}$. Currents are approximated by finite elements on the elements and integrals are evaluated using quadrature, allowing the approximations for φ and $\mathbf{A}$ to be written in the form of Eq. (6.1)

$$\varphi(\mathbf{x}) = \frac{1}{\varepsilon}\sum_j G(\mathbf{x} - \mathbf{x}_j)Q_j; \quad \mathbf{A}(\mathbf{x}) = \mu\sum_j G(\mathbf{x} - \mathbf{x}_j)\mathbf{I}_j \tag{6.13}$$

where Q_j and $\mathbf{I}_j$ are the charge and current associated with quadrature point j. Galerkin testing the finite element formulation of the zero tangential field conditions reduces the calculation to the matrix equation V = ZI, where the voltage vector V contains amplitudes from the incident electric field, the current vector I contain the unknown amplitudes of the surface currents and Z is the impedance matrix.

The FMM was extended to electromagnetic scattering (Engheta et al. 1992; Coifman et al. 1993a, 1993b) by replacing the spherical wave expansion by a plane wave expansion and this yielded $O(N^{3/2})$ complexity. Fast multipole methods for electromagnetic scattering have been the subject of intensive research over the past two decades. The electromagnetic MLFMA (Brandt 1991; Chew et al. 2001) further reduces the complexity of FMM to O(N log N) and is now widely used for high frequency electromagnetic scattering problems.

The P^3M described above, with a Helmholtz kernel rather than a Laplace one, can be used to evaluate the sums in Eq. (6.13). Another fast method for scattering is the Adaptive Integral Method (AIM). This uses a multipole expansion on a lattice, and like P^3M uses Fast Fourier Transforms (FFTs) to evaluate the long-range part of the interaction (Bleszynski et al. 1994, 1996). The operations count scaling for AIM (and for single-mesh P^3M) degrades to $O(N^{3/2}\log N)$ for surface distributions of sources. In both AIM and P^3M the mesh sizes are limited by the signal wavelength.

6.8 The Block-P3M Method

Unlike the standard MLFMA algorithm, Block-P^3M does not break down at low frequencies, although more recent developments of MLFMA have introduced means of overcoming this problem (Darve and Havé 2004). The building blocks for a Block-P^3M/MoM solver are MoM integration routines to evaluate sources and fields at quadrature points, a Krylov solver (van der Vorst 2000) and a Block-P^3M solver. These allow fast computation of the voltage excitation vector V and the product ZI in the iterative solution for the current vector $I = Z^{-1}$ V.

Block-P^3M for MoM uses a Helmholtz kernel with the singularity removed, and is used in the following steps for the fast computation of the ZI product:

- map MoM currents to the quadrature points;
- use Block-P^3M to find fields;
- add in singular integral corrections;
- map fields at quadrature points to the MoM basis functions;
- sum field contributions to obtain the ZI product vector.

Recursive interpolation decomposes the solve step by repeating the short-range/long-range split of the Green's function and the assign/solve steps on increasingly coarse spatial meshes. A key feature of Block-P^3M is to decompose the solve step by repeatedly grouping the mesh into blocks but now applying the FFT solve only to the non-empty subset of blocks.

Block decomposition splits 1D into 2D convolution sums (i.e., 3D to 6D in $\mathbb{R}^3$) and introduces inter-block propagators H. Given Eq. (6.10), where $P = QR$, we can set $p = q + Qr$ and rewrite it as a 2D convolution

$$\varphi_{q,r} = \sum_{r'=0}^{R-1} \sum_{q'=0}^{Q-1} G_{q-q',r-r'} \sigma_{q',r'} \tag{6.14}$$

The convolution over index q is not periodic. However, zero padding σ with $(1+w)Q$ zeroes (to give ρ) and setting $H_{q,r} = F_q G_{q,r}$ gives a periodic convolution of period $(2+w)Q$ that yields $M=(2+w)Q$ values of $\psi_{q,r}$ where $\varphi_{q,r} \equiv \psi_{q,r}$ for $q \in [0, Q-1]$.

F_q is a filtering function which is 1 for $|q| < Q$ and goes to zero over the interval $Q \le |q| < (1+w)Q$. If the padding width w is zero, then F is a rectangular window of width $2Q$, but there are advantages in taking $w \approx 1$ and using smooth transitions such as cosine bells to reduce spectral infilling.

The periodic convolution (period $M=(2+w)Q$) and its Fourier transform over index q are

$$\psi_{q,r} = \sum_{r'=0}^{R-1} \sum_{q'=0}^{M-1} H_{q-q',r-r'} \rho_{q',r'} \tag{6.15}$$

$$\psi_r^k = \sum_{r'=0}^{R-1} H_{r-r'}^k \rho_{r'}^k \tag{6.16}$$

where k is the harmonic index corresponding to q, and q indices are modulo M. Any block r where $\rho_{q,r}=0$ for all q is ignored.

Inspection of Eq. (6.16) shows that the original problem [Eq. (6.10)] has been recovered, but now for different harmonics k on the large blocks with indices r. These harmonics are split into three groups, depending on the range of H_r^k. If $|H_r^k|$ are below the k-space truncation amplitude they are ignored, if H_r^k are short range (i.e., non-zero only for small $|\mathbf{x}|$ so that direct evaluation is faster than FFT evaluation), Eq. (6.16) is evaluated by direct summation, otherwise a further level of decomposition is used, setting $R = ST$ and repeating the above sequence. At the final level, all remaining harmonics are short range. The computation of the potential is completed by repeatedly computing inverse FFTs (IFFTs) at each level and combining the results with the short-range contributions until the first level is reached.

In multilevel Block-P^{3}M, the assign and interpolate steps may be single level or multilevel (c.f. Recursive fast solvers above). The solve step goes through L levels of blocks computing harmonics from the smallest blocks at level 1 to the largest at level L, and then from largest to smallest computing field values.

For level 1 to L:

- For all harmonics at the current level where G is short range, directly evaluate the sum, Eq. (6.10);
- Terminate if all harmonics have been processed, otherwise, group sub-blocks into blocks of side Q, yielding the convolution sum, Eq. (6.14);
- Zero-pad and FFT the finer mesh index q for non-empty r blocks giving for each harmonic k Eq. (6.16);
- Set $r \to p$, $R \to P$, $\psi \to \varphi$, $H \to G$, $\rho \to \sigma$ to recover Eq. (6.10) and go to the next level.

For level L to 1:

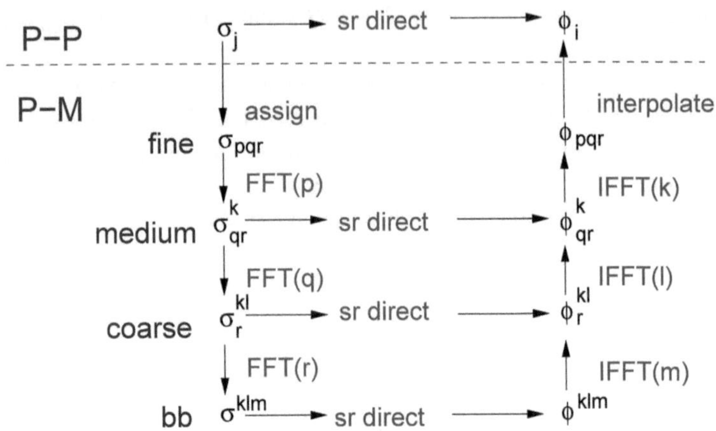

Fig. 6.8 Flow diagram for three-level Block-P^{3}M

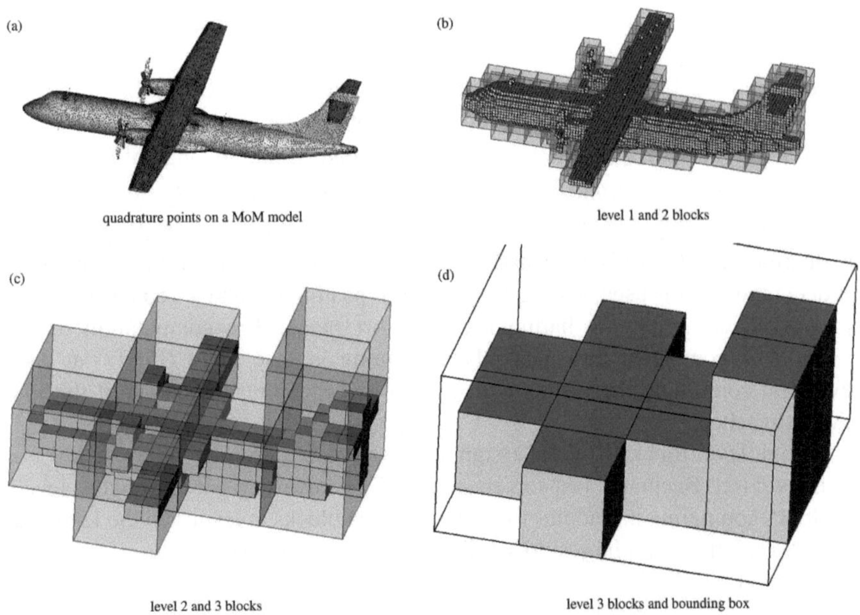

Fig. 6.9 A MoM model of an ATR aircraft for computing antenna interactions and the nested blocks used in computing the sums in Eq. (6.13) using three-level Block-P^{3}M

- Inverse FFT (IFFT) harmonics to get field values at the lower level;
- discard field values for padding index values;
- combine with directly summed terms at this level.

Figures 6.8 and 6.9 illustrate a three-level scheme for the mesh field calculation with single assign/interpolate levels. Figure 6.8 gives the flow diagram and Fig. 6.9

illustrates the steps for a real MoM model (Leclerc 2008). Sources at quadrature points (Fig. 6.9a) on the MoM model are assigned to the fine blocks (b) to give σ_{pqr}. These are grouped in the medium-level blocks [larger blocks in (b)], and the data are Fourier transformed to give σ^k_{qr} on the medium blocks. Zero-padding is used to avoid periodicity. The result is a mesh of values for each retained k on the finer mesh in (c).

For wavenumbers k where H^k_{qr} is short-range, the convolution is summed directly, and where it is long-range, the data are Fourier transformed to give σ^{kl}_r on the coarse mesh. The result is a mesh of values for each retained kl on the mesh in Fig. 6.9d. The same procedure as is used for lower levels is followed to get σ^{klm} on the bounding box.

The direct sum on the bounding box gives φ^{klm}. This is inverse transformed and combined with the directly calculated φ^{kl}_r. The process is repeated to get φ^k_{qr} and then φ_{pqr} and finally the result is interpolated to the test point positions.

The Block-P^3M algorithm overcomes the problem of the reduced performance of the original P^3M algorithm for non-uniform source distributions that are typically found in surface integral calculations. It gives O(N log N) complexity and O(N) storage for non-uniform distributions of sources and for any displacement-invariant Green's function. The algorithm is built on FFTs on uniform lattices and works equally well for Laplace and Helmholtz kernels. Unlike FMM, it does not rely on truncated divergent series, and so does not suffer breakdown when point separations become small compared to the wavelength.

The Block-P^3M algorithm is particularly attractive for electromagnetic scattering calculations using the method of moments. It has the same complexity and storage scaling as the MLFMA, but superior properties: MLFMA has a point-matched discontinuous approximation to G that is recursively extrapolated using a mixture of rectangular and polar meshes and suffers from low-frequency breakdown. In contrast, Block-P^3M has a least-squares fit, continuous (to the order of the splines used) approximation to G that is recursively interpolated on rectangular meshes, and does not suffer low-frequency breakdown.

6.9 Final Discussion

The fast fourier transform algorithm is a classical example of a divide-and-conquer algorithm, recursively splitting the $O(N^2)$ fourier sum into shorter sums until they are simple enough to be solved directly and then combined to give the complete fourier sum in O(N log N) operations. Its success relies on the harmonics being the Nth root of unity, and its importance comes from it ability to rapidly diagonalise cyclic matrices (convolution theorem).

Initially the FFT was used for periodic convolutions on a regular lattice, but as outlined in this paper it has been extended to encompass much more complex action-at-a-distance calculations. Small insights have led to big steps, and I find this reminiscent of Jim Dungey's approach as a PhD supervisor. Jim was not a self-aggrandising supervisor who told his team what to do, and indeed rarely put his name on the papers of the students he supervised. Instead, he dispensed 'pearls of wisdom'—key ideas—which often seemed unrelated to the problem in hand, but if you thought long and hard about them they would lead the way forward.

A key idea underlying all the applications considered here is the dual representation [Eqs. (6.2) and (6.3)]. Not only does this allow the FFT on a regular lattice to be applied to arbitrary particle distributions, but by serendipity its smoothing effect on short range forces suppresses collisional effects and made PM a practical approach to simulating collisionless Vlasov fluids. Zero padding, introduced to extend PM to collisionless models of galaxy evolution is also key in the Adaptive-P^3M and Block-P^3M algorithms. Introducing strict charge conservation in electromagnetic PM models has the important result that the action-at-a-distance calculation is needed only once in the initialisation of simulations.

The P^3M split of the force calculation into particle-particle and particle-mesh parts and using variational optimisation to get an accurate but computationally fast calculation is an additional divide-and-conquer step. It revolutionised the simulation of correlated systems with long-range forces. Another level of recursive refinement was added in Adaptive-P^3M to speed up strongly non-uniform distributions encountered in galaxy clustering simulations.

Although the main thrust of this article is related to particle methods, the fast convolution solver methods are equally applicable to boundary element method calculations in electromagnetics. Electrostatic and magnetostatic action-at-a-distance formulation involve the Laplace kernel and are scale free, so P^3M and recursive refinement methods are applicable. For the electromagnetic case, the Helmholtz kernel is not scale free, but surface distributions of sources can be efficiently handled using the recursive splitting of convolution sums into multidimensional sums introduce in Block-P^3M.

No doubt further ideas will be added in future to extend further the scope of FFT based methods for action-at-a-distance calculations. One aspect not discussed here, but is the focus of much study is the impact of multi-core, multi-threaded computer architectures. Divide-and-conquer algorithms are well suited to exploit these, as the splitting into distinct sub-problems allows efficient parallel execution on different processors, and smaller subproblems make for more efficient caching.

References

Appel, A.: An efficient program for many body simulation. SIAM J. Sci. Stat. Comput. **6**(1), 325–348 (1985)

Barnes, J., Hut, P.: A hierarchical O(N logN) force calculation algorithm. Nature **324**, 446–449 (1986)

Birdsall, C.K., Langdon, A.B.: Plasma Physics via Computer Simulation. Adam-Hilger, Bristol (1991)

Bleszynski, E., Bleszynski, M., Jaroszewicz, T.: A fast integral-equation solver for electromagnetic scattering problems. IEEE AP-S Int. Symp. Dig. **1**, 416–419 (1994)

Bleszynski, E., Bleszynski, M., Jaroszewicz, T.: AIM: Adaptive integral method for solving large-scale electromagnetic scattering and radiation problems. Radio Sci. **31**(5), 1225–1251 (1996)

Boris, J.P.: Relativistic plasma simulation-optimization of a hybrid code. In: Proceeding of Fourth Conference on Numerical Simulations of Plasmas, pp. 3–67. Naval Research Laboratory, Washington, DC (1970).

Bracewell, R.N.: The Fourier Transform and Its Applications. McGraw-Hill, New York, NY (1965)

Brandt, A.: Multi-level adaptive solutions to boundary-value problems. Math. Comput. **31**(138), 333–390 (1977)

Brandt, A.: Multilevel computations of integral transforms and particle interactions with oscillatory kernels. Comput. Phys. Commun. **65**(1–3), 24–38 (1991)

Brigham, E.O.: The Fast Fourier Transform and Its Applications. Prentice Hall, Englewood Cliffs, NJ (1988)

Chew, W., Jin, J.-M., Michielssen, E., Song, J. (eds.): Fast and Efficient Algorithms in Computational Electromagnetics. Artech House, Norwood, MA (2001)

Clemmow, P., Dougherty, J.: Electrodynamics of Particles and Plasmas. Addison-Wesley, Reading, MA (1969)

Coifman, R., Rokhlin, V., Wandzura, S.: The fast multipole method for the wave equation: a pedestrian description. IEEE Antennas Propag. Mag. **35**(3), 7–12 (1993a)

Coifman, R., Rokhlin, V., Wandzura, S.: The fast multipole method for electromagnetic scattering calculations. In: *IEEE Antennas and Propagation Society International Symposium*, vol. 1, pp. 48–51. IEEE, New York, NY, USA, 1993 International Symposium Digest Antennas and Propagation (Cat. No.93CH3289-6) (1993b)

Cooley, J.W., Lewis, P.A., Welch, P.D.: Historical notes on the fast Fourier transform. Proc. IEEE **55**(10), 1675–1677 (1967)

Couchman, H.: Mesh-refined P^3M: a fast adaptive N-body algorithm. Astrophys. J. **368**, L23–L26 (1991)

Darden, T., York, D., Pedersen, L.: Particle mesh Ewald: an Nlog(N) method for Ewald sums in large systems. J. Chem. Phys. **98**(12), 10089–10092 (1993)

Darve, E., Havé, P.: Efficient fast multipole method for low-frequency scattering. J. Comput. Phys. **197**(1), 341–363 (2004)

Dehnen, W.: A hierarchical O(N) force calculation algorithm. J. Comput. Phys. **179**(1), 27–42 (2002)

Eastwood, J., Hockney, R., Lawrence, D.: P3M3DP-The three-dimensional periodic particle-particle/particle-mesh program. Comput. Phys. Commun. **19**, 215–261 (1980)

Eastwood, J.W.: Consistency of fields and particle motion in the Speiser model of the current sheet. Planet. Space Sci. **20**(10), 1555–1568 (1972)

Eastwood, J.W.: The warm current sheet model, and its implications on the temporal behaviour of the geomagnetic tail. Planet. Space Sci. **22**(12), 1641–1668 (1974)

Eastwood, J.W.: Optimal P^3M algorithms for molecular dynamics simulation. In: Flooper, M.B. (ed.) Computational Methods in Classical and Quantum Physics, pp. 206–228. Advance Publications, London (1976)

Eastwood, J.W., Brownrigg, D.K.R.: Remarks on the solution of Poisson's equation for isolated systems. J. Comput. Phys. **32**, 24–28 (1979)

Eastwood, J.W.: The virtual particle electromagnetic particle-mesh method. Comput. Phys. Commun. **64**(2), 252–266 (1991)

Eastwood, J.W.: The block-P^3M algorithm. Comput. Phys. Commun. **179**(1), 46–50 (2008a)

Eastwood, J.W.: The block-P^3M algorithm for fast MoM calculations. Sci. Meas. Tech. **2**(6), 409–417 (2008b)

Eastwood, J.W., Arter, W., Brealey, N.J., Hockney, R.W.: Body-fitted electromagnetic PIC software for use on parallel computers. Comput. Phys. Commun. **87**(1), 155–178 (1995)

Eastwood, J.W., Hawkins, K.C., Hook, M.P.: The tapered MILO. IEEE Trans. Plasma Sci. **26**(3), 698–713 (1998)

Efstathiou, G., Eastwood, J.W.: On the clustering of particles in and expanding universe. Mon. Not. R. Astron. Soc. **194**, 503–526 (1981)

Engheta, N., Murphy, W., Rokhlin, V., Vassiliou, M.: The fast multipole method (FMM) for electromagnetic scattering problems. IEEE Trans. Antennas Propag. **40**(6), 634–641 (1992)

Frigo, M., Johnson, S.G.: FFTW: an adaptive software architecture for the FFT. In: *Acoustics, Speech and Signal Processing, 1998. Proceedings of the 1998 I.E. International Conference on*, vol. 3, pp. 1381–1384. IEEE (1998)

Gibson, W.C.: The Method of Moments in Electromagnetics. Chapman & Hall, Boca Raton, FL (2007)

Greengard, L., Rokhlin, V.: A fast algorithm for particle simulation. J. Comput. Phys. **73**(2), 325–348 (1987)

Harrington, R.F.: Field Computation by Moment Methods. Macmillan, New York (1968) (reprint: OUP, 1996)

Hockney, R.W.: A fast direct solution of Poisson's equation using Fourier analysis. J. ACM **12**(1), 95–113 (1965)

Hockney, R.W., Eastwood, J.W.: Computer Simulation Using Particles. McGraw-Hill, New York (1981). reprinted, Taylor & Francis Group, 1988

Langdon, A.B., Birdsall, C.K.: Theory of plasma simulation using finite-size particles. Phys. Fluids **13**, 2115 (1970)

Leclerc, D.: ATR, Toulouse, Private Communication (2008).

Lewis, H.R.: Energy-conserving numerical approximations for Vlasov plasmas. J. Comput. Phys. **6**, 136 (1970)

Marder, B.: A method for incorporating Gauss' law into electromagnetic PIC codes. J. Comput. Phys. **68**(1), 48–55 (1987)

Rokhlin, V.: Rapid solution of integral equations of classical potential theory. J. Comput. Phys. **60** (2), 187–207 (1985)

Stratton, J.A.: Electromagnetic Theory. McGraw-Hill, New York (1941). reprint: Wiley-IEEE Press, 2007

Swarztrauber, P.N.: The methods of cyclic reduction, Fourier analysis and the FACR algorithm for the discrete solution of Poisson's equation on a rectangle. SIAM Rev. **19**(3), 490–501 (1977)

Thacker, R., Couchman, H.: A parallel adaptive P^3M code with hierarchical particle reordering. Comput. Phys. Commun. **174**(7), 540–554 (2006)

van der Vorst, H.A.: Krylov subspace iteration. Comput. Sci. Eng. **2**(1), 32–37 (2000)

Chapter 7
Jim Dungey's Contributions to Magnetospheric ULF Waves and Field Line Resonances

W. Jeffrey Hughes

Abstract Jim Dungey introduced the concept of geomagnetic field line resonances in an obscure report published when he was a post-doc in 1954. This paper first describes how Dungey arrived at the idea of an Outer Atmospheric Cavity filled with a cold hydrogen plasma within which the dynamics were controlled by hydromagnetics and Alfvén waves, and how he showed that this cavity would support field line resonances. How Dungey's idea provided a ready explanation for the sinusoidal nature of magnetospheric ULF waves and led to the development of the rich field of ULF wave research are then briefly sketched.

7.1 Resonating Field Lines and Micropulsations

The idea that geomagnetic field lines might be capable of supporting a standing wave or resonance was first put forward by Jim Dungey in 1954 in an obscure report that is packed with important new ideas (Dungey 1954). This report was written during Jim's year as a post-doc at Penn State. The previous year Owen Storey (1953) had published his theory of whistler dispersion that depended on geomagnetic field lines being populated by a plasma (or at least electrons, ions then being required for charge neutrality) along their entire length. Storey used observations of whistler wave dispersion together with his new theory to deduce electron densities in the "outer atmosphere" as it was then called. Jim combined this new idea together with the established concept of the Chapman-Ferraro cavity (Chapman and Ferraro 1931a, b, 1932, 1933) to conceive of a large plasma-filled outer atmospheric cavity, which today we call the magnetosphere, within which hydromagnetics and Alfvén waves would control the dynamics.

W.J. Hughes (✉)
Department of Astronomy, Boston University, Boston, MA 02215, USA
e-mail: hughes@bu.edu

D. Southwood et al. (eds.), *Magnetospheric Plasma Physics: The Impact of Jim Dungey's Research*, Astrophysics and Space Science Proceedings 41,
DOI 10.1007/978-3-319-18359-6_7

Jim pictured the cavity filled with a hydrogen plasma in diffusive equilibrium with the upper F-region ionosphere, something like our modern concept of the plasmasphere, that filled flux tubes out to the outer boundary of the cavity, at least at low latitudes. The properties of the interplanetary medium were unknown and subject to considerable guesswork, but whatever they were, Jim realized that there would be a thin boundary containing a current that separated the outer atmospheric cavity and interplanetary space. Within this geomagnetic cavity, Jim calculated, magnetic energy density would be far greater than plasma thermal energy density. Using this approximation, the cold plasma approximation, he derived the equations for hydromagnetic waves in a rotationally symmetric magnetic field, such as a dipole, and found two modes corresponding roughly to the shear and fast modes in a uniform plasma which are coupled. He then went on to show that in the limits of both no azimuthal phase variation and very rapid azimuthal phase variation the modes decouple leading to the basic theory of field line resonances now found in textbooks (the first appearance being Dungey 1968, eqs. 22–26). Figure 7.1 illustrates these resonances. Jim understood that to first approximation the Earth and ionosphere would together act as a very good conductor, reflecting incident waves with little loss and effectively tying the ends of the field lines as the electric field must be very small. Given this boundary condition, a wave mode that is guided along the magnetic field, such as the shear Alfvén mode, will set up resonant oscillations or eigenmodes on a field line. The poloidal mode, illustrated on the right in Fig. 7.1, has field lines oscillating in the meridian plane and occurs for rapid azimuthal phase variation. In the toroidal mode which occurs for no azimuthal phase variation, field lines are displaced azimuthally, so that the whole magnetic shell oscillates. Table 7.1, taken from Jim's (Dungey 1954) report, shows his estimates of the fundamental resonant period of a field line versus the geomagnetic latitude of its foot-point. Given the almost wild guesses Jim had to make, these estimates are in remarkably good agreement with the values we would calculate and observe today.

In 1954 observations of magnetospheric ULF waves, or micropulsations as they were then called, were rudimentary. However Jim realized that his resonant field lines might well explain why these geomagnetic variations have a much more sinusoidal character than any other geomagnetic disturbance. As we will see below it took another 20 years for this to be clearly established observationally.

In another section of the 1954 report Jim proposed that Kelvin-Helmholtz waves excited on the outer boundary of the geomagnetic cavity might be a source of these waves, and derived an initial theory for this instability (see also Dungey 1955). However so little was known about the interplanetary medium, that he thought that the Earth's orbital velocity (30 km s^{-1}) through a stationary interplanetary gas rather than the solar wind might be the cause of the velocity shear.

The rather obscure Penn State report, parts of which were republished some years later as part of a contribution Jim wrote based on summer school lectures he presented (Dungey 1963a), laid the foundations for the theory of magnetospheric ULF waves and the whole subdiscipline that emerged. Jim continued to be active in this field for the rest of his career, making contributions to both wave excitation by wave particle interactions and to the effect of the ionosphere on ULF waves, which is the subject of the next section.

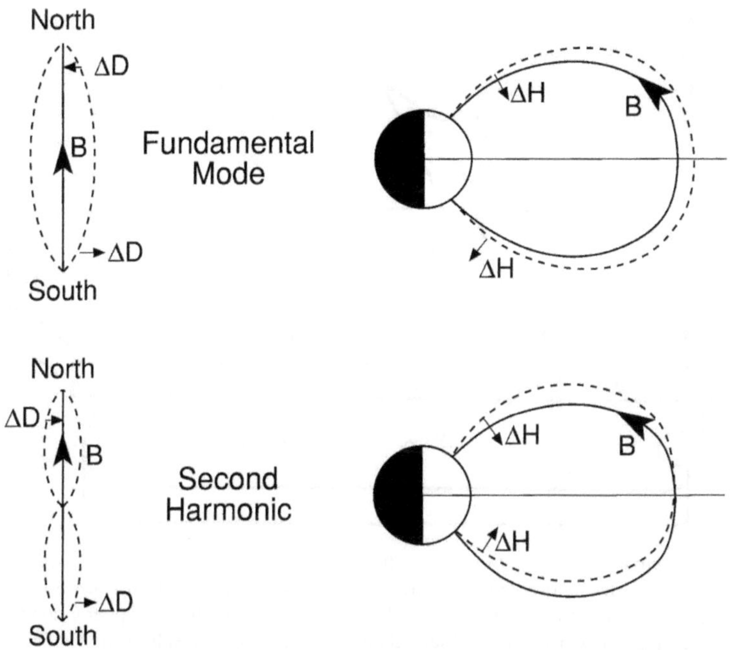

Fig. 7.1 Cartoons showing the oscillation of a geomagnetic field line in the fundamental (*top*) and second harmonic (*bottom*) of the toroidal mode (*left*) and poloidal mode (*right*). On the *left* we look towards the earth with the field line stretched out from north to south. The field line motion and the magnetic perturbations are both in the east-west or azimuthal direction (ΔD). On the *right* the field line is displaced within a meridian plane so both the motion and magnetic perturbations are in the radial direction (ΔH) (after Hughes 1994)

Table 7.1 Resonant period versus latitude

λ_0	τ
45°	10 s
55°	54 s
65°	11 min
70°	55 min

From Dungey (1954)

7.2 The Effect of the Ionosphere

Until the late 1960s, when spacecraft magnetometers in geostationary orbit began to observe them in space, the only way to observe a magnetospheric ULF wave was with a ground-based magnetometer. And for a long time after that ground-based observations provided the only way to observe the spatial variation of a ULF wave. As a result the ionosphere plays two important functions in the study of ULF waves: it provides the boundary condition at the ends of a resonating field line as

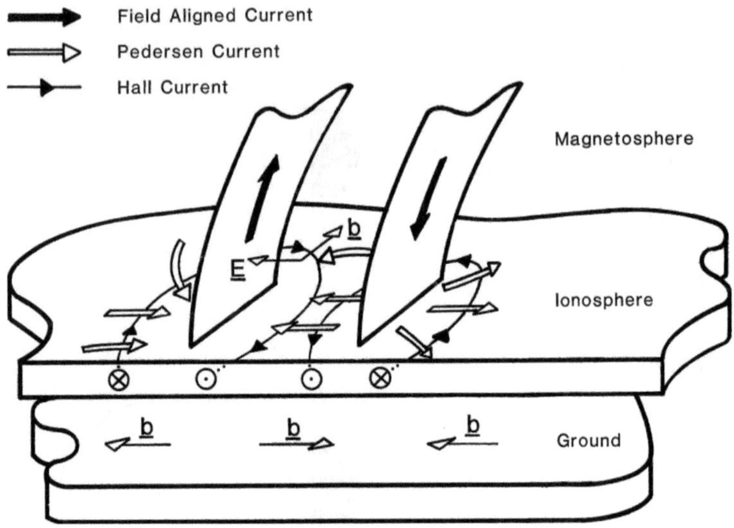

Fig. 7.2 A schematic representation of how the field aligned currents associated with a field line resonance close in the ionosphere via Pedersen currents. As these currents are solenoidal they create little or no magnetic perturbation on the ground provided the ionospheric conductivity is uniform. The magnetic perturbation observed on the ground is the result of the Hall currents that close on themselves within the ionosphere (after Southwood and Hughes 1983)

recognized by Jim in the 1954 Penn State Report, and it also modifies the wave signal before it is detected at ground level.

From the outset Jim realized that the solid earth, neutral atmosphere and ionosphere must be thought of as a whole system when considering their effect on resonating field lines. ULF waves which typically have periods of tens of seconds have wavelengths in free space and in the magnetosphere that are far larger than the height of the ionosphere, making this entire boundary effectively a thin sheet. In 1963 Jim published a short paper pointing this out and showing the importance of including the horizontal variation of the signal in the problem (Dungey 1963b). But he realized that the problem wasn't fully solved, and after I joined his group as a graduate student in 1971 he suggested that I work on it as part of my thesis project.

Although he understood that the anisotropic conductivity of the ionosphere would couple wave modes, Jim did not explore this effect in his 1963 paper. It turned out that the anisotropic conductivity when combined with a horizontal variation in the signal was the critical effect. If the ionosphere were simply a Pedersen conductor the field aligned currents of the transverse Alfvén wave would simply close via Pedersen currents in a solenoidal pattern, effectively screening the ULF wave magnetic signal from the ground (see Fig. 7.2). But the Hall conductivity creates another current that closes on itself (provided that the conductivity is spatially uniform) causing a different magnetic variation on the ground that is rotated through a right angle from the magnetospheric wave above

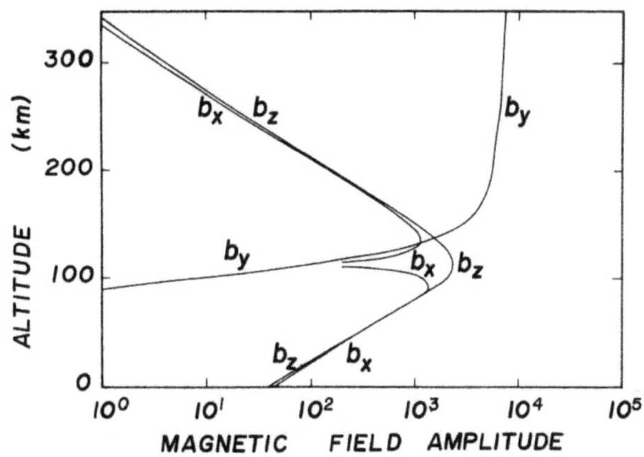

Fig. 7.3 The altitude variation of the components of the magnetic field of a monochromatic Alfvén wave incident on the ionosphere. See text for detailed explanation (after Hughes and Southwood 1976)

the ionosphere (Hughes 1974; Hughes and Southwood 1974). This effect is illustrated more quantitatively in Fig. 7.3, which is the result of a numerical integration of the wave equations through a realistic earth-atmosphere-ionosphere conductivity structure. The wave is monochromatic and has a horizontal variation, $k_\perp$, in the x-direction. The Alfvén mode in the magnetosphere is polarized with the magnetic perturbation, **b**, perpendicular to **k**, so in the y-direction. However, below the ionosphere where the non-conducting atmosphere precludes any vertical current, $(\text{curl } \mathbf{b})_z = 0$ so the magnetic perturbation has to be parallel to **k**, i.e., in the x-direction. The direction of the wave magnetic field changes in a thin region just above 100 km altitude. This is where the conductivity peaks in the E-region ionosphere and hence where the major currents flow.

It is characteristic of Jim that his name does not appear on the publications describing this work even though he provided much of the impetus and motivation behind it. He tended to have his students publish by themselves or just with other collaborators, which means that it is easy to underestimate his contributions to the development of magnetospheric physics when simply looking at the published record.

This ionospheric rotation of the wave polarization helped make sense of the ground-based observations coming from the magnetometers in Canada and elsewhere. By the end of the decade observations from the new ionospheric radars provided the first solid evidence of field line resonance structures at the foot of the field lines in the ionosphere (e.g., Walker et al. 1979). The radars observed the latitudinal variation of the wave amplitude and phase that had been predicted theoretically by Southwood (1974) and Chen and Hasagawa (1974) earlier in the decade.

Direct confirmation of the ionospheric rotation proved observationally challenging, and was further complicated by the fact that when spatial variation in the ionospheric conductivity is included, the rotation, though significant, may not be by exactly a right angle. A recent observation by Ponomarenko and Waters (2013) using rapidly sampled beams from two coherent scatter ionospheric radars, one located in Tasmania and the other in New Zealand, that intersect over a magnetometer located on Macquarie Island allowed them to directly measure the polarization of a Pi2 pulsation simultaneously in the ionosphere and on the ground, confirming the rotation.

7.3 Probing the Magnetosphere with Field Line Resonances

By the 1980s digital spectral data analysis and computer displays had advanced to the point, and spacecraft magnetometer data was clean enough, that Dungey's field line resonances could be dramatically and beautifully confirmed directly using spacecraft data. When the dayside magnetosphere is quiet enough that the signal is not hidden by larger disturbances, each geomagnetic field line can be seen to be ringing with multiple resonant harmonics. Figure 7.4, taken from Takahashi et al. (1984), shows dynamic spectra of the magnetic field observed by three spacecraft in geosynchronous orbit over a 24 h period (one orbit). Multiple harmonics equally spaced in frequency, in one case going up to the sixth harmonic, are clearly visible on the 10–100 mHz range, i.e., with periods between 10 and 100 s. corresponding to the Pc3 geomagnetic pulsations. The frequencies of the harmonics are set by the length of the field line and the variation of the Alfvén velocity along it. The frequencies slowly vary throughout the day as the spacecraft move onto field lines of different lengths and plasma densities. If the length and shape of the field line are known (or can be estimated using a geomagnetic model) the plasma mass density on that field line can be determined. As spacecraft instrumentation has improved, so has our ability to observe these almost omnipresent small amplitude harmonics. Magnetometer data from the recently launched Van Allan Probes have detected resonances up to the 11th harmonic (Craig Kletzing, GEM Workshop tutorial lecture, June 2013).

Using the frequencies of geomagnetic field line resonances, observed either from the ground or from space, to deduce magnetospheric plasma densities is now a standard practice. Figure 7.5 shows an example where this technique, sometimes referred to as magnetoseismology, is used. By combining the plasma mass density estimates with measurements of the electron number density, the mean ion mass is obtained providing clues to the ionic composition of the magnetospheric plasma. The data on the right are taken during the outbound and inbound legs of the orbit of the CRRES spacecraft that is illustrated on the left of the figure. The top panels show the frequency of the fundamental harmonic of the field line as a function of the spacecraft's L shell (a parameter that labels the geomagnetic field line on which the spacecraft is located using the distance from the center of the earth to the point on that field line furthest from the Earth). During both outbound and inbound legs the frequencies drop from about 6 to 2 mHz as the spacecraft moves onto longer field lines, going from $L \approx 4$ to $L \approx 6.5$. The center panels show the plasma mass densities (black dots) derived from these frequencies as well as the electron number

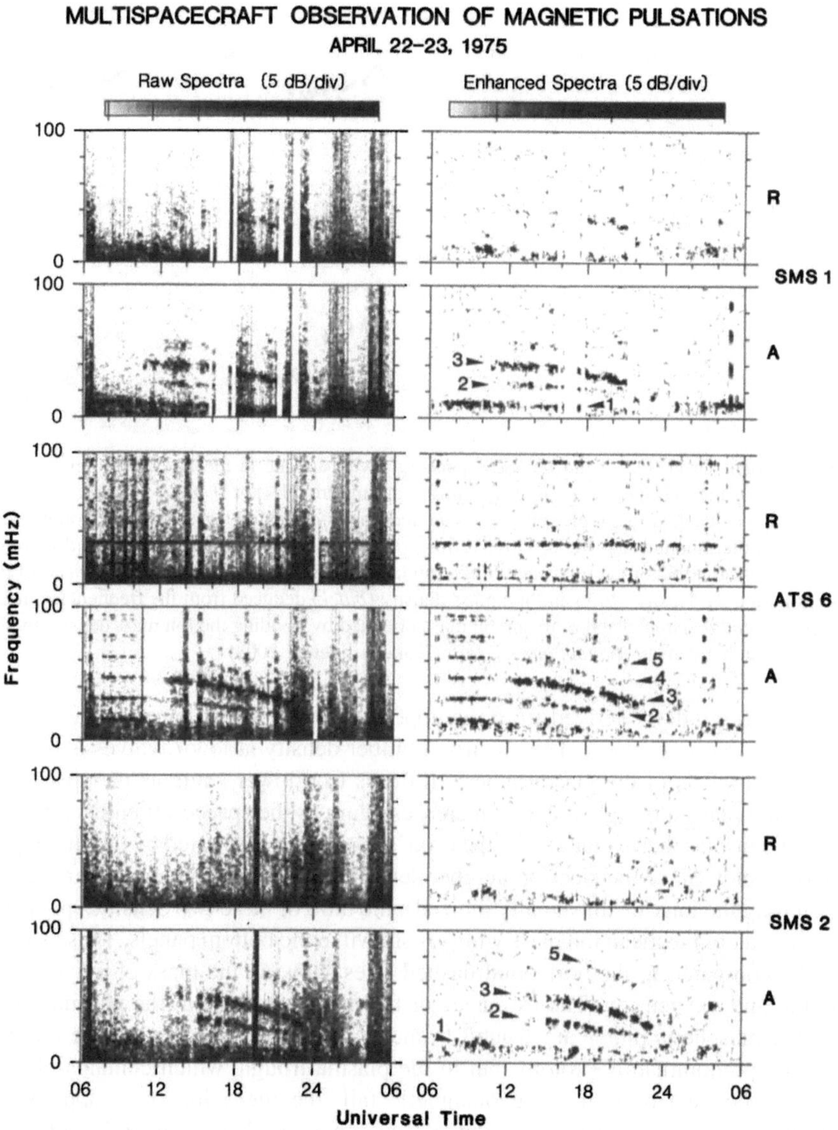

Fig. 7.4 Dynamic spectra of the variations observed in the radial (R) and azimuthal (A) components of the magnetic field by three geosynchronous spacecraft over a 24 h period. Local noon is near the center of the spectra. Multiple harmonics approximately equally spaced in frequency are evident especially in the azimuthal components. These are the signatures of field line resonant harmonic oscillations (after Takahashi et al. 1984)

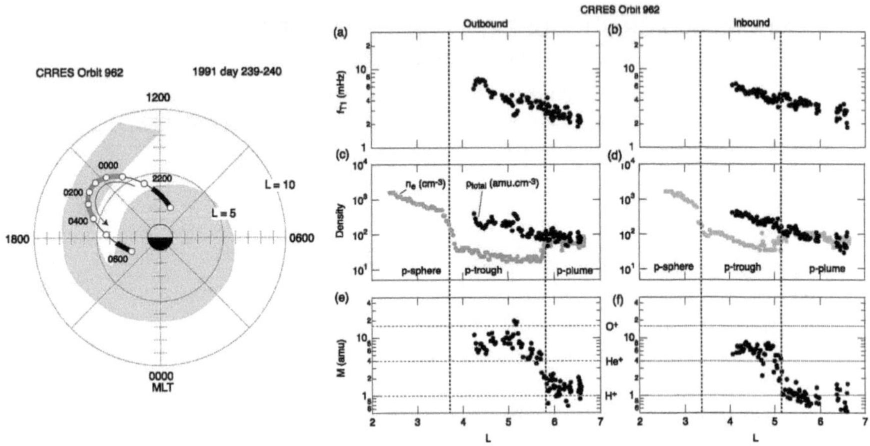

Fig. 7.5 *Left*: Orbit 962 of the CRRES spacecraft with apogee in the late afternoon. CRRES passed out of the plasmasphere (*shaded*) into the plasma trough (*not shaded*) then into a plasma plume shown attached to the plasmasphere (*shaded*) on its outbound leg, and back through these regions on the inbound leg. *Right*: Observations made during the outbound (*right*) and inbound (*left*) legs of orbit 962 of the frequency of the fundamental field line resonance (*top panels*) of the electron number density (*grey*) and ion mass density (*black*) deduced from the frequencies in the *top panel* (*middle panels*) and of mean ion mass obtained by dividing the ion mass density by the electron number density (*bottom panels*) (after Takahashi et al. 2008)

densities (grey dots) derived from the upper hybrid frequency measured using the CRRES wave instrument. The electron number density at low L valves is high as the spacecraft is in the plasmasphere, it drops to a lower value as the spacecraft enters the plasma trough and then increases again as the spacecraft enters a higher density plasmaspheric plume near the orbit's apogee, as illustrated by the shading in the orbit plot. Since the electron number density must equal the ion number density (assuming the ions are all signally ionized), the ratio of these two densities provides an estimate of the mean ion mass which is shown in the bottom panels. These values can be compared to the horizontal dashed lines showing the mass of a hydrogen, helium and oxygen ion. The estimates of the mean ion mass in the plasmasphere and plasmaspheric plume are around 1 amu or not much larger, indicating that the ions are predominantly protons, but in the plasma trough, which contains plasma that has convected from the geomagnetic tail, the mean ion mass approaches 10 amu, indicating a large fraction of oxygen ions. Thus Dungey's field line resonances can be used to show that the composition of these two plasma populations are quite different, indicating their different origins in spite of their proximity.

7.4 ULF Wave-Particle Resonances

Another important aspect of Dungey's work on ULF waves is his work on ULF resonant wave-particle interactions (Dungey 1963a, 1964; Southwood et al. 1969; Dungey and Southwood 1970). Jim recognized early that energetic particles trapped in the radiation belts could resonantly interact with ULF waves standing on a field line through the particles' bounce and drift motions (Dungey 1964), somewhat analogous to wave-particle gyroresonance. If the ULF wave variation in time and azimuth are represented as exp $i\ (\omega t - m\phi)$ then the important parameters are the ULF wave frequency, ω, and azimuthal wave number, m, and the particle's bounce and drift frequencies, ω_b and ω_d. Resonance occurs when $\omega - m\omega_\mathrm{b} = N\omega_\mathrm{d}$ where N is an integer (Dungey 1964). Jim recognized immediately that the symmetry of the wave mode was important. Resonance only occurs for N an odd integer if the wave electric field is antisymmetric about the equator, as in a second harmonic oscillation (see Fig. 7.1) and for N even if the wave is symmetric. Somewhat later Southwood and Kivelson (1982) graphically illustrated this using the diagram reproduced in Fig. 7.6 which illustrates the drift of bouncing ions through an antisymmetric wave. In this example resonance only occurs on the left where the ions drift one azimuthal wave length during each bounce, or $N = 1$, and not on the right where $N = 2$. Using a similar figure it is easy to show that symmetric modes

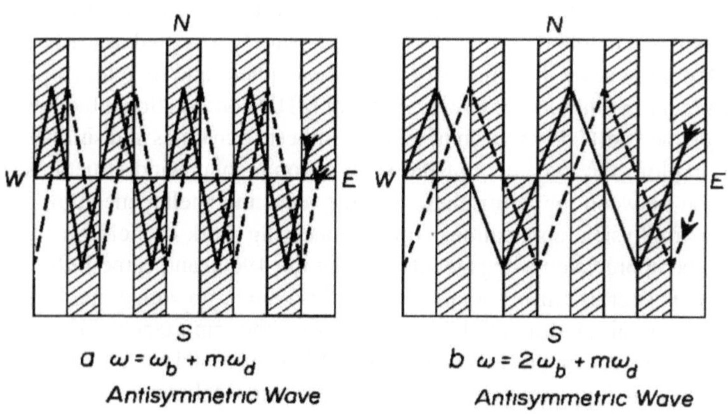

Fig. 7.6 Sketches showing the paths of trapped energetic ions that are both bouncing back and forth along field lines and drifting westward through a standing ULF wave. The *plane* depicts the drift or *L* shell. Field lines are *vertical straight lines* and the *horizontal line* is the equator. The ULF wave is a second harmonic (see Fig. 7.1) for which the direction of the azimuthal electric field (*shading*) is antisymmetric about the equator. On the *left* particles drift westward one azimuthal wavelength per bounce period, while on the *right* they drift two wavelengths per bounce. The particle on the *left* moving along the *solid line* remains in *shaded areas* so experiences only one direction of electric field hence is either continually accelerated or decelerated by the wave depending on the direction of the field. All other particles pass through equal amounts of *shaded* and *unshaded region* over the course of each bounce so receive no net acceleration (after Southwood and Kivelson 1982)

only resonate when N is even. Again, Dungey's initial insight was not appreciated by others until much later.

After ULF waves were first observed at geosynchronous orbit (Cummings et al. 1969) Jim thought that such sinusoidal waves could only be generated by a wave-particle mechanism and not by a hydromagnetic instability such as the Kelvin-Helmholtz instability (Dungey and Southwood 1970). Pursuing this track, Jim pushed to get good observational estimates of the value of the azimuthal wave number m. The first measurements of m made on the ground (Green 1976) and in space (Hughes et al. 1978) were both the result of his efforts.

Dungey's early work on ULF wave-particle resonance, which was largely ignored at the time, laid the foundation for much subsequent work, which continues today as we try to understand the connection between enhanced ULF wave activity and the enhancements observed in radiation belt particle fluxes during the passage of high speed solar wind streams.

7.5 Conclusion

Jim Dungey's insightful recognition in 1954 of the existence of an Outer Atmospheric Cavity filled with a cold plasma whose dynamics can be described using hydromagnetics was an idea way ahead if its time. At the time hydromagnetics was a new idea that few people had taken the time to understand, in sharp contrast to today when magnetohydrodynamics (MHD) is widely accepted as the best description of the large scale dynamics of the magnetosphere and its interaction with the interplanetary medium. Today several global MHD computational models compete to provide the "best" description of magnetospheric dynamics and similar codes are used to explore the dynamics of the solar corona, solar wind structures, and the interaction between the heliosphere and the local interstellar medium, as well as similar astrophysical applications. Jim's pioneering work on field line resonances began to be appreciated a decade later during the 1960s and came fully to the fore two decades later as our observational abilities caught up with Jim's foresight.

As with the initial idea of field line resonances, the importance of Dungey's early work on the effect of the ionosphere on ULF waves and on ULF wave particle resonances were both largely initially unappreciated. Only later did the importance of these ideas become apparent. As often happened with Dungey's ideas, the work continued to be developed at his urging by his students and others, usually without explicit acknowledgement of Dungey's contribution, so that by the time the importance of the idea became generally apparent, Dungey's leading role in its development was far from clear.

Although Dungey is best known for his work on magnetic reconnection and its control of magnetospheric dynamics (Dungey 1961), his earlier work on field line resonances provided the foundation for an entirely different field of magnetospheric physics, the study of the ULF waves that are ubiquitous. This is but another example of how Jim Dungey's pioneering work had a profound influence on almost all subsequent research in magnetospheric physics.

Acknowledgements I thank David Southwood and others for organizing the symposium to celebrate Jim Dungey's 90th birthday, and for inviting me to participate and to write this paper, and the Royal Astronomical Society for hosting the event. But most of all I thank Jim for the stimulation, help, example, and encouragement he gave me first as his graduate student and then throughout my career. He made it all possible.

References

Chapman, S., Ferraro, V.C.A.: A new theory of magnetic storms. Terr. Magn. Atmos. Electr. **36**, 77–97 (1931a)

Chapman, S., Ferraro, V.C.A.: A new theory of magnetic storms. Terr. Magn. Atmos. Electr. **36**, 171–186 (1931b)

Chapman, S., Ferraro, V.C.A.: A new theory of magnetic storms. Terr. Magn. Atmos. Electr. **37**, 147–156 (1932)

Chapman, S., Ferraro, V.C.A.: A new theory of magnetic storms. Part II. The main phase. Terr. Magn. Atmos. Electr. **38**, 79 (1933)

Chen, L., Hasagawa, A.: A theory of long period magnetic pulsations: 1. Steady state excitation of field line resonance. J. Geophys. Res. **79**, 1024–1032 (1974)

Cummings, W.D., O'Sullivan, R.J., Coleman Jr., P.J.: Standing Alfvén waves in the magnetosphere. J. Geophys. Res. **74**, 778–793 (1969)

Dungey, J.W.: Electrodynamics of the outer atmosphere. Scientific Report 69, Ionosphere Research Laboratory, Pennsylvania State University, University Park, PA (1954)

Dungey, J.W.: Electrodynamics of the outer atmosphere. In: Proceedings of the Ionosphere Conference, p. 255. The Physical Society, London (1955)

Dungey, J.W.: Interplanetary magnetic field and the auroral zones. Phys. Rev. Lett. **6**, 47–49 (1961)

Dungey, J.W.: The structure of the exosphere or adventures in velocity space. In: Dewitt, C., Hieblot, J., Lebeau, A. (eds.) Geophysics: The Earth's Environment, pp. 503–550. Gordon and Breach, New York (1963)

Dungey, J.W.: Hydromagnetic waves and the ionosphere. In: Proceedings of the International Conference on the Ionosphere, pp. 230–232. Institute of Physics, London (1963b)

Dungey, J.W.: Effects of electromagnetic perturbations on particles trapped in the radiation Belts. Space Sci. Rev. **4**, 199–222 (1964)

Dungey, J.W.: Hydromagnetic waves. In: Matsushita, S., Campbell, W.H. (eds.) Physics of Geomagnetic Phenomena, p. 913. Academic Press, New York (1968)

Dungey, J.W., Southwood, D.J.: Ultra low frequency waves in the magnetosphere. Space Sci. Rev. **10**, 672–688 (1970)

Green, C.A.: The longitudinal phase variation of mid-latitude Pc 3-4 micropulsations. Planet. Space Sci. **24**, 79–85 (1976)

Hughes, W.J.: The effect of the atmosphere and ionosphere on long period magnetospheric micropulsations. Planet. Space Sci. **22**, 1157–1172 (1974)

Hughes, W.J.: Magnetospheric ULF waves: a tutorial with a historical perspective. In: Engebretson, M.J., Takahashi, K., Scholer, M. (eds.) Solar Wind Sources of Magnetospheric Ultra-Low-Frequency Waves. American Geophysical Union Geophysical Monograph Series, vol. 81, pp. 1–11. American Geophysical Union, Washington, DC (1994)

Hughes, W.J., Southwood, D.J.: Effect of atmosphere and ionosphere on magnetospheric micropulsation signals. Nature **248**, 493–495 (1974)

Hughes, W.J., Southwood, D.J.: The screening of micropulsation signals by the atmosphere and ionosphere. J. Geophys. Res. **81**, 3234–3240 (1976)

Hughes, W.J., McPherron, R.L., Barfield, J.N.: Geomagnetic pulsations observed simultaneously on three geostationary satellites. J. Geophys. Res. **83**, 1109–1116 (1978)
Ponomarenko, P.V., Waters, C.L.: Transition of Pi2 ULF wave polarization structure from the ionosphere to the ground. Geophys. Res. Lett. **40**, 1474–1478 (2013). doi:10.1002/grl.50271
Southwood, D.J.: Some features of field line resonances in the magnetosphere. Planet. Space Sci. **22**, 483–491 (1974)
Southwood, D.J., Dungey, J.W., Etherington, R.J.: Bounce resonant interaction between pulsations and trapped particles. Planet. Space Sci. **17**, 349–361 (1969)
Southwood, D.J., Hughes, W.J.: Theory of hydromagnetic waves in the magnetosphere. Space Sci. Rev. **35**, 301–366 (1983)
Southwood, D.J., Kivelson, M.K.: Charged particle behavior in low-frequency geomagnetic pulsations 2. Graphical approach. J. Geophys. Res. **87**, 1707–1710 (1982)
Storey, L.R.O.: An investigation of whistling atmospherics. Philos. Trans. R. Soc. Lond. **246**, 113–141 (1953)
Takahashi, K., McPherron, R.L., Hughes, W.J.: Multispacecraft observations of the harmonic structure of Pc3-4 magnetic pulsations. J. Geophys. Res. **89**, 6758–6774 (1984)
Takahashi, K., Ohtani, S.-I., Denton, R.E., Hughes, W.J., Anderson, R.R.: Ion composition in the plasma trough and plasma plume derived from a combined release and radiation effects satellite magnetoseismic study. J. Geophys. Res. **113**, A12203 (2008). doi:10.1029/2008JA013248
Walker, A.D.M., Greenwald, R.A., Stuart, W.F., Green, C.A.: STARE auroral radar observations of Pc 5 geomagnetic pulsations. J. Geophys. Res. **84**, 3373–3388 (1979)

Chapter 8
The Science of the Cluster Mission

Matthew G.G.T. Taylor, C. Philippe Escoubet, Harri Laakso, Arnaud Masson, Mike Hapgood, Trevor Dimbylow, Jürgen Volpp, Silvia Sangiorgi, and Melvyn L. Goldstein

Abstract In 1966, in the concluding part of his inaugural lecture at Imperial College London, Jim Dungey discussed the future of magnetospheric physics, in particular indicating that progress in the field required "bunches" of satellites. Indeed, the previous year Dungey had submitted a proposal to the European Space Agency's predecessor ESRO (European Space Research Organisation) proposing the launch of bunches of spacecraft into the magnetosphere. However it was not until 2000, following the successful 1982 proposal led by G. Haerendel, that the first four spacecraft mission, Cluster, was initiated. This paper provides a select few highlights of the Cluster mission related specifically to some objectives presented in the 1960s by Dungey. In addition, we will indicate future prospects for Cluster, in particular coordination of a number of multi-spacecraft missions—Cluster, THEMIS, Van Allen Probes and Swarm, approaching "bunches of bunches" of satellites.

8.1 Introduction

This paper was presented as part of the celebration of Professor J. W. Dungey's 90th Birthday in January 2013 and focuses on the pioneering ideas of Professor Dungey in the realm of multiple satellite missions. Following a brief overview of the current mission status, in light of the 90th birthday celebrations, we relate some of the key

M.G.G.T. Taylor (✉) • C.P. Escoubet • H. Laakso • A. Masson
ESA/ESTEC, D-SRE, Keplerlaan 1, 2200 AG Noordwijk, The Netherlands
e-mail: mtaylor@esa.int

M. Hapgood • T. Dimbylow
RAL Space/STFC, Harwell, Oxford, UK

J. Volpp • S. Sangiorgi
ESA/ESOC, Darmstadt, Germany

M.L. Goldstein
Code 672, NASA/GSFC, Greenbelt, MD 20771, USA

D. Southwood et al. (eds.), *Magnetospheric Plasma Physics: The Impact of Jim Dungey's Research*, Astrophysics and Space Science Proceedings 41,
DOI 10.1007/978-3-319-18359-6_8

ideas put forward in the 1960s by Professor Dungey, in terms of the capability of "bunches of satellites", to a selection of scientific achievements of the current Cluster mission. Finally we look forward to the future of Cluster and collaboration with other missions, and the potential of "bunches of bunches" of satellites.

In March 1965, J. W. Dungey submitted a proposal (Appendix, taken from ESRO-4897 1966) to the European Space Agency's predecessor, ESRO (European Space Research Organisation), through the British National Committee on Space Research. The Tetrahedral Observatory Probe System (TOPS) was suggested for the Second Large ESRO project (SLEP) (Ulivi 2006) and called for the launch of several identically instrumented satellites in orbits that enabled them to move as a close group. This "bunch" of satellites would remove the ambiguity of temporal vs. spatial variations, with four providing the minimum number by which one could distinguish between curvature or planarity of sheet-like structures. Simultaneous magnetic field measurements from these four spacecraft would also allow the evaluation of the curl of the magnetic field, providing direct measurement of current systems. In the following year, concluding his inaugural lecture on the Magnetosphere, at Imperial College, Dungey discussed multiple satellites:

> Looking to the future I believe that progress requires bunches of satellites, though these are as yet in no published program.
>
> When one comes to study waves, bunches of satellites are also needed from several point of view. First one wants to know the geometry of the waves and second their direction of propagation. For any magnetic disturbance it would be extremely useful to obtain the curl of the magnetic field because this tells one the current.
>
> Unfortunately very few people yet appreciate the need for satellite bunches and, since satellites are being launched singly, the scientific returns are less than they could be.

However, it was not until the 1980s that such a mission concept really began to take shape. ESSAIM, meaning swarm of bees in French, was a project presented to CNES in 1981 at a meeting held in "les Arcs" in the French Alps (A. Roux, private communication). CNES recommended the formation of a definition team, which was chaired by A. Roux and M. Blanc, with the support of R. Pellat. The mission concept involved an equatorial orbit covering the plasma sheet, magnetopause and bow shock, with a tetrahedral spacecraft configuration. At the same time a German–Swedish concept, led by B. Hultqvist, G. Haerendel and G. Paschmann, of two spacecraft in the exterior cusp regions was being studied.

In November 1982, the key aspects of these studies were combined. The proposal, led by Haerendel (with Dungey listed as a contributor), was entitled "Cluster: Study in three dimensions of plasma turbulence and small-scale structure" (Haerendel et al. 1982) and was submitted in response to a call for new scientific missions in the ESA Horizon 2000 programme. At the time, focus from NASA was on coordinated global (large scale) investigations of the magnetosphere/solar wind system, in collaboration with the USSR and Japan. By contrast the Cluster mission was a mother and three closely separated daughter spacecraft:

> . . . directed towards the in-situ study of certain processes and phenomena that are ubiquitous in cosmic plasmas. The location of this study is the outer magnetosphere, because it is

> the closest object of this kind. The mission is designed to focus on small-scale structures and macroscopic (MHD) turbulence which arise in many places in the magnetosphere. Such regions are: the bow shock, the magnetosheath (i.e., the region of shocked solar wind plasma), the magnetopause, the cusp regions, the boundary layers, and the plasma sheet of the tail.
>
> The purpose of the CLUSTER mission is to perform in-situ 3D measurements, by means of four simple spacecraft appropriately located in space.
>
> . . .such a system is the only means by which localized or turbulent plasma processes can be investigated or understood.

Following an assessment and Phase-A study, in February 1986 Cluster, alongside the Solar and Heliospheric Observatory (SOHO) mission, was selected by the ESA Science Programme Committee for implementation, the two missions becoming the first Cornerstone of the new Horizon 2000 programme. Soon after financial constraints led to the adoption of four identical spacecraft, more in line with the "bunch" discussed by Dungey in 1966. After 10 years of design, development and testing, the four spacecraft were placed on top of the Ariane-5 for its maiden launch. Unfortunately, following a catastrophic launch failure, all four spacecraft were lost. Finally in April 1997, following a great deal of work by the community examining a variety of revivals of the mission concept, including a single spacecraft 'Phoenix' concept and the original Cluster, a 'Phoenix' along side three smaller satellites, the SPC approved the Cluster-II mission, a full revival of the four identical spacecraft concept. Fast-forwarding to 2000, the pairs of spacecraft were successfully launched by Soyuz in July and August from the Baikonur Cosmodrome (Escoubet et al. 2001).

8.2 Cluster Today

To target the required science regions, the Cluster spacecraft were put into a ($4 \times 19.6\ R_E$) polar orbit (period ~56 h). At the time of writing, Cluster has far exceeded its nominal 2-year mission lifetime, and has a currently approved mission extension up to the end of 2014. We expect to propose for a further mission extension up to end of 2016.

As the mission has continued, proposals for extension have been driven by new science objectives rather than a continuation of the original mission goals. Each time, the choice of objectives has been influenced by the mission's growing collection of new scientific discoveries and operational experience, and has taken advantage of the capability of changing the spacecraft separations to make measurements on new scales, relevant to new science goals.

Further opportunities to perform unique science have been presented by the changing inclination and perigee height of the evolving Cluster orbit (arising through lunar–solar gravitational perturbations), which has provided access to

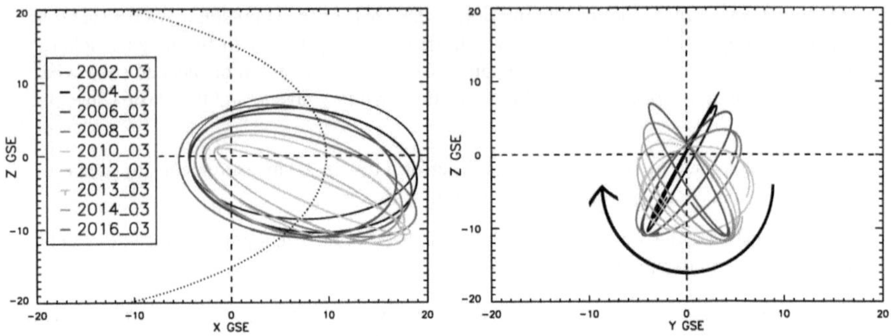

Fig. 8.1 A broad sample of Cluster orbits from 2002 up to 2016 in Geocentric Solar Ecliptic (GSE) coordinates. The *dotted line* on the *left hand plot* indicates the nominal magnetopause position. The plot clearly demonstrates the large evolution of the orbit, in particular in the *left-hand* Z–X plot, the *line* of apsides has evolved such that the apogee has moved further below the ecliptic until ~2012 where it has begun to climb back up. We note that the orbit in 2014–2016 has a similar but mirror image tilt (in the YZ plane) to 2006–2008. In the *right-hand plot* the evolution of the tilt of the orbit plane can clearly be seen, such that the ascending node has moved away from perigee towards apogee. The impact of this is that Cluster moves in the positive Z direction at apogee after ~2012

new regions of near-Earth space, such as the auroral acceleration region and the subsolar point. This evolution is exemplified in Fig. 8.1, where we show representative orbits from across the mission timeline, up to 2015/2016, a period currently being proposed for a further mission extension.

During the course of the mission, the relative distance between the four spacecraft has been varied to form a nearly perfect tetrahedral configuration at 100, 250, 600, 2,000, 5,000 and 10,000 km inter-spacecraft separation targeted to study scientifically interesting regions at different scales, as indicated in Haerendel et al. (1982), and alluded to in the TOPS proposal:

> Variation of the satellite separation would lead to information over a range of structure sizes

Figure 8.2 indicates the range of configuration changes over the period of the mission lifetime. In more recent years, since summer 2005, the mission has implemented multi-scale operations. In this configuration, three of the spacecraft (Cluster 1, 2 and 3) form a large triangle, while Cluster 3 and 4, whose orbits are very similar, can be drifted with respect to one another and their separation can be varied from a few 10 s of km to 10,000 km. In addition, new science has been possible through the synergies afforded by collaborations with other missions. Most recently, during the current extension, Cluster has implemented a Guest Investigator (GI) programme, through which the scientific community was invited to propose their own science objectives and to request specific spacecraft and instrument operations needed to deliver that science. This was the first time such an activity has been carried out in the field of in-situ space plasma research. As indicated in

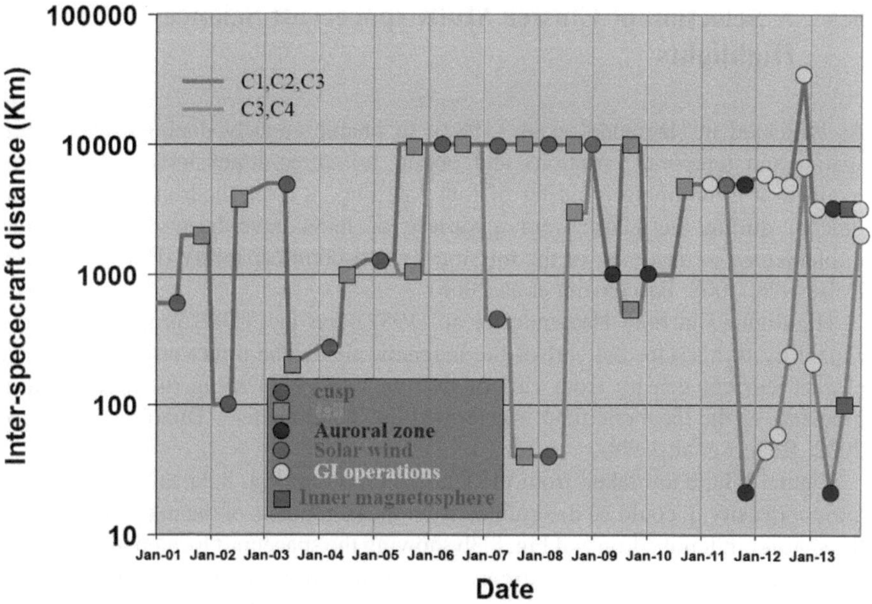

Fig. 8.2 Interspacecraft separation strategy. The different *coloured shapes* indicate different science targets. Recent points marked in *yellow* represent GI observations, which have dominated operations (Color figure online)

Fig. 8.2, alongside new targets such as the auroral acceleration region, these GI operations have dominated recent activities.

Another unique activity has been the implementation of the Cluster Active Archive (CAA). The CAA provides an unprecedented legacy of the Cluster mission and is responsible for the validation, storage and online availability of the entire set of Cluster calibrated high-resolution data and other allied products, in a standard format and with a complete set of metadata in machine readable format. The CAA has been operational since February 2006 and presently has over 1,700 active users downloading an average of 1 TB of data per month. At present located at ESTEC, ESA's centre in the Netherlands, in the near future the Archive will relocate to ESAC in Spain, to reside with ESA's other permanent archives.

Cluster continues to produce an impressive volume of science output, presently standing (as of January 2013) at 1,790 refereed publications [including publications from the Double Star mission (Liu et al. 2005)] and 70 theses, of which 59 are Ph.D.

8.3 A Selection of Cluster Multi-spacecraft Science Highlights

As indicated in Haerendel et al. (1982), to unambiguously distinguish between spatial and temporal variations and obtain full three-dimensional information, requires at a minimum a cluster of four satellites in a tetrahedron configuration. Indeed, during the Cluster era a variety of tools have been developed and implemented to make use of the mission's multi-point capability (Paschmann and Daly 1998, 2008; Paschmann et al. 2005).

Highlighted in both Haerendel et al. (1982) and the TOPS proposal was the capability to measure the curl of the magnetic field. The direct estimation of the electric current density from curl of the magnetic field using measured spatial gradients of the magnetic field was termed the "Curlometer" (Dunlop et al. 1988, 2002; Robert et al. 1998).

Figure 8.3a, b are taken from the Cluster proposal, Fig. 8.3a showing how the current density, **j**, could be determined from measurements of the magnetic field by the spacecraft tetrahedron and Fig. 8.3b showing the spacecraft located in the region of the tail neutral sheet, the current sheet in the central part of the magnetotail. Cluster has facilitated the measurement of current sheet thicknesses of a few R_E (Shen et al. 2008). Current density measurements have ranged between 10 and 30 nA m^{-2} (Sergeev et al. 2003; Runov et al. 2003, 2005, 2006; Davey et al. 2012), although for large spacecraft separations, these values could be underestimated (Forsyth et al. 2011). Nakamura et al. (2008), with the spacecraft separated by only 200 km, derived a strong neutral sheet current above 180 nA m^{-2} and a thickness of only a few 100 km, demonstrating a thinning and strengthening of the current sheet

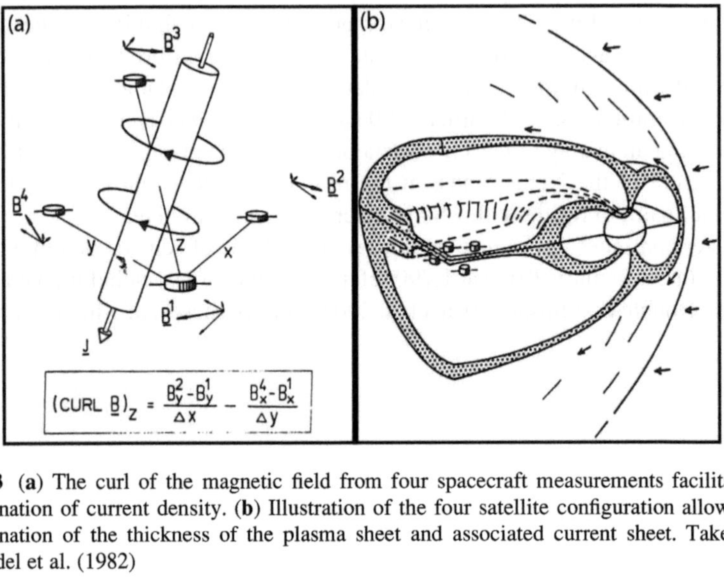

Fig. 8.3 (**a**) The curl of the magnetic field from four spacecraft measurements facilitates the determination of current density. (**b**) Illustration of the four satellite configuration allowing the determination of the thickness of the plasma sheet and associated current sheet. Taken from Haerendel et al. (1982)

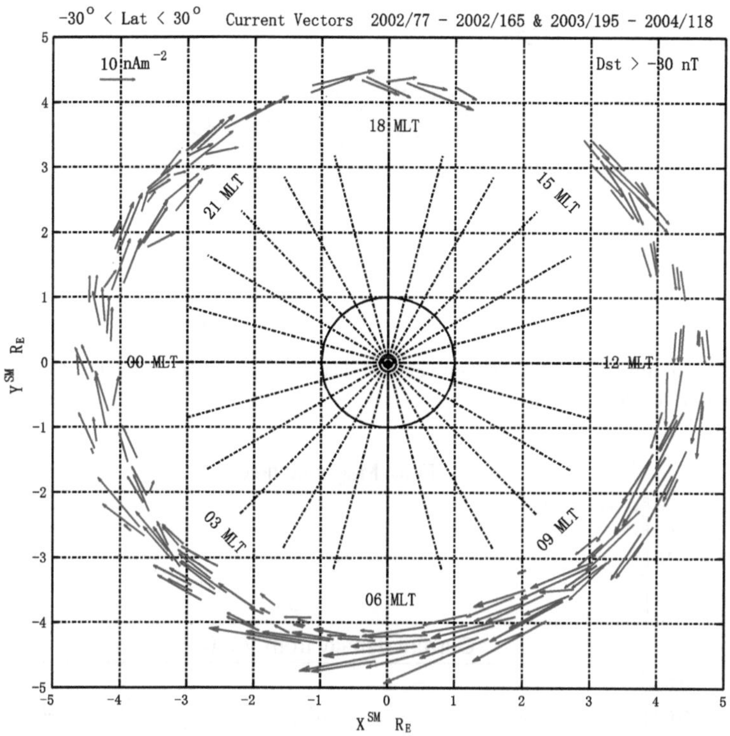

Fig. 8.4 The distribution of ring current vectors in the equatorial plane in SM coordinates for Cluster perigee crossings between 18 March 2002 (77/2002, orbit number 265) and 14 June 2002 (165/2002, orbit number 302) and between 14 July 2003 (195/2003) and 27 April 2004 (118/2004). Taken from Zhang et al. (2011)

during substorm activity. Davey et al. (2012) also showed enhanced current sheet current densities during storm time periods (with enhanced ring current indicated by SYM-H).

The curlometer has also been used in the inner magnetosphere, to directly measure, for the first time, the ring current itself. Vallat et al. (2005) used perigee data from February to June 2002 to provide a measure of a partial westward ring current at local times ranging from 17 to 1 MLT. The inclination of the orbit also enabled the study of the latitudinal extent and variation of the current density, ranging from $-65°$ to $65°$ in latitude and directed almost fully perpendicular to the magnetic field near the equator and becoming more field aligned at higher latitudes. Zhang et al. (2011) extended this study to cover all MLT (during non-storm periods with Dst > -30 nT) and showed a clear asymmetric distribution, with an average current density ranged from 9 to 27 nA m^{-2} with enhancements (by a factor 2 compared to other regions) in the range 5–11 MLT and a reduced magnitude for 12–17 MLT (Fig. 8.4). This asymmetry was suggested to be due to region-2 field aligned currents. Most recently, Grimald et al. (2012) have investigated the limitations of the curlometer technique for low altitude Cluster orbits.

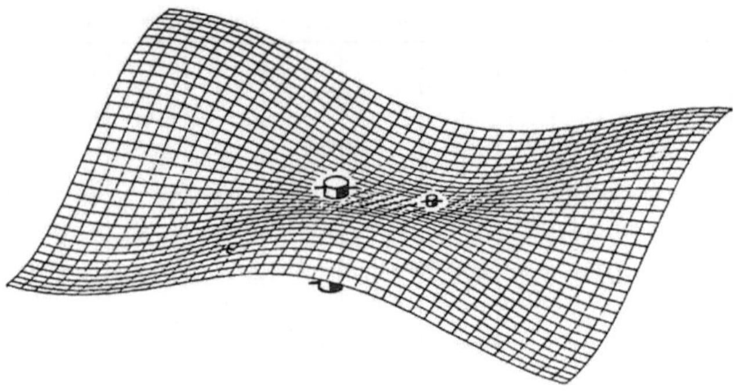

Fig. 8.5 Illustration of the magnetopause current sheet boundary characterized by four spacecraft. Taken from Haerendel et al. (1982)

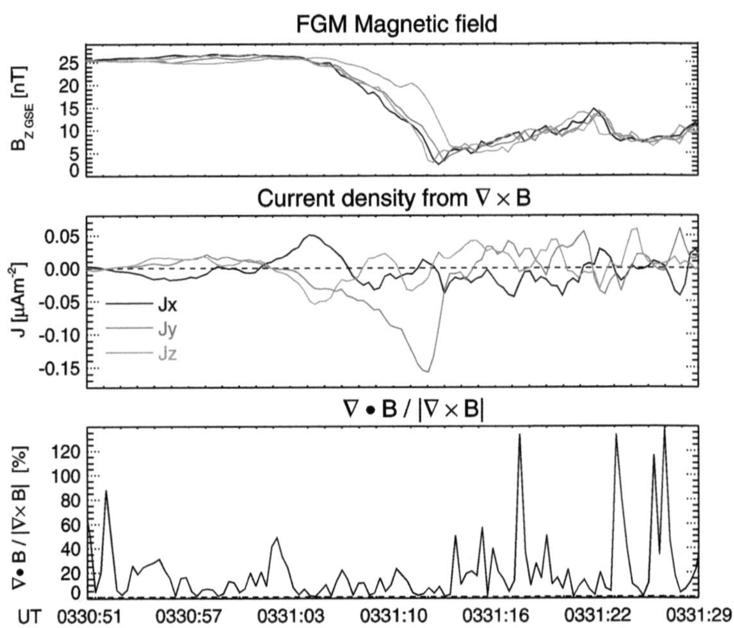

Fig. 8.6 Overview of a magnetopause crossing on 2 March 2002, around 0331 UT. *Panels* (from the *top*) show the GSE z component of the magnetic field, the GSE components of the current density and $|\nabla \cdot B| / |\nabla \times B| \ll 1$ as an indicator of the uncertainty of the current estimation [reproduced from Haaland et al. (2004) by permission of American Geophysical Union]

Separating the Earth's magnetospheric magnetic field from the magnetosheath field is the thin current layer forming the magnetopause (Fig. 8.5). The curlometer has allowed the direct measurement of the current density from the curl of the magnetic field (Dunlop et al. 2002). Using spacecraft separations of 100 km, Haaland et al. (2004) calculated a magnetopause current density of ~0.15 nA m^{-2} during a dayside crossing (Fig. 8.6). An indication of the uncertainty in the current density determination related to the employment of a linear approximation to determine the gradients in the magnetic field, can be obtained from the ratio of the apparent divergence of the magnetic field to the magnitude of the curl of the magnetic field, where it is necessary to have $|\nabla \cdot \mathrm{B}| / |\nabla \times \mathrm{B}| \ll 1$.

Motivated by the curlometer technique, Gurgiolo et al. (2010) used the four electron instruments on Cluster to determine the vorticity of the solar wind plasma in the free-flowing wind and in the vicinity of the foreshock. They found that much of the measured vorticity was caused by changes in the flow direction of the return population in the case of the foreshock (either reflected or leakage from the magnetosheath) and strahl electron populations outside of the foreshock, as they couple to changes in the magnetic field orientation. That, in turn, resulted in deflections in the total bulk velocity, leading to the production of vorticity. The technique was also used to examine vorticity in the inner plasma sheet. With the four Cluster spacecraft positioned in a tetrahedral configuration, it was possible to estimate directly the electron fluid vorticity, including three time periods when Cluster passed through a site of magnetic reconnection. Enhancements in vorticity were seen in association with each crossing of the ion diffusion region, similar to what had been seen in numerical simulations of magnetic reconnection.

Four spacecraft measurements can also be used to infer the normal and speed of a discontinuity (Russell et al. 1983), using the so-called 'timing method'. A number of different variations of this method have been developed, combining both single and multi-spacecraft techniques. These include the Discontinuity Analyzer (DA) developed by Dunlop and Woodward (1998) which utilizes the variance of the magnetic field measurements at the separate spacecraft of the tetrahedron, the Constant Thickness Approach (CTA) and Constant Velocity Approach (CVA) which assume constant thicknesses or velocities of the boundary (Haaland et al. 2004). Using a minimum variance analysis on the current vectors determined by the curlometer technique, Haaland et al. (2004) were able to derive an orientation of the magnetopause current sheet and also its thickness, of around 200 km. Dunlop et al. (2001) demonstrated application of the DA technique on the magnetopause, while Pedersen et al. (2001) used electron density estimates from spacecraft potential to determine the dynamic evolution of the magnetopause. Rezeau et al. (2001) demonstrated the local curvature of the magnetopause using observations of small-scale fluctuations of the magnetic field at each spacecraft.

Owen et al. (2004) reported periodic surface waves and using four spacecraft measurements of the electron plasma, determined the wavelength and direction and speed of propagation of the waves for the first time. The leading edge of the waves was shown to be steeper than the trailing edge, consistent with Kelvin-Helmholtz (KH) waves, although the waves were found to be stable to the instability criterion locally. Hasegawa et al. (2004) reported observations of large-scale vortices (40,000 km) on the flank of the magnetopause, a result of KH driven waves, and suggested that these giant swirls could be a viable mechanism for plasma transfer across the magnetopause. Nykyri et al. (2006) showed this process to be feasible via a combination of the KH instability and magnetic reconnection of high shear magnetic fields in the rolled up vortices. Multi-scale aspects of these structures were later revealed by Hasegawa et al. (2009), indicating reconnection on the trailing edge of the waves. The global nature of such waves can be examined by combining observations from a number of satellites. Foullon et al. (2008) reported the evolution of KH waves on the magnetopause using Geotail with Cluster, showing evidence for an inverse dependence between the interplanetary magnetic field (IMF) direction and wavelength at the flank and pointing to a connection to geomagnetic pulsation periods. Using a Cluster-Geotail conjunction, Nishino et al. (2010) revealed similar scale size wave structures on both dawn and dusk flanks of the magnetopause. Most recently, Hwang et al. (2012) have shown evidence for high-latitude KH waves driven during periods of strong westward IMF conditions.

Observations of wave-like boundary behavior from Cluster has not been limited to the magnetopause. Sergeev et al. (2004) reported on large-scale waves propagating in the magnetotail current sheet, with scale sizes 1,500–10,000 km and speed of 57–145 km s^{-1}. Zhang et al. (2005) revealed the macroscopic scale of these features by combining four spacecraft Cluster observations in the magnetotail at 16 R_E, with coherent signatures at the Double Star 1 spacecraft at 11 R_E.

Haerendel et al. (1982) indicated the capability of Cluster in probing the morphology of reconnection, as illustrated in Fig. 8.7a. Using four point measurements, Louarn et al. (2004) reported the three dimensional topology of the magnetopause resulting from reconnection at multiple sites. Xiao et al. (2006) presented the first observations of the magnetic null—the core region of the reconnection process. Using the four point magnetic field measurements, the latter authors used the Poincaré index, based on a linear extrapolation of the magnetic field gradient within the spacecraft tetrahedron, to identify the magnetic null point at the heart of magnetic reconnection in the magnetotail. Using the same technique, Xiao et al. (2007) identified a 'null line' connecting two magnetic reconnection sites and He et al. (2008a, b) presented a detailed analysis of the local magnetic structure around the null line and also uncovered evidence of electron trapping in the null region itself. This electron trapping may ultimately lead to the formation of energetic electron beams, a well-known but poorly understood consequence of reconnection. Such observations have also been made on the dayside high latitude magnetopause by Dunlop et al. (2009) (Fig. 8.7b), with the addition of observations at lower latitude by Double Star 1 enabling a large scale picture of reconnection on the dayside. Such multi-mission observations, with the launch of more magnetospheric probes such as Double Star and THEMIS (Angelopoulos 2008), have become commonplace. Dunlop et al. (2011) reported multi-mission measurements of the dayside magnetopause, using the unique configuration and collection of spacecraft during the April–July 2007 period, shortly after the launch of THEMIS and before the demise of Double Star 1. A number of conjunctions using the ten spacecraft were presented which look at the extent of the x-line, multiple x lines and the response of the magnetopause to changes in the IMF.

The four spacecraft have also been used to explore Ohm's law at small scales, where electric fields are thought to play a vital role in the microphysics of reconnection. The electric fields are governed by the generalized Ohm's law, which includes the frozen-in convection of magnetic field and plasma, as well as the effects of the (Hall) $\mathbf{j} \times \mathbf{B}$ force, which becomes important when the scales approach the ion inertial length and ions become demagnetized. At even smaller

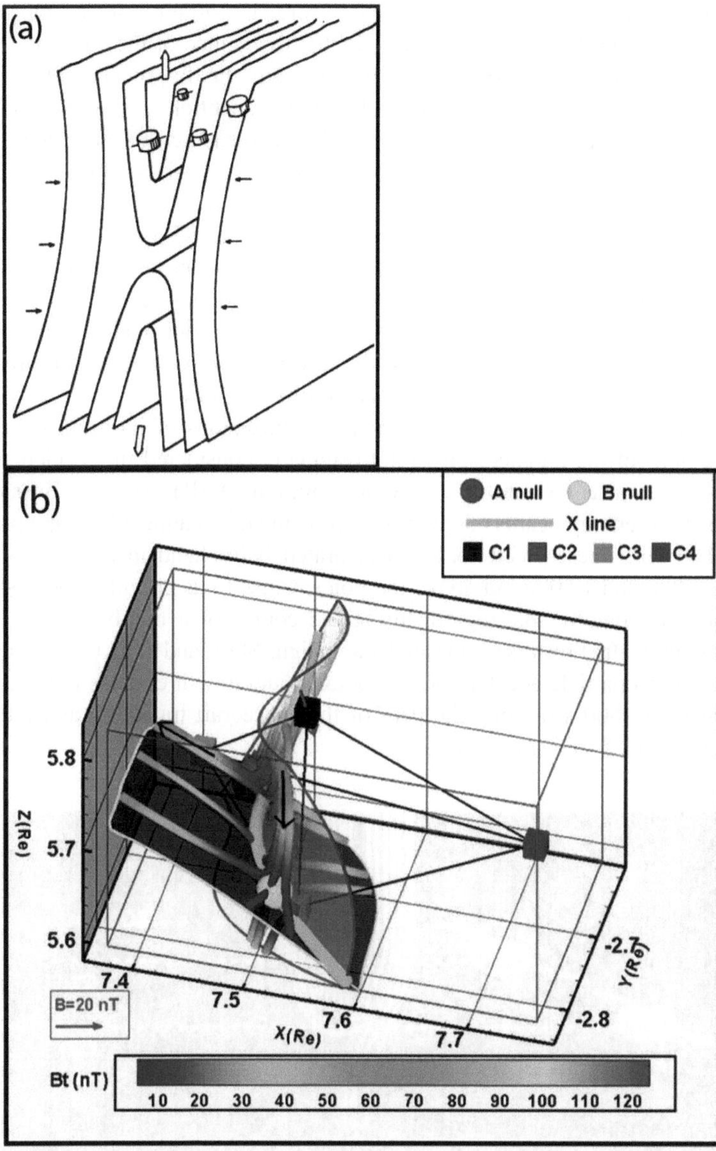

Fig. 8.7 (**a**) Illustration of a large-scale reconnection event. Taken from Haerendel et al. (1982). (**b**) Illustration of a magnetic null pair configuration, reconstructed using the method of He et al. (2008a). The A null is partially hidden by the fans formed by the field line geometry (shown by the *black* surface and half-transparent *white* surface), and the null lies just in front of the spacecraft configuration. Cluster 2 is close to the B null but is also partially hidden. The X line is denoted with a *curve* through A–B, while magnetic field lines converge along the fan surface (in *white*) to approach the A null and then travel out along the spine (marked in *black arrows*) of the A null. Field lines also converge along the spine of the B null and then diverge out along its fan surface (in *black*) [taken from Dunlop et al. (2009) copyright 2009 by The American Physical Society]

scales, Ohm's Law includes two terms that become important at scales approaching the electron inertial length. One of these is the divergence of the electron pressure tensor (the other is the electron inertia). In particular, the divergence of the electron pressure tensor is related to the onset of fast reconnection. Using carefully calibrated data, Henderson et al. (2006) were able to measure this divergence in the plasma sheet near a site of magnetic reconnection. They showed that the electric fields normal to the neutral sheet from the Hall and electron pressure divergence had opposite signs—the Hall term, larger by a factor of ~5.3, pointed towards the neutral sheet while the divergence of the electron pressure term pointed away, in agreement with numerical simulations (also see Henderson et al. 2008). Simulations predict that the region in which the divergence term alone supports the reconnection electric field is highly localised around the x-line and requires much higher resolution sampling, beyond the capability of Cluster.

Most recently, due to lunar solar gravity perturbations of its orbit, Cluster has visited regions of the magnetosphere not originally considered in its science goals. One such location is the auroral acceleration region (AAR). Auroral emissions are caused by high-energy beams of electrons hitting the upper atmosphere at altitudes as low as 100 km. These electrons are accelerated beforehand in the AAR, located between 4,000 and 12,000 km above the poles. Cluster has provided the first multipoint measurements in this region during the course of a number of campaigns. Reporting on the first results of one such campaign, Marklund et al. (2011) described how the auroral particle accelerator is structured and how it evolves in time, in one particular case ~800 km along the path of the spacecraft trajectory and remaining stable for at least 5 min (Fig. 8.8).

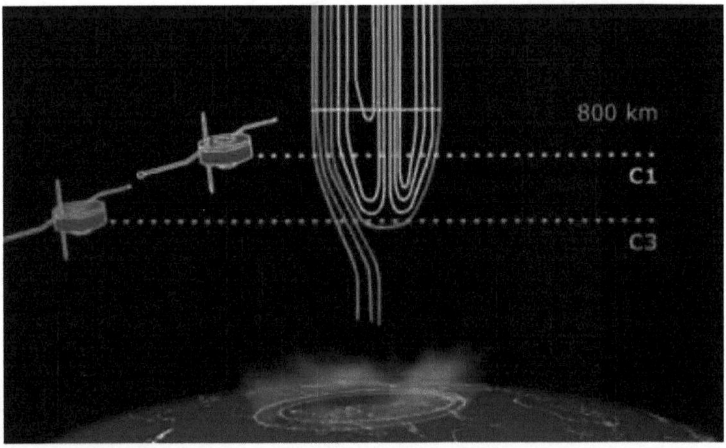

Fig. 8.8 Measurements made by the Cluster C3 and C1 spacecraft revealed the 2-D morphology and altitude distribution of the acceleration (electric) potential in the AAR, shown here by equipotential contours derived by combining data from both satellites. The pattern consists of two broad, *U-shaped* potential structures at the higher level (*pale green*) and a narrower *S-shaped* potential structure located below (*pale blue*). For the first time it was possible to constrain the size and longevity of the electric fields in the acceleration regions. The data showed that the electric field structures measured at least 800 km across and remained stable for at least 5 min. Image courtesy of ESA (Color figure online)

8.4 The Era of 'Bunches of Bunches' of Satellites

Since its launch in 2000, Cluster, which joined the Polar (Acuña et al. 1995) (1996–2008) and Geotail (Nishida 1994) (1993–date) magnetospheric spacecraft in orbit around the Earth, has in turn been joined by Double Star (2003–2009), THEMIS (launched 2007) and most recently the Van Allen probes in 2012. This flotilla of spacecraft has provided unprecedented coverage of the magnetospheric system in recent years and this is due to continue, with Cluster currently extended until the end of 2014. During this period we will be presented with exciting multi-mission conjunctions and configurations, as indicated in Fig. 8.9. In particular, Cluster will contribute to the science of the Van Allen probes mission in combination of THEMIS by providing additional measurements within the equatorial region with THEMIS and in additional by monitoring higher latitude phenomena associated with radiation belt processes, which no other current magnetospheric mission has the capability to do. This high latitude capability will greatly benefit the BARREL (BAlloon Array for RBSP Relativistic Electron Losses) experiment (Millan et al. 2011), where up to 20 balloons were launched from the Antarctic in January and February 2013, which measure the particle precipitation (another campaign will be carried out in 2014). Cluster will provide high latitude measurements of wave activity on field lines connecting BARREL and the Van Allen probes.

A key part of the current extension involves the collaboration with ESA's Earth Observation mission, Swarm, due for launch in 2013. The objective of this three-satellite mission is to provide the best ever survey of the internal geomagnetic field and the first global representation of its variations on time scales from an hour to several years. It is well known that (in particular at high latitude regions) strong currents flowing along geomagnetic field lines connect the magnetosphere and the ionosphere. The magnetic fields created by these currents constitute the biggest noise factor for the Swarm mission, and need to be monitored (and eventually eliminated from the Swarm data) as accurately as possible. At the same time, space physicists have not yet understood the internal structure of these current systems, in particular their temporal development and density distribution. This creates an optimum win-win collaboration, where one scientist's noise is another scientist's data. Currently, activities are focusing on determining ionospheric conductances and convection maps from the satellite data. This will provide a better understanding of ionospheric-magnetospheric coupling. In turn, this can help to significantly improve the quality of current density estimates, which rely today on model corrections.

A proposal was submitted in 2012 to extend the Cluster mission further to the end of 2016. Key aspects of the extension are examining the declining phase of the current solar cycle taking advantage of the unprecedented satellite coverage of the magnetosphere, but in particular providing a large database of Cluster data spanning more than a whole solar cycle. This will yield a unique comparative measurements of the magnetosphere under different solar wind driving conditions (Fig. 8.10).

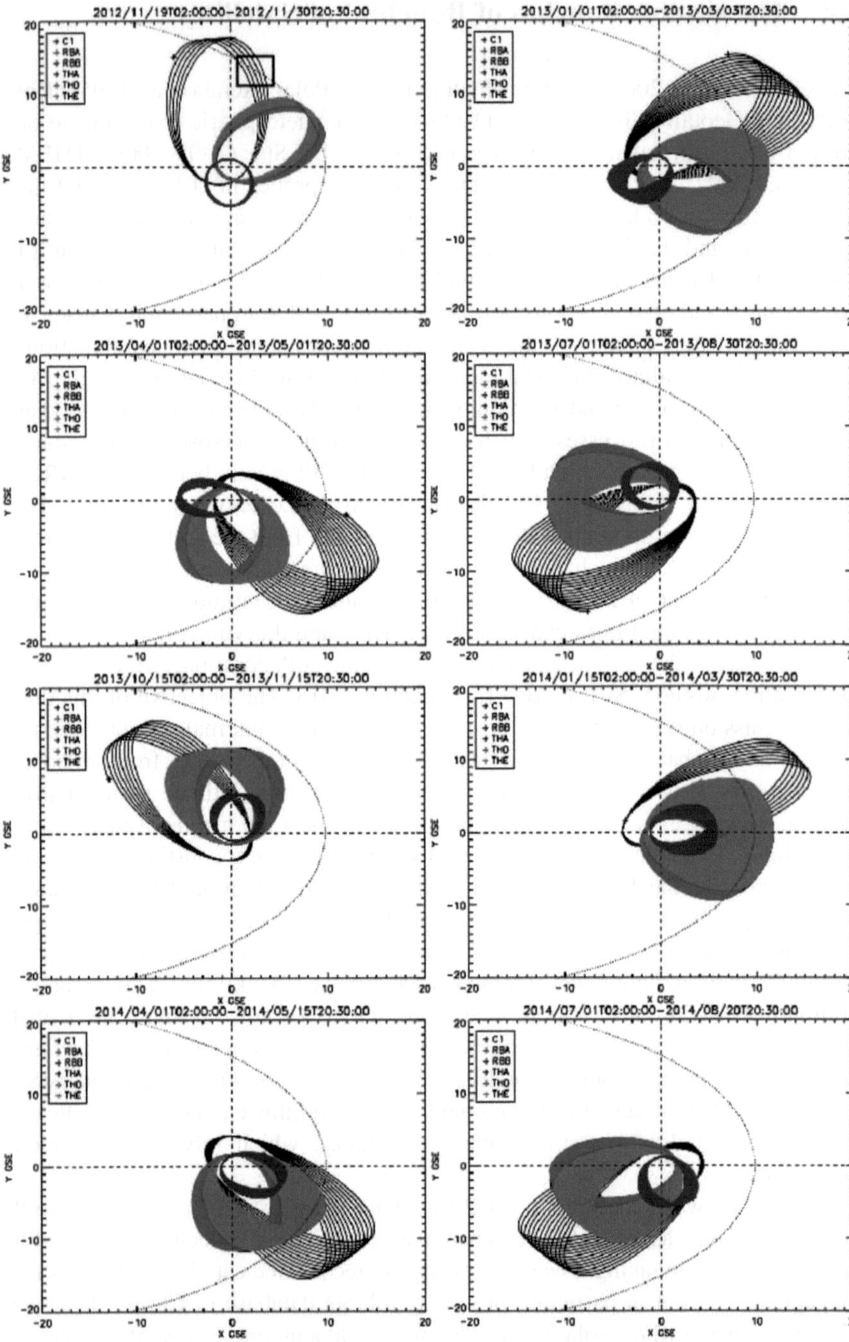

Fig. 8.9 Composite plot of Cluster, THEMIS and Van Allen probes 2012–2014, targeting the following broad science questions: dawn-dusk differences (Autumn 2012), particle loss at the dayside magnetopause (winter 2012–2013), dawnside waves and electron drift (spring 2013), ion injection, energization, and drift (summer 2013), duskside electromagnetic ion cyclotron waves

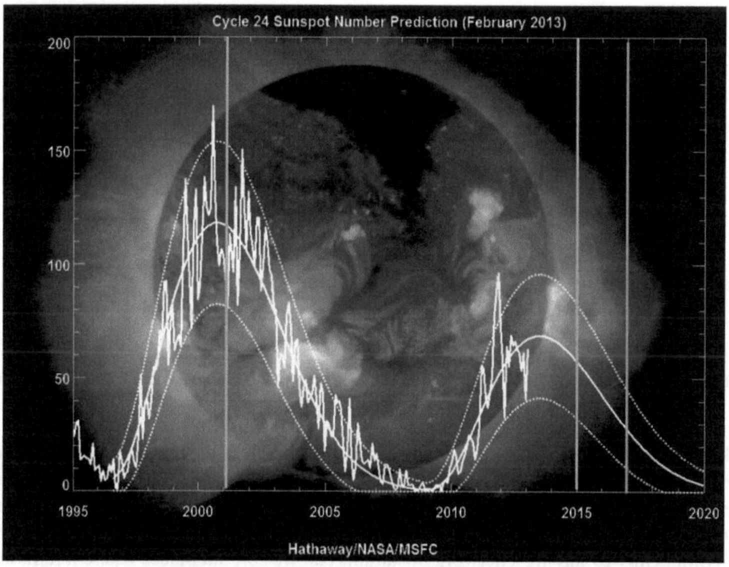

Fig. 8.10 Solar cycle 24 prediction. Solar maximum is predicted in Summer-Autumn 2013. The *yellow vertical lines* bound the current Cluster mission timeline. The *blue line* indicates the bound of the proposed extension. Image courtesy of NASA (Color figure online)

Following on from the first announcement of opportunity in 2010, we plan to make another call for Guest Investigators in 2014. Members of the community will for the second time be invited to submit proposals for how the Cluster instrument and spacecraft complement should be operated for specific science studies. 2015 will also see the launch by NASA of Magnetospheric Multiscale (MMS) mission (Sharma and Curtis 2005). This mission builds on the success of Cluster by surpassing its spatial and temporal resolution, targeting the core of reconnection, the electron diffusion region. Cluster and MMS will provide a micro-macro-scope combination of observations, with large separations between the two missions providing a macroscopic picture of plasma processes, and for the first time, two tetrahedra that can investigate small-scale phenomena simultaneously. We note that there are plans to reconfigure the three THEMIS spacecraft such that they are configured around the MMS spacecraft. In addition, in September 2015 and 2016 the GEOTAIL spacecraft will also be in the magnetotail. In such a situation, the combination of these spacecraft will provide an unprecedented opportunity to examine the cross scale coupling of plasma phenomena. As noted by Haerendel et al. (1982):

Fig. 8.9 (continued) (autumn 2013); effects of solar wind pressure variations (winter 2013–2014); ultralow frequency waves (spring 2014) and electron injection (summer 2014). The *rectangle* in the *upper left plot* indicates the inbound magnetopause crossing by Cluster

> The magnetosphere is our closest example of interacting cosmic plasmas. Apart from the interest in its overall configuration, dynamics, mass and energy balance, there is a more general interest in the fundamental processes of these interactions. They have microscopic and small-scale macroscopic aspects.

8.5 Summary

This paper has provided a sample of the scientific output of the Cluster mission in the scope of comments, predictions, and proposals made by Jim Dungey in the 1960s. In particular we have emphasized those related to boundary waves and measurements of the curl of the magnetic field, alluding to the benefit of 'bunches' of satellites. Borrowing again from Dungey's inaugural lecture:

> Looking to the future I believe that progress requires bunches of satellites, though these are as yet in no published program. One is continually conscious of this need for reasons which have a direct analogue on the ground. Observations from just one observatory, whether stationary or moving, can only tell us for instance whether it is raining at one place at any given time. Only by having a network of stations can we sort out places where it rains all the time, times when it rains everywhere and belts of rain which move across the country.

We have now reached a period whereby one can describe "bunches of bunches" of satellites in orbit around the Earth, providing a very powerful capability to unravel the ubiquitous plasma physics phenomena.

Cluster has a critical part in this continued exploration.

Acknowledgements The authors would like to thank all PIs and their teams who provided the Cluster data, and the JSOC and ESOC teams for their very efficient operation of the Cluster spacecraft. MGGTT would also like to thank D. J. Southwood for the invitation to be part of Professor Dungey's birthday celebrations.

Appendix

Copy of part of ESRO-4897 1966. **TOPS (Tetrahedral Observatory Probe System) submitted by Jim W. Dungey to the European Space Agency's predecessor, ESRO (European Space Research Organisation), through the British National Committee on Space Research.**

COS/24
Addendum 1
Paris, 5 April 1966

Suggestion for Second Large ESRO Project *)

Tetrahedral Observatory Probe System

by J.W. Dungey, Imperial College, London

It is proposed to launch several (say four) satellites either simultaneously or sequentially into similar orbits with identical periods, so that the satellites move as a close group. The satellites should have identical instrumentation, the system being designed to measure spatial gradients of magnetic fields and particle intensities. Perigee must be sufficiently high so that the differential change in their periods due to drag should not produce appreciable separation for several months. The system would be most useful for orbits with apogees lying a little beyond the magnetospheric boundary. It would be an advantage if the separation distance varied over the orbit, the maximum envisaged being a few thousand kilometres. Structures of different sizes could then be effectively investigated. Assuming a bunch of four satellites, their orbits should not be coplanar.

1. Scientific Objectives

(i) Particle Measurements

Measurements of energetic electrons fequently show "spikes' and with a single satellite there is always an ambiguity as to whether these represent variations in time or space. A bunch of satellites would remove this ambiguity and would further reveal the space-time structure of the particle distribution, the orientation of a plane sheet, a third would detect either curvature of motion of the sheet, but could not distinguish between them, a fourth would probably indicate whether curvature or motion were the more plausible. Variation of the satellite separation would lead to information over a range of structure sizes.

(ii) Steady Magnetic Fields

Simultaneous data from four satellites make it possible to evaluate curl $\underline{B}$ and time averaging should give useful information on the large scale current system. This is the information required for solution of the ring current problem; it is impossible to locate the current by means of measurements from a single spacecraft. The local current density is determined by curl $\underline{B}$.

*) This proposal was submitted to ESRO through the British National Committee on Space Research in March 1965.

P/4519.

COS/24
Addendum 1
Page 2.

(iii) Waves

Waves are expected with a wide range of wavelength down to a few kilometres at VLF. The nature of the information obtained depends on the ratio of satellite separation to wavelength. For long wavelengths the system would be used to determine the propagation velocity by correlation analysis and to compare waveforms. Also for rather long wavelengths the current density could be obtained as for the steady field and this would be useful in building a hydromagnetic model. Assuming the occurrence of periodic waves, the system would be most effective when the wavelength is only a few times the satellite separation. If the disturbance is represented by the superposition of plan waves, a least squares fit of the data from four satellites could be made to a superposition of three elliptically polarised waves or four plane polarised waves. For short wavelengths the signals may be totally different at the different satellites. There is evidence that VLF is propagated in quite narrow "ducts". Recording of the amplitude and polarisation at four points should yield much semiquantitative information on this problem. Satellite separations of up to a few thousand kilometres would be well suited to the study of these problems.

2. Instrumentation

For magnetic measurements flux gates have proved satisfactory and it is estimated that the disturbances to be measured would exceed the noise for periods longer than one second. The likely method for measuring VLF would be by coils. Observations on Injun suggest that waves would easily be detected down to 100 c/s. This leaves a gap centred on 10 c/s in which the amplitude to be expected is not known. Theoretical work at Imperial College is in progress to estimate what amplitude at 10 c/s would be needed for such disturbances to have an important effect on particle distributions. Satellite observations in this frequency range are under consideration in the US and so it is possible that data will be available before the design of T.O.P.S. would be frozen.

Particle measurements should be chosen for particles which are known to show spikes, probably electrons of 1 KeV to 1 MeV. These are the measurements for which the system was primarily suggested, but optional extras could be measurements of soft protons, electrons harder than 1 MeV and electron density.

Instruments used should all have been proven on previous satellites.

References

Acuña, M.H., Ogilvie, K.W., Baker, D.N., Curtis, S.A., Fairfield, D.H., Mish, W.H.: The global geospace science program and its investigations. Space Sci. Rev. **71**, 5 (1995). Reprinted in The Global Geospace Mission, ed. By C.T. Russell, Kluwer Academic Publishers

Angelopoulos, V.: The THEMIS mission. Space Sci. Rev. **141**, 5 (2008). doi:10.1007/s11214-008-9336-1

ESRO-4897: Report on the Thirteenth Meeting of ad Hoc Working Group COS on Cosmic Rays and Trapped Radiation, B. Peters (Chairman), COS/24 (1966)

Davey, E.A., M. Lester, S.E. Milan, R.C. Fear, and C. Forsyth (2012), The orientation and current density of the magnetotail current sheet: A statistical study of the effect of geomagnetic conditions, J. Geophys. Res., 117, A07217, doi:10.1029/2012JA017715

Dunlop, M.W., Southwood, D.J., Glassmeier, K.-H., Neubauer, F.M.: Analysis of multipoint magnetometer data. Adv. Space Res. **8**, 273 (1988)

Dunlop, M.W., Woodward, T.I.: Multi-spacecraft discontinuity analysis: orientation and motion. In: Paschmann, G., Daly, P.W. (eds.) Analysis Methods for Multi-Spacecraft Data, p. 271. ESA, Noordwijk (1998)

Dunlop, M.W., Balogh, A., Cargill, P., Elphic, R.C., Fornaçon, K.-H., Georgescu, E., Sedgemore-Schulthess, F., the FGM team : Cluster observes the Earths magnetopause: coordinated four-point magnetic field measurements. Ann. Geophys. **19**, 1449 (2001)

Dunlop, M.W., Balogh, A., Glassmeier, K.-H., Robert, P.: Four-point Cluster application of magnetic field analysis tools: the Curlometer. J. Geophys. Res. **107**, 23 (2002)

Dunlop, M.W., Zhang, Q.-H., Xiao, C.-J., He, J.-S., Pu, Z., Fear, R.C., Shen, C., Escoubet, C.P.: Reconnection at high latitudes: antiparallel merging. Phys. Rev. Lett. **102**, 075005 (2009)

Dunlop, M.W., Zhang, Q.-H., Bogdanova, Y.V., Trattner, K.J., Pu, Z., Hasegawa, H., Berchem, J., Taylor, M.G.G.T., Volwerk, M., Eastwood, J.P., Lavraud, B., Shen, C., Shi, J.-K., Wang, J., Constantinescu, D., Fazakerley, A.N., Frey, H., Sibeck, D., Escoubet, P., Wild, J.A., Liu, Z.X., Carr, C.: Magnetopause reconnection across wide local time. Ann. Geophys. **29**, 1683 (2011). doi:10.5194/angeo-29-1683-2011

Escoubet, C.P., Fehringer, M., Goldstein, M.: The Cluster mission. Ann. Geophys. **19**, 1197 (2001)

Forsyth, C., Lester, M., Fazakerley, A.N., Owen, C.J., Walsh, A.P.: On the effect of line current width and relative position on the multi-spacecraft curlometer technique. Planet. Space Sci. **59**, 598 (2011). doi:10.1016/j.pss.2009.12.007

Foullon, C., Farrugia, C.J., Fazakerley, A.N., Owen, C.J., Gratton, F.T., Torbert, R.B.: Evolution of Kelvin-Helmholtz activity on the dusk flank magnetopause. J. Geophys. Res. **113**, A11203 (2008). doi:10.1029/2008JA013175

Grimald, S., Dandouras, I., Robert, P., Lucek, E.: Study of the applicability of the curlometer technique with the four Cluster spacecraft in regions close to earth. Ann. Geophys. **30**, 597 (2012). doi:10.5194/angeo-30-597-2012

Gurgiolo, C., Goldstein, M.L., Vinas, A.F., and Fazakerley, A.N.: First measurements of electron vorticity in the foreshock and solar wind, Ann. Geophys., 28, 2187–2200, doi:10.5194/angeo-28-2187-2010, 2010

Haaland, S., Sonnerup, B.U.Ö., Dunlop, M.W., Georgescu, E., Paschmann, G., Klecker, B., Vaivads, A.: Orientation and motion of a discontinuity from Cluster curlometer capability: minimum variance of current density. Geophys. Res. Lett. **31**(10), L10804 (2004). doi:10.1029/2004GL020001

Haerendel, G., Roux, A., Blanc, M., Paschmann, G., Bryant, D., Korth, A., Hultqvist, B.: Cluster, study in three dimensions of plasma turbulence and small-scale structure. Mission proposal, 1982

Hasegawa, H., Fujimoto, M., Phan, T.D., Rème, H., Balogh, A., Dunlop, M.W., Hashimoto, C., TanDokoro, R.: Transport of solar wind into earth's magnetosphere through rolled-up Kelvin-Helmholtz vortices. Nature **430**, 755 (2004)

Hasegawa, H., Retinò, A., Vaivads, A., Khotyaintsev, Y., André, M., Nakamura, T.K.M., Teh, W.-L., Sonnerup, B.U.Ö., Schwartz, S.J., Seki, Y., Fujimoto, M., Saito, Y., Rème, H., Canu, P.: Kelvin-Helmholtz waves at the earth's magnetopause: multiscale development and associated reconnection. J. Geophys. Res. **114**, A12207 (2009). doi:10.1029/2009JA014042

He, J.-S., Tu, C.-Y., Tian, H., Xiao, C.-J., Wang, X.-G., Pu, Z.-Y., Ma, Z.-W., Dunlop, M.W., Zhao, H., Zhou, G.-P., Wang, J.-X., Fu, S.-Y., Liu, Z.-X., Zong, Q.-G., Glassmeier, K.-H.,

Rème, H., Dandouras, I., Escoubet, C.P.: A magnetic null geometry reconstructed from Cluster spacecraft observations. J. Geophys. Res. **113**, A05205 (2008a). doi:10.1029/2007JA012609
He, J.-S., Zong, Q.-G., Deng, X.-H., Tu, C.-Y., Xiao, C.-J., Wang, X.-G., Ma, Z.-W., Pu, Z.-Y., Lucek, E., Pedersen, A., Fazakerley, A.N., Cornilleau-Wehrlin, N., Dunlop, M.W., Tian, H., Yao, S., Tan, B., Fu, S.-Y., Glassmeier, K.-H., Rème, H., Dandouras, I., Escoubet, C.P.: Electron trapping around a magnetic null. Geophys. Res. Lett. **35**, L14104 (2008b)
Henderson, P.D., Owen, C.J., Lahiff, A.D., Alexeev, I.V., Fazakerley, A.N., Lucek, E., Reme, H.: Cluster PEACE observations of electron pressure tensor divergence in the magnetotail. Geophys. Res. Lett. **33**(22), L22106 (2006). doi:10.1029/2006GL027868
Henderson, P.D., Owen, C.J., Lahiff, A.D., Alexeev, I.V., Fazakerley, A.N., Yin, L., Walsh, A.P., Lucek, E., Réme, H.: The relationship between j $\times$ B and ∇.Pe in the magnetotail plasma sheet: Cluster observations. J. Geophys. Res. **113** (2008). doi:10.1029/2007JA012697
Hwang, K.-J., M.L. Goldstein, M.M. Kuznetsova, Y. Wang, A.F. Viñas, and D.G. Sibeck (2012), The first in-situ observation of Kelvin-Helmholtz waves at high-latitude magnetopause during strongly dawnward interplanetary magnetic field conditions, J. Geophys. Res., 117, A08233, doi:10.1029/2011JA017256
Liu, Z.X., Escoubet, C.P., Pu, Z., Laakso, H., Shi, J.K., Shen, C., Hapgood, M.: The double star mission. Ann. Geophys. **23**, 2707 (2005). SRef-ID: 1432-0576/ag/2005-23-2707
Louarn, P., Fedorov, A., Budnik, E., Fruit, G., Sauvaud, J.A., Harvey, C.C., Dandouras, I., Rème, H., Dunlop, M., Balogh, A.: Cluster observations of complex 3D magnetic structures at the magnetopause. Geophys. Res. Lett. **31**(19), L19805 (2004). doi:10.1029/2004GL020625
Nakamura, R., Baumjohann, W., Fujimoto, M., Asano, Y., Runov, A., Owen, C.J., Fazakerley, A.N., Klecker, B., Rème, H., Lucek, E.A., Andre, M., Khotyaintsev, Y.: Cluster observations of an ion-scale current sheet in the magnetotail under the presence of a guide field. J. Geophys. Res. **113**, A07S16 (2008). doi:10.1029/2007JA012760
Nishida, A.: The GEOTAIL mission. Geophys. Res. Lett. **21**, 2871–2873 (1994)
Nishino, M.N., Hasegawa, H., Fujimoto, M., Saito, Y., Mukai, T., Dandouras, I., Rème, H., Retinò, A., Nakamura, R., Lucek, E., Schwartz, S.J.: A case study of Kelvin-Helmholtz vortices on both flanks of the earth's magnetotail. Planet. Space Sci. **59**, 502–509 (2010). doi:10.1016/j.pss.2010.03.011
Nykyri, K., Otto, A., Lavraud, B., Mouikis, C., Kistler, L.M., Balogh, A., Rème, H.: Cluster observations of reconnection due to the Kelvin-Helmholtz instability at the dawnside magnetospheric flank. Ann. Geophys. **24**, 2619 (2006). http://direct.sref.org/1432-0576/ag/2006-24-2619
Marklund, G.T., Sadeghi, S., Karlsson, T., Lindqvist, P.-A., Nilsson, H., Forsyth, C., Fazakerley, A., Lucek, E.A., Pickett, J.: Altitude distribution of the auroral acceleration potential determined from Cluster satellite data at different heights. Phys. Rev. Lett. **106**, 055002 (2011)
Millan, R.M., the BARREL Team: Understanding relativistic electron losses with BARREL. J. Atmos. Sol. Terr. Phys. (2011). doi:10.1016/j.jastp.2011.01.006
Owen, C.J., Taylor, M.G.G.T., Krauklis, I.C., Fazakerley, A.N., Dunlop, M.W., Bosqued, J.M.: Cluster observations of surface waves on the dawn flank magnetopause. Ann. Geophys. **22**, 971–983 (2004)
Paschmann, G., Daly, P.W. (eds.): Analysis methods for multi-spacecraft data. ISSI Scientific Report, SR-001 (1998)
Paschmann, G., Schwartz, S.J., Escoubet, C.P., Haaland, S. (eds.): Outer magnetospheric boundaries: Cluster results. Space Science Series of ISSI, 20 (2005)
Paschmann, G., Daly, P.W. (eds.): Multi-spacecraft analysis methods revisited. ISSI Scientific Report, SR-008 (2008)
Pedersen, A., Décréau, P., Escoubet, C.-P., Gustafsson, G., Laakso, H., Lindqvist, P.-A., Lybekk, B., Masson, A., Mozer, F., Vaivads, A.: Four-point high time resolution information on electron densities by the electric field experiments (EFW) on Cluster. Ann. Geophys. **19**, 1483 (2001)

Rezeau, L., Sahraoui, F., d'Humières, E., Belmont, G., Chust, T., Cornilleau-Wehrlin, N., Mellul, L., Alexandrova, O., Lucek, E., Robert, P., Décréau, P., Canu, P., Dandouras, I.: A case study of low-frequency waves at the magnetopause. Ann. Geophys. **19**, 1463 (2001)

Robert, P., Dunlop, M.W., Roux, A., Chanteur, G.: Accuracy of current density determination. In: Paschmann, G., Daly, P.W. (eds.), Analysis Methods for Multi-Spacecraft Data, ISSI Scientific Report, SR-001, 395 (1998).

Runov A., R. Nakamura, W. Baumjohann, R. A. Treumann, T. L. Zhang, M. Volwerk, Z. Voros, A. Balogh, K.-H. Glassmeier, B. Klecker, H. Rème, and L. Kistler, Current sheet structure near magnetic X-line observed by Cluster, Geophys. Res. Lett., 30, No. 11, 1579, doi:10.1029/2002GL016730, 2003

Runov, A., Sergeev, V.A., Nakamura, R., Baumjohann, W., Zhang, T.L., Asano, Y., Volwerk, M., Vörös, Z., Balogh, A., Rème, H.: Reconstruction of the magnetotail current sheet structure using multi-point Cluster measurements. Planet. Space Sci. **53**, 237 (2005). doi:10.1016/j.pss.2004.09.049

Runov, A., et al.: Local structure of the magnetotail current sheet: 2001 Cluster observations. Ann. Geophys. **24**, 247 (2006). doi:10.5194/angeo-24-247-2006

Russell, C.T., Mellott, M.M., Smith, E.J., King, J.H.: Multiple spacecraft observations of interplanetary shocks: four spacecraft determination of shock normals. J. Geophys. Res. **88**, 4739 (1983)

Sergeev, V., et al.: Current sheet flapping motion and structure observed by Cluster. Geophys. Res. Lett. **30**(6), 1327 (2003). doi:10.1029/2002GL016500

Sergeev, V., Runov, A., Baumjohann, W., Nakamura, R., Zhang, T.L., Balogh, A., Louarnd, P., Sauvaud, J.-A., Rème, H.: Orientation and propagation of current sheet oscillations. Geophys. Res. Lett. **31**, L05807 (2004). doi:10.1029/2003GL019346

Sharma, A.S., Curtis, S.A.: Magnetospheric multiscale mission. In: Sharma, A.S., Kaw, P. (eds.) Nonequilibrium Phenomena in Plasmas. Astrophysics and Space Science Library, vol. 321, p. 179. Springer, New York (2005)

Shen, C., Rong, Z.J., Li, X., Dunlop, M., Liu, Z.X., Malova, H.V., Lucek, E., Carr, C.: Magnetic configurations of the tilted current sheets in magnetotail. Ann. Geophys. **26**, 3525 (2008). doi:10.5194/angeo-26-3525-2008

Ulivi, P.: ESRO and the deep space: European planetary exploration planning before ESA. J. Br. Interplanet. Soc. **59**, 204 (2006)

Vallat, C., Dandouras, I., Dunlop, M., Balogh, A., Lucek, E., Parks, G.K., Wilber, M., Roelof, E.C., Chanteur, G., Rème, H.: First current density measurements in the ring current region using simultaneous multi-spacecraft CLUSTER-FGM data. Ann. Geophys. **23**, 1849 (2005). doi:10.5194/angeo-23-1849-2005

Xiao, C.J., Wang, X.G., Pu, Z.Y., Zhao, H., Wang, J.X., Ma, Z.W., Fu, S.Y., Kivelson, M.G., Liu, Z.X., Zong, Q.G., Glassmeier, K.H., Balogh, A., Korth, A., Reme, H., Escoubet, C.P.: In situ evidence for the structure of the magnetic null in a 3D reconnection event in the Earth's magnetotail. Nat. Phys. **2**, 478 (2006)

Xiao, C.J., Wang, X.G., Pu, Z.Y., Ma, Z.W., Zhao, H., Zhou, G.P., Wang, J.X., Kivelson, M.G., Fu, S.Y., Liu, Z.X., Zong, Q.G., Dunlop, M.W., Glassmeier, K.-H., Lucek, E., Rème, H., Dandouras, I., Escoubet, C.P.: Satellite observations of separator line geometry of three-dimensional magnetic reconnection. Nat. Phys. **3**, 609 (2007)

Zhang, T.L., Nakamura, R., Volwerk, M., Runov, A., Baumjohann, W., Eichelberger, H.U., Carr, C., Balogh, A., Sergeev, V., Shi, J.K., Fornacon, K.-H.: Double Star/Cluster observation of neutral sheet oscillations on 5 August 2004. Ann. Geophys. **23**, 2909 (2005). doi:10.5194/angeo-23-2909-2005

Zhang, Q.-H., Dunlop, M.W., Lockwood, M., Holme, R., Kamide, Y., Baumjohann, W., Liu, R.-Y., Yang, H.-G., Woodfield, E.E., Hu, H.-Q., Zhang, B.-C., Liu, S.-L.: The distribution of the ring current: Cluster observations. Ann. Geophys. **29**, 1655 (2011). doi:10.5194/angeo-29-1655-2011

Chapter 9
Observing Magnetic Reconnection: The Influence of Jim Dungey

Jonathan P. Eastwood

Abstract As part of the Festspiel celebrating the 90th birthday of Prof. Jim Dungey, this paper reviews his influence on experimental efforts to observe reconnection in space plasmas. Jim has influenced this area of research in two key ways: firstly, in the development of theories of magnetic reconnection and secondly, in being an early and vocal advocate of the need for multi-point observations. The advent of multi-point missions such as Cluster and THEMIS in the past decade has enabled considerable progress, and we illustrate with examples how multi-point data and techniques have indeed improved our understanding of how reconnection works.

9.1 Introduction

On 10th January 2013, a Festspiel was held at the Royal Astronomical Society in London in celebration of Professor Jim Dungey's 90th birthday. During the meeting, I gave a presentation discussing the influence of Jim Dungey on our understanding of magnetic reconnection, from the point of view of experimental observations. This paper is based on the remarks I made at the meeting: rather than present a comprehensive review of the field of reconnection research, my aim is to highlight some of the ways in which Jim has influenced the study of magnetic reconnection. In particular I will focus more on magnetic reconnection as a plasma process, rather than solar wind—magnetosphere coupling more generally. More information about the effect of reconnection on the gross configuration and dynamics of the magnetosphere is given by Milan (chapter "Sun et Lumière: Solar Wind-Magnetosphere Coupling as Deduced from Ionospheric Flows and Polar Auroras") and Cowley (chapter "Dungey's Reconnection Model of the Earth's Magnetosphere: The First 40 Years").

J.P. Eastwood (✉)
The Blackett Laboratory, Imperial College London, London SW7 2AZ, UK
e-mail: jonathan.eastwood@imperial.ac.uk

D. Southwood et al. (eds.), *Magnetospheric Plasma Physics: The Impact of Jim Dungey's Research*, Astrophysics and Space Science Proceedings 41,
DOI 10.1007/978-3-319-18359-6_9

Whilst Jim was primarily a theorist rather than an experimentalist, his role in the experimental study of reconnection in space plasmas cannot be understated, for two reasons:

The first is that he has played perhaps *the* primary role in developing our understanding of both what reconnection is and how it applies to the Earth's magnetosphere, as well as solar wind driven magnetospheres more generally. Whilst Jim's most celebrated paper is his 1961 Letter to the Physical Review (Dungey 1961) (see chapters "Dungey's Reconnection Model of the Earth's Magnetosphere: The First 40 Years" and "Sun et Lumière: Solar Wind-Magnetosphere Coupling as Deduced from Ionospheric Flows and Polar Auroras"), we take the opportunity here to highlight two other papers. The first is his 1953 article (Dungey 1953), which provides tremendous insight, at a very early stage in the development of reconnection theory, into what reconnection actually does. The second is his 1994 review article (Dungey 1994), which finishes by identifying the exploration of differential ion-electron motion in reconnection, stating "I am hopeful that recent activity represents the onset of rapid progress". This was an extremely prescient observation, since the most exciting advances in reconnection research in the last decade have indeed been in this area.

The second reason is Jim's early identification of the need for multi-point satellite measurements, with the general case being made during his inaugural lecture at Imperial College (Dungey 1966), and a specific proposal for a four satellite mission, TOPS (Tetrahedral Observatory Probe System), being made to ESRO (European Space Research Organisation, the forerunner of ESA) in 1965. The four-satellite Cluster mission was launched in 2000 and data from Cluster (see chapter "The Science of the Cluster Mission") has been central to recent significant developments in our experimental understanding of magnetic reconnection. In fact, Jim identified the need for "bunches of satellites", and whilst a specific magnetospheric constellation mission has not yet been launched, researchers have combined data from multiple (often multi-)satellite missions to study reconnection much as Jim envisaged.

In Sect. 9.2, we discuss Jim's influence on the development of reconnection theory, bookended by his 1953 and 1994 papers. In Sect. 9.3, we discuss the importance of multi-point measurements, and the contributions Jim made to cement this idea in the community. In Sect. 9.4, we show how multi-point measurements and techniques have been used to study reconnection, illustrated using some representative examples, and conclusions are presented in Sect. 9.5.

9.2 Dungey's Influence on the Development of Magnetic Reconnection Theory

Magnetic reconnection is one of the most important concepts in plasma physics. If a plasma can be considered ideal (that is, the conductivity is effectively infinite and $\mathbf{E} + \mathbf{v} \times \mathbf{B} = \mathbf{0}$), then the magnetic field behaves as if it is 'frozen' into the plasma (Dungey 1963). This means that magnetic field lines can be used to prescribe the 'connectivity' of the plasma, since two plasma parcels frozen to the same magnetic field line remain connected as the plasma evolves. Furthermore, this idea can be used to qualitatively predict aspects of the plasma's motion because the magnetic field lines act as if they have a tension and exert a pressure. This was described by Jim as a 'priceless jewel', citing for example the formation of the Parker spiral in the interplanetary magnetic field as something whose shape can be predicted simply by considering frozen in field behaviour (Dungey 1994).

Two ideal plasmas cannot interpenetrate as described by the frozen-in field theorem, and so in practice a thin boundary layer (current sheet) forms to separate them, as shown in Fig. 9.1a. If the two plasmas are pushed together (for example by applying a uniform out-of-plane electric field), energy can be stored in the compressed magnetic fields. If the current sheet becomes sufficiently thin, then magnetic reconnection may occur, changing the topology of the plasma, accelerating jets, and releasing the stored magnetic energy, as shown in Fig. 9.1b. The system reconfigures itself, with an X-line separating the plasma flowing in from the top and

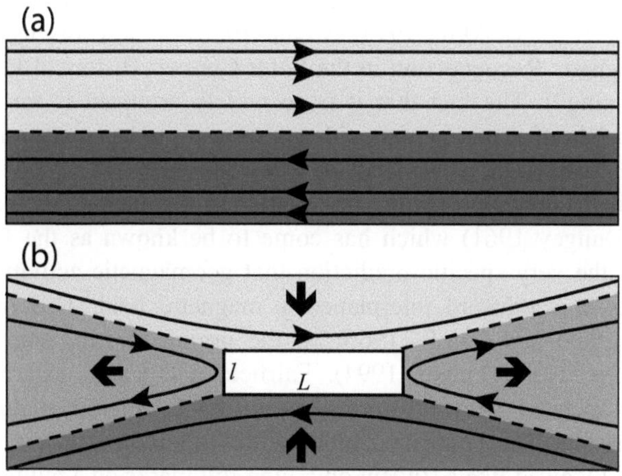

Fig. 9.1 (**a**) A current sheet forms to separate two regions of ideal plasma. (**b**) Breakdown of the current sheet at a point leads to the formation of an X line geometry centred on a diffusion region (*white box*) where the plasma is no longer ideal and magnetic field lines cannot be ascribed an identity (although of course the magnetic field exists). The diffusion region has height l and width L where $l < L$. Plasma flowing in from the *top* and *bottom* is able to interpenetrate and mix, and is then ejected to the *left* and *right* in jets along the current sheet. The plasma jets are heated, and the plasma in the outflow is once more frozen to the magnetic field

bottom from oppositely-directed jets of hot plasma flowing out along the current sheet. The X-line is centred on the diffusion region where the frozen-in field theorem breaks down and the plasma is demagnetized. In this region the magnetic field lines do not behave as if they are frozen to the plasma: whilst the magnetic field exists, field line identity is now meaningless here. The two plasmas mix, and then exit the diffusion region to the left and right. As the plasma exits the diffusion region, the plasma becomes frozen to the magnetic field once again. Whilst the consequences of reconnection are large scale, it is the diffusion region that plays a crucial role, because it is here that the plasma decouples from the magnetic field allowing the whole process to proceed.

The early development of magnetic reconnection theory is discussed in more detail by Cargill (chapter "Magnetic Reconnection in the Solar Corona: Historical Perspective and Modern Thinking") but here we highlight the fact that the key physics concepts which underpin magnetic reconnection were clearly stated in Jim's 1953 paper, preceding the more well known papers by Sweet (1958) and Parker (1957). In particular he noted three key points which summarize exactly the physical picture of reconnection:

- "The lines of force ... can be regarded as being broken and rejoined",
- "The total length of the lines of force decreases in the process, and it follows that the energy of the field decreases",
- "The discharge extends in the direction perpendicular to the paper [i.e. out of the plane of Fig. 9.1] up to a distance where the change in field is considerable."

Even though the concept of magnetic reconnection is now widely accepted, it is important to note that its acceptance met with significant resistance for a prolonged period of time (e.g., Heikkila 1973; Alfvén 1976), as also discussed by Cargill (chapter "Magnetic Reconnection in the Solar Corona: Historical Perspective and Modern Thinking"). The fact that it *is* so widely accepted is for one principle reason: the weight of experimental evidence shows it to be an accurate representation of plasma behaviour, as we shall now describe.

Applied to the magnetosphere, Jim predicted the pattern of magnetospheric convection (Dungey 1961) which has come to be known as the Dungey cycle. This leads to the very specific prediction that geomagnetic activity is correlated with intervals of southward interplanetary magnetic field. This prediction was tested by Fairfield and Cahill (1966), at the urging of Jim, who had acted as Fairfield's supervisor (Dungey 1994). Fairfield and Cahill examined the IMF orientation associated with substorms, identified by their signature in ground magnetometer data. This confirmed, albeit with limited data, the predicted correlation, "a result I viewed like a student obtaining full marks on a multiple choice test of 10 questions with three choices for each" (the choices being northward IMF, southward IMF or indeterminate) (Dungey 1994).

Although this evidence was consistent with reconnection, it did not directly prove its existence and in the late 1970s, Sonnerup (1979) commented that "even in the most recent literature opinions about the cosmic occurrence of the process range from full acceptance (Vasyliunas 1975) to outright rejection (Alfvén 1976)." This is even in the context of flow reversal data from the magnetotail showing behaviour consistent with reconnection occurring there (e.g., Hones Jr. 1976). The 'smoking gun', a reconnection jet confined to the magnetopause current sheet, was not discovered until spacecraft with fast plasma measurements (able of constructing a moment of the distribution on timescales of the order of a few seconds) were available. With the advent of the International Sun Earth Explorer (ISEE) mission in the late 1970s, reconnection jets were indeed observed (Paschmann et al. 1979).

Perhaps one of the most important advances that has occurred in the past 15–20 years has been the rapid development of theories explaining reconnection in the context of Hall physics. Jim accurately predicted this in the final discussion section ('The Future Has Started') of his 1994 paper, writing "I am hopeful that recent activity (Cai and Ding 1994; Mandt et al. 1994) represents the onset of rapid progress ... a feature which is yet to appear occurs when the electrons are magnetized but the ions not..." In fact, as shown in Fig. 9.2, the more massive ions decouple from the magnetic field in the ion diffusion region which surrounds the small electron diffusion region where the electrons decouple and the plasma reconnects. Differential motion between the ions and the still-magnetized and drifting electrons creates Hall currents in the plane of reconnection. These currents are associated with out-of-plane magnetic fields which forms a quadrupole pattern (Sonnerup 1979).

This also explains why reconnection is fast. In Sweet-Parker reconnection, although plasma flows out at the Alfvén speed, the reconnection rate $R \sim (l/L) \cdot V_A$ is very small because l/L is small (see Fig. 9.1), and this appears to makes reconnection irrelevant as a physical process. In Hall reconnection, the fact that the electrons are magnetized and the ions not means that the outflow from the electron diffusion region is whistler-like. In fact, if the outflow width l is small, then the associated wavenumber k is large, and since the whistler speed scales with k ($\omega \sim k^2$), the aspect ratio is no longer relevant for calculations of the reconnection rate. As a result, the reconnection rate is fast (Mandt et al. 1994). The GEM

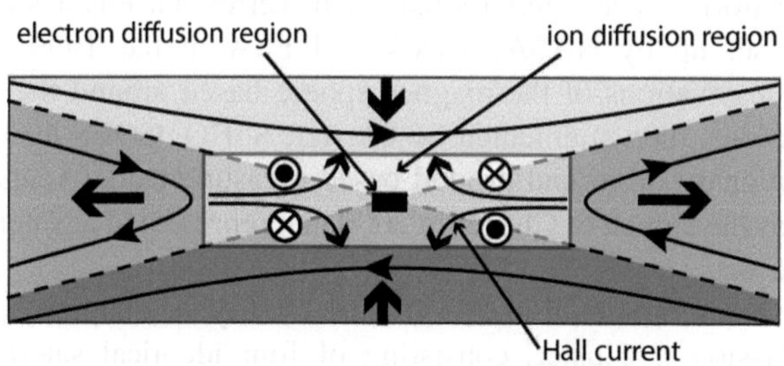

Fig. 9.2 Differential ion-electron motion, ultimately due to their different masses, leads to the formation of a two-scale diffusion region. The heavier ions demagnetize in the ion diffusion region, whereas the electrons decouple from the magnetic field, and reconnection finally occurs, in the electron diffusion region. The different response of the ions and electrons in the ion diffusion region leads to the formation of in-plane Hall currents and a characteristic quadrupole pattern in the out-of-plane magnetic field

modelling challenge subsequently showed that Hall physics is the necessary ingredient for fast reconnection (Birn et al. 2001; Drake et al. 2008).

The first observational evidence for Hall reconnection physics subsequently came from measurements of Hall currents (Fujimoto et al. 1997), and then measurements of the Hall magnetic field and electric field confirming predictions (Øieroset et al. 2001; Mozer et al. 2002). This coincided with the launch of Cluster, and so set the stage for an explosion in reconnection research making use of multi-satellite data and techniques, as described below in Sect. 9.4. However, the possibilities of multi-satellite data had been envisaged decades earlier by Jim, as we now describe.

9.3 Bunches of Spacecraft

Jim Dungey was an early proponent of multi-satellite missions for studying space plasma physics. In particular, he discussed the concept of 'bunches' of satellites in his inaugural lecture (Dungey 1966), eloquently making the case for multi-point measurements. Referring to the direct analogy of ground measurements, he noted that "observations from just one observatory, whether stationary or moving, can only tell us for instance whether it is raining at one place at any given time. Only by having a network of stations can we sort out places where it rains all the time, times when it rains everywhere and belts of rain which move across the country." A bunch of at least four spacecraft is required to study a plasma in three dimensions, and in March 1965 Jim proposed the TOPS (Tetrahedral Observatory Probe System) mission to ESRO, as discussed by Taylor (chapter "The Science of the Cluster Mission").

The first multi-satellite missions, ISEE (Ogilvie et al. 1977) and the Active Magnetospheric Particle Tracer Explorers (AMPTE) (Acuña et al. 1985) missions were launched in the late 1970s and 1980s, both specifically designed as three-satellite observatories. The International Solar Terrestrial Physics (ISTP) science initiative was set up by NASA, JAXA and ESA in the 1990s to coordinate simultaneous observations of the magnetosphere based around the Polar, Geotail and Wind missions, then augmented by Cluster, SoHO, GOES and LANL spacecraft in geostationary orbit, and ground based measurements (Acuña et al. 1995). However, it was the launch of Cluster in 2000 that represented a watershed in multi-spacecraft analysis.

As discussed in more detail in this volume by Taylor (chapter "The Science of the Cluster Mission"), Cluster, consisting of four identical satellites (Escoubet et al. 2001) has transformed our understanding of magnetospheric physics and the physics of collisionless plasmas. Theoretically, the four satellite tetrahedron of Cluster (whose scale has varied from 100s to 1,000s of km) can be used to investigate several different types of plasma structure, as illustrated in Fig. 9.3, with three main techniques—the discontinuity analyser, the k-filtering/wave

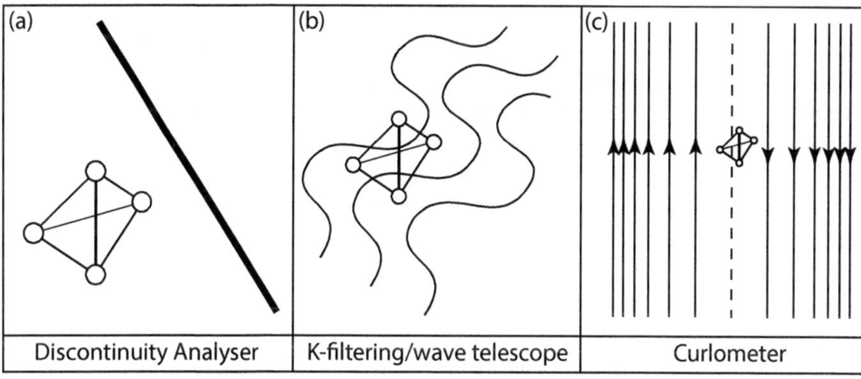

Fig. 9.3 The principal cluster multi-spacecraft analysis techniques: (**a**) discontinuity analyzer (tetrahedron scale > structure scale); (**b**) k-filtering/wave telescope (tetrahedron scale ~ structure scale) and (**c**) curlometer (tetrahedron scale < structure scale)

telescope and the curlometer (Dunlop et al. 1988). Examples of the use of these techniques is given in Sect. 9.4, but here brief descriptions are given:

- Discontinuity Analyzer: if the boundary is thin, and is measured at a specific time by each satellite, then its orientation and normal velocity can be found. In combination with single satellite techniques (e.g. minimum variance analysis), boundary curvature and acceleration can be probed (Dunlop and Woodward 1998). The first tests of this technique applied to the magnetopause were used to show the stability of the magnetopause normal, and also demonstrate the acceleration of the magnetopause over the satellites (Dunlop et al. 2002a)
- Waves-telescope/k-filtering: For waves with wavelengths larger than the spacecraft separation (to avoid aliasing), the power and phase information from a Fourier analysis of data from the different spacecraft can be combined to determine the existence of superposed propagating waves, and calculate their frequency and k-vectors (Pinçon and Motschmann 1998). This technique was first applied to the magnetosheath and solar wind, demonstrating its utility in the analysis of space plasma ULF waves (Glassmeier et al. 2001).
- Curlometer (current density): For structure whose scale size is much bigger than the tetrahedron, the four satellite magnetic field measurements can be combined to make a linear estimate of the curl of the magnetic field and thus the current density (Dunlop et al. 1988; Khurana et al. 1996; Robert et al. 1998). This technique was also first validated and tested in detail using magnetopause data (Dunlop et al. 2002b)

THEMIS (Time History of Events and Macroscale Interactions during Substorms), which launched in 2007, is a multi-satellite mission, consisting of five spacecraft, or probes, labelled P1–P5, together with an extensive set of ground based measurements (Angelopoulos 2008). Whereas the Cluster satellites are relatively closely spaced, THEMIS consists of distributed multi-point measurements. The main goal of the THEMIS mission was to establish the sequence of events that occurred in the magnetotail during a geomagnetic substorm and in its initial season of tail observations, the satellites were deployed so that P1 had an apogee ~ 30 Re (Earth radii), P2 ~ 19 Re, P3 and P4 ~ 12 Re and P5 ~ 10 Re. In 2009, P1 and P2 were removed from Earth orbit, and following a complex series of orbital manoeuvres, were placed in lunar orbit in mid-2011, forming the two-satellite ARTEMIS (Acceleration, Reconnection, Turbulence & Electrodynamics of Moon's Interaction with the Sun) mission which includes amongst its mission goals a two-spacecraft study of the magnetotail at 40–60 Re downtail (Angelopoulos 2011; Sibeck et al. 2011).

Looking to the future, at the time of writing the next major multi-satellite mission is Magnetospheric Multi-Scale (Burch and Drake 2009), a four satellite NASA mission which launched in 2015. MMS is a reconnection focused mission, and is designed to study the physics of the electron diffusion region with very high time resolution particle and field measurements. Beyond MMS, various ideas for future multi-satellite missions have been studied. The 12 satellite Cross-Scale concept was proposed to ESA (Schwartz et al. 2009), but was ultimately not selected. NASA has studied a magnetospheric constellation mission concept 'DRACO' (Spence et al. 2001, 2004), based on the use of the NASA ST-5 microsatellite (Slavin et al. 2008). DRACO was originally envisaged as a 50–100 satellite mission, but more recently, a cut-down 36 satellite magnetospheric constellation mission proposal has been developed and submitted by Kepko and Le in a white paper (http://www8.nationalacademies.org/SSBSurvey/publicviewHeliophysics.aspx) to the 2010 NRC decadal survey in solar and space physics (Natl. Res. Council 2012), which quoted the cost at $0.775 Bn.

This demonstrates that a space mission with 100+ satellites, two orders of magnitude larger than any science mission previously flown, is economically unfeasible in the traditional science mission model. The baseline satellite design must be orders-of-magnitude smaller and lighter, so that many can be launched together, and simple, so that they are easy to produce in a production-line process. With these constraints, a CubeSat (see e.g. www.cubesat.org) or similar small-satellite platform is thus the ideal starting point for such a mission.

In 2012, Imperial College London and Satellite Services SSBV received a grant from the UK Space Agency to perform a preparatory study of CENTINEL, a magnetospheric space weather constellation mission consisting of 100+ CubeSats that would monitor the Earth's magnetotail and the onset and evolution of geomagnetic storms. Guided by the science requirements, we examined available instrumentation, possible orbits (an example is shown in Fig. 9.4), deployment scenarios, the radiation environment and communications. In the context of possible pathways to implementation, and commercial space weather considerations, it was recommended that a pathfinder mission carrying appropriate instrumentation is flown demonstrating key technologies and most importantly, the ability of CubeSats to operate and return science quality data from beyond low Earth orbit.

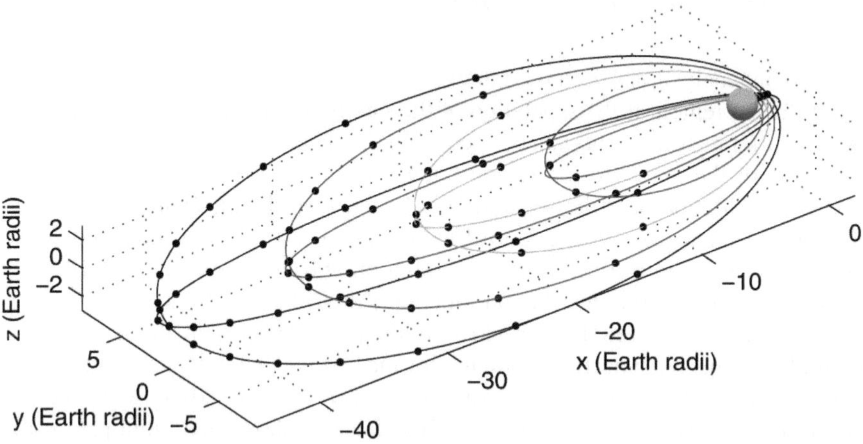

Fig. 9.4 A possible CENTINEL orbit configuration. This configuration consists of 74 CubeSats arranged in two orbital planes, each containing four nested orbits. Such a configuration would fill the Earth's magnetotail over the region where energy is stored and subsequently released during geomagnetic storms and substorms. The full CENTINEL mission would consist of two or three such sets of nested orbits, separated in local time, so as to ensure that the magnetotail would be continuously observed

CENTINEL, or a mission like it, offers the prospect of ultimately fulfilling Jim's vision of bunches of satellites studying the magnetosphere, and is possible because of a new generation of miniaturised instruments. For example, Imperial College has developed a new generation of light-weight and low power magnetometers based on magneto-resistive (MR) technology (Brown et al. 2012). A magnetometer using MR technology, MAGIC, was launched on the CINEMA CubeSat in September 2012, operating and returning data from low Earth orbit.

9.4 Using Multi-point Measurements to Study Reconnection

Here we examine a selection of interesting recent results which highlight the use of multi-spacecraft analysis to study reconnection and its consequences. We do not attempt to provide a comprehensive and exhaustive review of the literature, rather illustrations of how multi-spacecraft analysis can and has been used to examine and reveal the physics of reconnection in the spirit of Jim's original idea of 'bunches of satellites'. For more comprehensive reviews of reconnection observations and theory, we refer the reader to the following recent reviews (Paschmann 2008; Vaivads et al. 2009; Fuselier and Lewis 2011; Gosling 2012; Paschmann et al. 2013).

9.4.1 Structure and Boundaries

Multi-point analysis has been used in two distinct ways to study the structure of the ion diffusion region. If the separation of the satellites is relatively small, then four point Cluster measurements can be used to do timing analysis and thus establish the thickness of the ion diffusion region boundaries to be a few ion inertial lengths (Vaivads et al. 2004). If the separation is larger, then the different satellites pass through different parts of the ion diffusion region, directly revealing the spatial structure of e.g. the Hall fields (Eastwood et al. 2007) and spatial changes in the structure of the electron distribution function (Chen et al. 2008).

A key result derived from multi-satellite measurements is the existence of multiple X-line reconnection both in the magnetotail current sheet and at the magnetopause. In different analyses, Hasegawa et al. (2010) and Øieroset et al. (2011) used multi-point THEMIS data to show that multiple X-line reconnection leading to flux transfer event (FTE) formation was occurring on the dayside magnetopause. In the magnetotail, Eastwood et al. (2005) used multi-point Cluster data to show that a correlated southward/northward field and tailward/Earthward flow reversal was in fact caused by the Earthward motion of an island sandwiched by two X-lines rather than a tailward retreating X-line.

9.4.2 Waves

Jets produced by magnetic reconnection contain turbulent fluctuations and wave analysis techniques can be used to study their properties. For anti-parallel reconnection, and relatively low beta, analysis shows that the waves are consistent with parallel-propagating whistler waves (Eastwood et al. 2009). Further studies showed that if the plasma beta is large then waves propagate obliquely (Huang et al. 2010), and if there is a guide field then kinetic Alfven waves are observed (Chaston et al. 2009; Huang et al. 2012).

9.4.3 Current Density

In one of the earliest Cluster analyses of the ion diffusion region Runov et al. (2003) used four-satellite techniques to measure both the current density and the curvature of the magnetic field. Similarly Vaivads et al. (2004) used timing analysis in combination with differentiation of the magnetic field time series to determine the current density and thus show the presence of the Hall electric field by comparing $\mathbf{E}$ to $\mathbf{J} \times \mathbf{B}$. The curlometer technique has also been used to examine the Generalized Ohm's law: Henderson et al. (2006, 2008) compared four-point analysis of the electron plasma data with $\mathbf{J} \times \mathbf{B}$, finding that on the scale of the

Cluster tetrahedron this was generally anti-correlated with div P_e. Finally, multi-point Cluster data has been used to examine the current structure inside magnetotail flux ropes (e.g., Slavin et al., 2003).

9.4.4 Null Points

In three dimensional reconnection, the concept of an X-line is generalized to the idea of a separator line terminated by two magnetic null points (Lau and Finn 1990). Magnetic Nulls can in principle be detected using multipoint observations (Greene 1992), and this technique was used on Cluster data in the magnetotail to make the first attempt at identifying nulls in space plasma data (Xiao et al. 2006, 2007). This was subsequently extended to the magnetopause, (Dunlop et al. 2009) where Cluster was thought to have surrounded a magnetic null at the X line whilst at the same time the Double Star TC-1 satellite monitored input conditions in the magnetosheath.

9.4.5 Large Scale Structure and Dynamics

Thus far we have considered analyses that made use of measurements from relatively closely separated satellites (less than 1 Re). We now present a few illustrative examples where multiple spacecraft were used to determine the large scale structure and dynamics generated by reconnection.

On the magnetopause the length of the X-line has been studied first by (Phan et al. 2000) using two spacecraft observations and then by (Dunlop et al. 2011a, b), who used THEMIS, Cluster and Double Star TC-1 (i.e. ten satellites) to show extended magnetic reconnection occurring across the magnetopause. Flux transfer events are another common feature of magnetopause reconnection whose source could be multiple X-line reconnection, patchy reconnection, or bursty reconnection at a single elongated X line. Fear et al. (2009) used Cluster, THEMIS, Double Star TC-1 and SuperDARN radar data to trace the production of FTEs going both northward and southward from a subsolar reconnection X-line, providing new insights into their production. Subsequently Eastwood et al. (2012) used THEMIS and ARTEMIS data to show the survival of FTE flux ropes far along the tail magnetopause, in a manner consistent with their production along an extended X-line on the dayside magnetopause.

In the magnetotail, as mentioned earlier the primary goal of THEMIS was to determine the chain of events leading to a magnetospheric substorm. Angelopoulos et al. (2008) used the five THEMIS satellites to show the chain of events leading to a magnetospheric substorm was initiated by magnetic reconnection in the magnetotail. The onset of magnetic reconnection is thought to lead to the formation of dipolarization fronts (Sitnov et al. 2009), which have also been observed in multi-satellite data. In the example analysed by Runov et al. (2009), the dipolarization front was traced moving towards the Earth past all five THEMIS satellites in sequence, rather like a weather front propagating through the magnetotail in a manner exactly analogous to the idea that Jim proposed in his inaugural lecture and was quoted earlier.

9.4.6 Solar Wind Reconnection

The discovery of reconnection in solar wind current sheets provides a very different laboratory for reconnection research, because it can operate with essentially open boundary conditions (Gosling 2012). As predicted by Jim, X-lines can grow to be extremely long. For example, Wind, ACE and Cluster were used to identify a reconnection X-line extending for 100s of Earth radii in the solar wind (Phan et al. 2006). In further studies Gosling et al. used ACE, Wind, Geotail and the two STEREO spacecraft to identify an X-line extending 4.26×10^6 km, or 668 Re (Gosling et al. 2007).

9.5 Discussion and Conclusions

Magnetic reconnection is a process fundamental to plasma physics, and it is now widely accepted that reconnection plays a key role controlling the interaction of the solar wind with the Earth's magnetosphere. Jim Dungey played a central role both in developing the theory of magnetic reconnection, and applying that theory to magnetospheric physics: experimental observations, especially over the past 30+ years, have demonstrated the correctness of his predictions. As predicted by Jim, considerable advances in recent years have occurred because of the increased availability of multipoint data, and the launch of specifically multipoint missions such as Cluster and THEMIS.

In the context of reconnection, multi-point measurements have been fundamentally important to understanding structure: for example, the structure of the ion diffusion region, and the larger-scale structure of reconnection both on the magnetopause and in the magnetotail. These measurements have shown beyond doubt that collisionless reconnection does occur in near Earth space, and that many of its properties are consistent with the predictions of theory, especially theory based on Hall physics. In fact, the routine use of multi-point analysis now so common, and so firmly entrenched, that for a generation of space plasma physicists (including the author), the thought of not having multi-point measurements is hard to imagine. It is therefore appropriate that Jim, as an early advocate for multi-point measurements, should see many of his ideas tested experimentally by a community that has fully embraced multi-point missions.

It is also noteworthy that the next large NASA heliophysics mission, MMS, is a four satellite tetrahedron entirely dedicated to the experimental study of magnetic reconnection. The MMS tetrahedron will be separated on very small (electron) scales, so that it can reveal the structure and physics that controls the electron diffusion region. This demonstrates the journey reconnection has taken over the past 60 years to now being one of the central pillars of modern collisionless plasma physics. Finally, new technological advances are now enabling extremely small satellites, which may finally enable constellation-class missions, fulfilling Jim's wish to fill the magnetosphere with 'bunches' of satellites and provide new ways to observe, measure and study magnetic reconnection.

Acknowledgements I would like to thank David Southwood for organising the FestSchrift and inviting me to speak and write this paper, as well for his comments on an early version of the manuscript. I am also grateful to Peter Cargill for his comments on the manuscript during the writing process. The CENTINEL mission study, from which Fig. 9.3 is taken, was performed by myself and Peter Fox (Imperial College London) and Max Pastena and James Barrington-Brown (SSBV Space and Ground Systems). This work was supported by an STFC Advanced Fellowship, award ST/G00725X/1 at ICL.

References

Acuña, M.H., Ousley, G.W., McEntire, R.W., Bryant, D.A., Paschmann, G.: Editorial: AMPTE-mission overview. IEEE Txrans. Geosci. Remote Sens. **GE-23**(3), 175–176 (1985)

Acuña, M.H., Ogilvie, K.W., Baker, D.N., Curtis, S.A., Fairfield, D.H., MISH, W.H.: The global geospace program and its investigations. Space Sci. Rev. **71**, 5–21 (1995)

Alfvén, H.: On frozen-in field lines and field line reconnection. J. Geophys. Res. **81**, 4019–4021 (1976)

Angelopoulos, V.: The THEMIS mission. Space Sci. Rev. **141**, 5–34 (2008)

Angelopoulos, V.: The ARTEMIS mission. Space Sci. Rev. **165**, 3–25 (2011). doi:10.1007/s11214-010-9687-2

Angelopoulos, V., McFadden, J.P., Larson, D., Carlson, C.W., Mende, S.B., Frey, H., Phan, T., Sibeck, D.G., Glassmeier, K.-H., Auster, U., Donovan, E., Mann, I.R., Rae, J., Russell, C.T., Runov, A., Zhou, X.-Z., Kepko, L.: Tail reconnection triggering substorm onset. Science **321**, 931–935 (2008)

Birn, J., Drake, J.F., Shay, M.A., Rogers, B.N., Denton, R.E., Hesse, M., Kuznetsova, M., Ma, Z. W., Bhattacharjee, A., Otto, A., Pritchett, P.L.: Geospace environmental modeling (GEM) magnetic reconnection challenge. J. Geophys. Res. **106**(A3), 3715–3719 (2001)

Brown, P., Beek, T., Carr, C., O'Brien, H., Cupido, E., Oddy, T., Horbury, T.S.: Magnetoresistive magnetometer for space science applications. Meas. Sci. Technol. **23**, 025902 (2012). doi:10.1088/0957-0233/23/2/025902

Burch, J.L., Drake, J.F.: Reconnecting magnetic fields. Am. Sci. **97**, 392–399 (2009)

Cai, H.J., Ding, D.Q., Lee, L.C.: Momentum transport near a magnetic X line in collisionless reconnection. J. Geophys. Res. **99**(1), 35–42 (1994)

Chaston, C.C., Johnson, J.R., Wilber, M., Acuña, M.H., Goldstein, M.L., Reme, H.: Kinetic Alfvén wave turbulence and transport through a reconnection diffusion region. Phys. Rev. Lett. **102**, 015001 (2009)

Chen, L.-J., Bessho, N., Lefebvre, B., Vaith, H., Fazakerley, A.N., Bhattacharjee, A., Puhl-Quinn, P.A., Runov, A., Khotyaintsev, Y., Vaivads, A., Georgescu, E., Torbert, R.B.: Evidence of an

extended electron current sheet and its neighboring magnetic island during magnetotail reconnection. J. Geophys. Res. **113**, A12213 (2008). doi:10.1029/2008JA013385
Drake, J.F., Shay, M.A., Swisdak, M.: The Hall fields and fast magnetic reconnection. Phys. Plasmas **15**, 042306 (2008)
Dungey, J.W.: Conditions for the occurrence of electrical discharges in astrophysical systems. Philos. Mag. **44**(354), 725–738 (1953)
Dungey, J.W.: Interplanetary magnetic field and the auroral zones. Phys. Rev. Lett. **6**(2), 47–48 (1961)
Dungey, J.W.: The structure of the exosphere or adventures in velocity space. In: DeWitt, C., Hieblot, J., Lebeau, A. (eds.) Geophysique Exterieure/Geophysics the Earth's Environment, pp. 504–550. Gordon and Breach/Science Publishers, New York (1963)
Dungey, J.W.: The Magnetosphere, Imperial College of Science and Technology. University of London, London (1966)
Dungey, J.W.: Memories, maxims and motives. J. Geophys. Res. **99**(10), 19189–19197 (1994)
Dunlop, M.W., Woodward, T.I.: Multi-spacecraft discontinuity analysis: orientation and motion. In: Paschmann, G., Daly, P.W. (eds.) Analysis Methods for Multi-Spacecraft Data, pp. 271–305. International Space Science Institute, Bern (1998)
Dunlop, M.W., Southwood, D.J., Glassmeier, K.-H., Neubauer, F.M.: Analysis of multipoint magnetometer data. Adv. Space Res. **8**, 273–277 (1988)
Dunlop, M.W., Balogh, A., Glassmeier, K.-H. Four-point Cluster application of magnetic field analysis tools: the discontinuity analyzer. J. Geophys. Res. **107**(A11) (2002a). doi: 10.1029/2001JA0050089
Dunlop, M.W., Balogh, A., Glassmeier, K.-H., Robert, P.: Four-point Cluster application of magnetic field analysis tools: the Curlometer. J. Geophys. Res. **107**(A11), 1384 (2002b). doi:10.1029/2001JA0050088
Dunlop, M.W., Zhang, Q.-H., Xiao, C.J., He, J.-S., Pu, Z., Fear, R.C., Shen, C., Escoubet, C.P.: Reconnection at high latitudes: antiparallel merging. Phys. Rev. Lett. **102**, 075005 (2009)
Dunlop, M.W., Zhang, Q.-H., Bogdanova, Y.V., Lockwood, M., Pu, Z., Hasegawa, H., Wang, J., Taylor, M.G.G.T., Berchem, J., Lavraud, B., Eastwood, J., Volwerk, M., Shen, C., Shi, J.-K., Constantinescu, D., Frey, H., Fazakerley, A.N., Sibeck, D., Escoubet, P., Wild, J.A., Liu, Z.X.: Extended magnetic reconnection across the dayside magnetopause. Phys. Rev. Lett. **107**, 025004 (2011a)
Dunlop, M.W., Zhang, Q.-H., Bogdanova, Y.V., Trattner, K.J., Pu, Z., Hasegawa, H., Berchem, J., Taylor, M.G.G.T., Volwerk, M., Eastwood, J.P., Lavraud, B., Shen, C., Shi, J.-K., Wang, J., Constantinescu, D., Fazakerley, A.N., Frey, H., Sibeck, D., Escoubet, P., Wild, J.A., Liu, Z.X., Carr, C.: Magnetopause reconnection across wide local time. Ann. Geophys. **29**, 1683–1697 (2011b)
Eastwood, J.P., Sibeck, D.G., Slavin, J.A., Goldstein, M.L., Lavraud, B., Sitnov, M., Imber, S., Balogh, A., Lucek, E.A., Dandouras, I.: Observations of multiple X-line structure in the Earth's magnetotail current sheet: a Cluster case study. Geophys. Res. Lett. **32**, L11105 (2005). doi:10.1029/2005GL022509
Eastwood, J.P., Phan, T.-D., Mozer, F.S., Shay, M.A., Fujimoto, M., Retinò, A., Hesse, M., Balogh, A., Lucek, E.A., Dandouras, I.: Multi-point observations of the Hall electromagnetic field and secondary island formation during magnetic reconnection. J. Geophys. Res. **112**, A06235 (2007). doi:10.1029/2006JA012158
Eastwood, J.P., Phan, T.D., Bale, S.D., Tjulin, A.: Observations of turbulence generated by magnetic reconnection. Phys. Rev. Lett. **102**(3), 035001 (2009)
Eastwood, J.P., Phan, T.D., Fear, R.C., Sibeck, D.G., Angelopoulos, V., Øieroset, M., Shay, M.A.: Survival of flux transfer event (FTE) flux ropes far along the tail magnetopause. J. Geophys. Res. **117**, A08222 (2012). doi:10.1029/2012JA017722
Escoubet, C.P., Fehringer, M., Goldstein, M.L.: The Cluster mission. Ann. Geophys. **19**, 1197–1200 (2001)

Fairfield, D.H., Cahill, L.J.: Transition region magnetic field and polar magnetic disturbances. J. Geophys. Res. **71**, 155–169 (1966)

Fear, R.C., Milan, S.E., Fazakerley, A.N., Fornacon, K.-H., Carr, C.M., Dandouras, I.: Simultaneous observations of flux transfer events by THEMIS, Cluster, Double Star and SuperDARN: acceleration of FTEs. J. Geophys. Res. **114**, A10213 (2009). doi:10.1029/2009JA014310

Fujimoto, M., Nakamura, M.S., Shinohara, I., Nagai, T., Mukai, T., Saito, Y., Yamamoto, T., Kokubun, S.: Observations of earthward streaming electrons at the trailing boundary of a plasmoid. Geophys. Res. Lett. **24**(22), 2893–2896 (1997)

Fuselier, S.A., Lewis, W.S.: Properties of near-earth magnetic reconnection from in-situ observations. Space Sci. Rev. **160**, 95–121 (2011)

Glassmeier, K.-H., Motschmann, U., Dunlop, M.W., Balogh, A., Carr, C.M., Mussman, G., Fornacon, K.-H., Schweda, K., Vogt, J., Georgescu, E., Buchert, S.: Cluster as a wave telescope—first results from the fluxgate magnetometer. Ann. Geophys. **19**, 1439–1447 (2001)

Gosling, J.T.: Magnetic reconnection in the solar wind. Space Sci. Rev. **172**, 187–200 (2012)

Gosling, J.T., Eriksson, S., Blush, L.M., Phan, T.D., Luhmann, J.G., McComas, D.J., Skoug, R.M., Acuña, M.H., Russell, C.T., Simunac, K.D.: Five spacecraft observations of oppositely directed exhaust jets from a magnetic reconnection X-line extending $>$ 4.26 x 106 km in the solar wind at 1 AU. Geophys. Res. Lett. **34**, 108 (2007). doi:10.1029/2007GL031492

Greene, J.M.: Locating three-dimensional roots by a bisection method. J. Comp. Phys. **98**, 194–198 (1992)

Hasegawa, H., Wang, J., Dunlop, M.W., Pu, Z.Y., Zhang, Q.-H., Lavraud, B., Taylor, M.G.G.T., Constantinescu, O.D., Berchem, J., Angelopoulos, V., McFadden, J.P., Frey, H.U., Panov, E.V., Volwerk, M., Bogdanova, Y.V.: Evidence for a flux transfer event generated by multiple X-line reconnection at the magnetopause. Geophys. Res. Lett. **37**, L16101 (2010). doi:10.1029/2010GL044219

Heikkila, W.J.: Critique of fluid theory of magnetospheric phenomena. Astrophys. Space Sci. **23**, 261–268 (1973)

Henderson, P.D., Owen, C.J., Lahiff, A.D., Alexeev, I.V., Fazakerley, A.N., Lucek, E.A., Rème, H.: Cluster PEACE observations of electron pressure tensor divergence in the magnetotail. Geophys. Res. Lett. **33**, L22106 (2006). doi:10.1029/2006GL027868

Henderson, P.D., Owen, C.J., Lahiff, A.D., Alexeev, I.V., Fazakerley, A.N., Yin, L., Walsh, A.P., Lucek, E., Rème, H.: The relationship between j x B and div P_e in the magnetotail plasma sheet: Cluster observations. J. Geophys. Res. **113**, A07S31 (2008). doi:10.1029/2007JA012697

Hones Jr., E.W.: Observations in the Earth's magnetotail relating to magnetic merging. Sol. Phys. **47**, 101–113 (1976)

Huang, S.Y., Zhou, M., Sahraoui, F., Deng, X.H., Pang, Y., Yuan, Z.G., Wei, Q., Wang, J.F., Zhou, X.M.: Wave properties in the magnetic reconnection diffusion region with high β: application of the k-filtering method to Cluster multispacecraft data. J. Geophys. Res. **115**, A12211 (2010). doi:10.1029/2010JA015335

Huang, S.Y., Zhou, M., Sahraoui, F., Vaivads, A., Deng, X.H., André, M., He, J.S., Fu, H.S., Li, H.M., Yuan, Z.G., Wang, D.D.: Observations of turbulence within reconnection jet in the presence of guide field. Geophys. Res. Lett. **39**, L11104 (2012). doi:10.1029/2012GL0522210

Khurana, K.K., Kepko, E.L., Kivelson, M.G., Elphic, R.C.: Accurate determination of magnetic field gradients from four-point vector measurements—II: use of natural constraints on vector data obtained from four spinning spacecraft. IEEE Trans. Magn. **32**(5), 5193–5205 (1996)

Lau, Y.-T., Finn, J.M.: Three-dimensional kinematic reconnection in the presence of field nulls and closed field lines. Astrophys. J. **350**, 672–691 (1990)

Mandt, M.E., Denton, R.E., Drake, J.F.: Transition to whistler mediated magnetic reconnection. Geophys. Res. Lett. **21**(1), 73–76 (1994)

Mozer, F.S., Bale, S.D., Phan, T.D.: Evidence of diffusion regions at a subsolar magnetopause crossing. Phys. Rev. Lett. **89**(1), 015002 (2002)

Natl. Res. Council: Solar and Space Physics: A Science for a Technological Society. National Research Council, Washington, DC (2012)

Ogilvie, K.W., von Rosenvinge, T., Durney, A.C.: International sun-earth explorer: a three-spacecraft program. Science **198**(4313), 131–138 (1977)

Øieroset, M., Phan, T.D., Fujimoto, M., Lin, R.P., Lepping, R.P.: In situ detection of collisionless reconnection in the Earth's magnetotail. Nature **412**, 414–417 (2001)

Øieroset, M., Phan, T.D., Eastwood, J.P., Fujimoto, M., Daughton, W., Shay, M.A., Angelopoulos, V., Mozer, F.S., McFadden, J.P., Larson, D.E., Glassmeier, K.-H.: Direct evidence for a three-dimensional magnetic flux rope flanked by two active magnetic reconnection X lines at Earth's magnetopause. Phys. Rev. Lett. **107**, 165007 (2011)

Parker, E.N.: Sweet's mechanism for merging magnetic fields in conducting fluids. J. Geophys. Res. **62**, 509–520 (1957)

Paschmann, G.: Recent in-situ observations of magnetic reconnection in near-Earth space. Geophys. Res. Lett. **35**, L19109 (2008). doi:10.1029/2008GL035297

Paschmann, G., Sonnerup, B.U.Ö., Papamastorakis, I., Sckopke, N., Haerendel, G., Bame, S.J., Asbridge, J.R., Gosling, J.T., Russell, C.T., Elphic, R.C.: Plasma acceleration at the Earth's magnetopause: evidence for reconnection. Nature **282**, 243–246 (1979)

Paschmann, G., Øieroset, M., Phan, T.: In-situ observations of reconnection in space. Space Sci. Rev. **178**, 385–417 (2013). doi:10.1007/s11214-012-9957-2

Phan, T.D., Kistler, L.M., Klecker, B., Haerendel, G., Paschmann, G., Sonnerup, B.U.Ö., Baumjohann, W., Bavassano-Cattaneo, M.B., Carlson, C.W., DiLellis, A.M., Fornacon, K.-H., Frank, L.A., Fujimoto, M., Georgescu, E., Kokubun, S., Möbius, E., Mukai, T., Øieroset, M., Paterson, W.R., Rème, H.: Extended magnetic reconnection at the Earth's magnetopause from detection of bi-directional jets. Nature **404**, 848–850 (2000)

Phan, T.D., Gosling, J.T., Davis, M.S., Skoug, R.M., Oieroset, M., Lin, R.P., Lepping, R.P., McComas, D.J., Smith, C.W., Reme, H., Balogh, A.: A magnetic reconnection X-line extending more than 390 Earth radii in the solar wind. Nature **439**(7073), 175–178 (2006)

Pinçon, J.L., Motschmann, U.: Multi-spacecraft filtering: general framework. In: Paschmann, G., Daly, P.W. (eds.) Analysis Methods for Multi-Spacecraft Data, pp. 65–78. International Space Science Institute, Bern (1998)

Robert, P., Roux, A., Harvey, C.C., Dunlop, M.W., Daly, P.W., Glassmeier, K.-H.: Tetrahedron geometric factors. In: Paschmann, G., Daly, P.W. (eds.) Analysis Methods for Multi-Spacecraft Data, pp. 323–348. International Space Science Institute, Bern (1998)

Runov, A., Nakamura, R., Baumjohann, W., Treumann, R.A., Zhang, T.L., Volwerk, M., Vörös, Z., Balogh, A., Glassmeier, K.-H., Klecker, B., Rème, H., Kistler, L.M.: Current sheet structure near magnetic X-line observed by Cluster. Geophys. Res. Lett. **30**(11), 1579 (2003). doi:10.1029/2002GL016730

Runov, A., Angelopoulos, V., Sitnov, M.I., Sergeev, V.A., Bonnell, J., McFadden, J., Larson, D., Glassmeier, K.-H., Auster, U.: THEMIS observations of an earthward-propagation dipolarization front. Geophys. Res. Lett. **36**, L14106 (2009). doi:10.1029/2009GL038980

Schwartz, S.J., Horbury, T., Owen, C.J., Baumjohann, W., Nakamura, R., Canu, P., Roux, A., Sahraoui, F., Louarn, P., Sauvaud, J.A., Pinçon, J.L., Vaivads, A., Marcucci, M.F., Anastasiadis, A., Fujimoto, M., Escoubet, P., Taylor, M., Eckersley, S., Allouis, E., Perkinson, M.-C.: Cross-scale: multi-scale coupling in space plasmas. Exp. Astron. **23**, 1001–1015 (2009)

Sibeck, D.G., Angelopoulos, V., Brain, D.A., Delory, G.T., Eastwood, J.P., Farrell, W.M., Grimm, R.E., Halekas, J.S., Hasegawa, H., Hellinger, P., Khurana, K.K., Lillis, R.J., Øieroset, M., Phan, T.-D., Raeder, J., Russell, C.T., Schriver, D., Slavin, J.A., Travnicek, P.M., Weygand, J.M.: ARTEMIS science objectives. Space Sci. Rev. **165**, 59–91 (2011). doi:10.1007/s11214-011-9777-9

Sitnov, M.I., Swisdak, M., Divin, A.V.: Dipolarization fronts as a signature of transient reconnection in the magnetotail. J. Geophys. Res. **114**, A04202 (2009). doi:10.1029/2008JA013980

Slavin, J.A., Lepping, R.P., Gjerloev, J., Goldstein, M.L., Fairfield, D.H., Acuña, M.H., Balogh, A., Dunlop, M.W., Kivelson, M.G., Khurana, K.K., Fazakerley, A.N., Owen, C.J., Rème, H.,

Bosqued, J.M.: Cluster current density measurements within a magnetic flux rope in the plasma sheet. Geophys. Res. Lett. **30**(7), 1362 (2003). doi:10.1029/2002GL016411

Slavin, J.A., Le, G., Strangeway, R.J., Wang, Y., Boardsen, S.A., Moldwin, M.B., Spence, H.E.: Space technology 5 multi-point measurements of near-Earth magnetic fields: initial results. Geophys. Res. Lett. **35**, L02107 (2008). doi:10.1029/2007GL031728

Sonnerup, B.U.Ö.: Magnetic field reconnection. In: Lanzerotti, L.T., Kennel, C.F., Parker, E.N. (eds.) Solar System Plasma Physics, vol. III, pp. 47–108. North-Holland, Amsterdam (1979)

Spence, H.E., Moore, T.E., DiJoseph, M., Buchanan, R., Anderson, B.J., Angelopoulos, V., Baumjohann, W., Borovsky, J., Carovillano, R., Craven, P., Fennell, J.F., Goodrich, C.C., Hesse, M., Li, H., Lynch, K.A., Panetta, P.V., Raeder, J., Reeves, G., Sibeck, D., Siscoe, G.L., Tsyganenko, N.A., Vondrak, R., Slavin, J., Spann, J.: The magnetospheric constellation mission dynamic response and coupling observatory (DRACO) understanding the global dynamics of the structured magnetotail. NASA, Greenbelt, MD (2001)

Spence, H.E., Moore, T.E., Klimas, A.J., Anderson, B.J., Angelopoulos, V., Baumjohann, W., Borovsky, J., Carovillano, R., Craven, P., Fennell, J.F., Goodrich, C.C., Hesse, M., Li, X., Lynch, K.A., Raeder, J., Reeves, G., Sibeck, D., Siscoe, G.L., Tsyganenko, N.A., Ford, K., Vondrak, R., Slavin, J., Hoeksema, J.T., Peterson, W.K.: The magnetospheric constellation (MC) global dynamics of the structured magnetotail. NASA, Washington, DC (2004)

Sweet, P.A.: The neutral point theory of solar flares. In: Lehnert, B. (ed.) Electromagnetic Phenomena in Cosmical Physics, pp. 123–134. Cambridge University Press, Cambridge (1958)

Vaivads, A., Khotyaintsev, Y., André, M., Retinò, A., Buchert, S., Rogers, B.N., Décréau, P., Paschmann, G., Phan, T.D.: Structure of the magnetic reconnection diffusion region from four-spacecraft observations. Phys. Rev. Lett. **93**(10), 105001 (2004)

Vaivads, A., Retinò, A., André, M.: Magnetic reconnection in space plasma. Plasma Phys. Controlled Fusion **51**, 124016 (2009)

Vasyliunas, V.M.: Theoretical models of magnetic field line merging. Rev. Geophys. **13**, 303–336 (1975)

Xiao, C.J., Wang, X.G., Pu, Z.Y., Zhao, H., Wang, J.X., Ma, Z.W., Fu, S.Y., Kivelson, M.G., Liu, Z.X., Zong, Q.G., Glassmeier, K.-H., Balogh, A., Korth, A., Reme, H., Escoubet, C.P.: In situ evidence for the structure of the magnetic null in a 3D reconnection event in the Earth's magnetotail. Nat. Phys. **2**, 478–483 (2006)

Xiao, C.J., Wang, X.G., Pu, Z.Y., Ma, Z.W., Zhao, H., Zhou, G.P., Wang, J.X., Kivelson, M.G., Fu, S.Y., Liu, Z.X., Zong, Q.G., Dunlop, M.W., Glassmeier, K.-H., Lucek, E., Reme, H., Dandouras, I., Escoubet, C.P.: Satellite observations of separator-line geometry of three-dimensional magnetic reconnection. Nat. Phys. **3**, 609–613 (2007)

Bosqued, J.M.: Cluster current density measurements within a magnetic flux rope in the plasma sheet. Geophys. Res. Lett. 30(7), 1162 (2003). doi:10.1029/2002GL016411
Slavin, J.A., Le, G., Strangeway, R.J., Wang, Y., Boardsen, S.A., Moldwin, M.B., Spence, H.E.: Space Technology 5 multi-point measurements of near-Earth magnetic fields: initial results. Geophys. Res. Lett. 35, L02107 (2008). doi:10.1029/2007GL031728
Sonnerup, B.U.Ö.: Magnetic field reconnection. In: Lanzerotti, L.J., Kennel, C.F., Parker, E.N. (eds.) Solar System Plasma Physics, vol. III, pp. 45–108. North Holland, Amsterdam (1979)

Chapter 10
Adventures in Parameter Space: Reconnection and the Magnetospheres of the Solar System

Margaret Galland Kivelson

Abstract The key to a deep understanding of the physics of a magnetosphere was provided by Jim Dungey in 1961 in a brief but highly influential paper (~3 citations per word to this time) that is referenced repeatedly in this volume. Jim was thinking about more than one system when he proposed the reconnecting magnetosphere; he was trying to understand both the terrestrial environment and the dynamics of the Sun. His ideas apply not only to those two systems but also to others of which nothing was known in 1961. From the early 1970s to the present time, spacecraft have been exploring the magnetospheres of other planets as well as the magnetosphere of the moon, Ganymede. Today the Voyager spacecraft are approaching the boundary of the largest solar system magnetosphere, the heliosphere. Understanding the structure and dynamics of these remote systems requires consideration of various parameters that govern such features as their size and shape and their interaction with the flowing plasmas in which they are embedded. So, taking a lead from another remarkable paper of Jim's that explains many ideas important to space plasmas by inviting the reader to participate in "Adventures in Velocity Space," this work provides an introduction to the range of parameter space that characterizes the diverse magnetospheres of the solar system and illustrates how Jim's perceptive work fits into the interpretation of other magnetospheric systems.

M.G. Kivelson (✉)
Department of Earth, Planetary, and Space Sciences, University of California, Los Angeles, CA 90095-1567, USA

Department of Atmospheric, Oceanic and Space Sciences, University of Michigan, Ann Arbor, MI 48109-2143, USA
e-mail: mkivelson@igpp.ucla.edu

D. Southwood et al. (eds.), *Magnetospheric Plasma Physics: The Impact of Jim Dungey's Research*, Astrophysics and Space Science Proceedings 41,
DOI 10.1007/978-3-319-18359-6_10

10.1 Introduction

A magnetosphere forms when a magnetized body is embedded in a plasma that we characterize by its flow velocity, **u**, plasma thermal pressure and mass density, p_{th} and ρ, and magnetic field, $\mathbf{B}_{ext}$. The properties of the central body that contribute to establishing magnetospheric properties include its size, characterized by its mean radius (R_o), surface magnetic field (B_o), angular velocity, $\mathbf{\Omega}$, and the density of neutral gas in its surroundings (possible sources include the atmosphere, rings, and moons) that serves as a source of trapped magnetospheric plasma. Dear to the heart of a physicist and critical to understanding the structure and the dynamics of the system are the dimensionless parameters formed from these quantities, such as β the ratio of thermal to magnetic pressure, and the Alfvén and sonic Mach numbers of the upstream flow ($M_A = u/\mathrm{v}_A$, $M_S = u/c_S$, where $\mathrm{v}_A = B/(2\mu_o\rho)^{1/2}$ is the Alfvén speed, $c_S = (\gamma p_{th}/\rho)^{1/2}$ is the sound speed, and γ is the adiabatic index). Other significant parameters are the internal and external pressure at the magnetopause, whose approximate equality sets the standoff distance of the nose of the magnetosphere, and, as will be clear in the discussion of the magnetospheres of Jupiter and Saturn, the ratio of rotational energy to magnetic and thermal energy of the magnetospheric plasma.

Dungey solved the problem of how reconnection with the solar wind accounts for properties of the global magnetosphere of Earth and controls ionospheric convection in 1961 and, shortly thereafter, went on to describe almost all critical aspects of magnetospheric dynamics (Dungey 1961, 1963), but it took time for his ideas to be accepted by all (well, most) of his colleagues. Over the decades, spacecraft exploration has provided ever more evidence that magnetic reconnection drives the dynamics of the terrestrial magnetosphere. But a physicist regards the explanation of a single example of a phenomenon as a first step. What would the magnetosphere look like and how would it work if we could change the properties of the upstream plasma and the internal parameters? Unlike laboratory scientists, we cannot twiddle with dials to control experimental conditions, but we are fortunate that other accessible systems have their dials set differently, taking us some distance down the experimental path. As early as 1973, the year in which Pioneer 10 flew by Jupiter, spacecraft began to explore planetary magnetospheres, systems that form in ranges of parameter space rarely or never encountered in Earth's immediate environment (see Table 10.1). The differences in external flow parameters (and also internal parameters) account in part for spatial scales that vary by a factor of 10^3 as revealed by six spacecraft to Jupiter, with another en route at the time of this writing, three to Saturn, one to Uranus and Neptune, two to Mercury, and one to Ganymede. Figure 10.1 shows sketches of the structure and indicates the scales of the magnetospheres of these bodies. There is also a representation of the largest solar system magnetosphere, the cavity that the solar wind carves out of the local interstellar medium (LISM) referred to as the heliosphere. The Voyager 1 and 2 spacecraft, on escape trajectories from the Sun and now well beyond 100 AU, are exploring its upstream boundary regions. Our sample of eight systems forms the basis for the study of comparative magnetospheres, and provides

Table 10.1 Variation with distance from the Sun (Jupiter) of the parameters of the external flow that contribute to establishing the spatial scales of solar system magnetospheres

Planet	R(AU) [or R(R_J)]	P_{ext} (nPa)	M_S	M_A	β	Spiral angles
Mercury	0.31	26.5000	5.5	3.9	0.5	17
	0.47	11.0000	6.1	5.7	0.9	25
Venus	0.72	5.0000	6.6	7.9	1.4	36
Earth	1.00	2.5000	7.2	9.4	1.7	45
Mars	1.52	1.1000	7.9	11.1	2.0	57
Jupiter	5.20	0.0920	10.2	13.0	1.6	79
Saturn	9.60	0.0270	11.6	13.3	1.3	84
Uranus	19.10	0.0069	13.3	13.3	1.0	87
Neptune	30.20	0.0027	14.6	13.3	0.8	88
Ganymede	15 R_J	1	0.5	0.3	0.4–1.6	N/A
Heliosphere	~100	7×10^{-5}	2.8	0.8	0.3	N/A

Table is adapted from Fujimoto et al. (2007) who references Slavin and Holzer (1981). Additions (last 2 lines) are from Jia et al. (2008) and Swisdak et al. (2010). R is the distance of the planet from the Sun or the distance from the Sun to the heliopause in AU or the distance from Jupiter in R_J; P_{ext} is the dynamic pressure of the external flowing plasma, M_S, M_A, and β refer to properties of the external plasma; the spiral angle refers to the angle between ± the mean orientation of the IMF and the radius vector from the Sun

evidence that not all magnetospheres are alike. For example, the sketched field lines of the planetary magnetospheres emerge from the polar regions but, not far from the surface, bend significantly towards alignment with the direction of the external flow (see, for example, the sketch of Mercury's magnetosphere). Polar field lines from Ganymede bend little; they fill an almost cylindrical region centered on Ganymede's dipole axis and transverse to the flow direction. Below we will consider the parameters responsible for this dramatic difference in structure.

Change of scale size over a large range does more than increase the volume carved out of the external plasma. It changes magnetospheric dynamics in many ways. For example, it takes of order minutes for the solar wind to flow from Mercury's upstream magnetopause to its distant neutral line, whereas it takes of order an hour for the solar wind to flow over the equivalent segment of the terrestrial magnetosphere and days for it to flow by the equivalent portion of the jovian magnetosphere. In considering the role of reconnection as introduced by Dungey (1961), one typically imagines that the interplanetary magnetic field (IMF) is capable of remaining in an orientation that favors reconnection over times long enough for the reconnected flux tubes to drape over the entire magnetotail. However, the sign of the north-south component of the IMF, B_z. fluctuates. D'Amicis et al. (2006) report that the time in minutes between reversals satisfies $\tau(t) = A\, t^{-\alpha}$ $\exp(-t/T_c)$ with α ranging between 1.33 and 1.56 and $T_c \geq 105$ min. This relation implies that although it is not unusual for the IMF to remain in an orientation favorable to reconnection for tens of minutes, times long enough for reconnected flux tubes to drape over the entirety of Mercury's or Earth's magnetospheres, it is rare for the orientation to remain favorable to reconnection for a time long enough

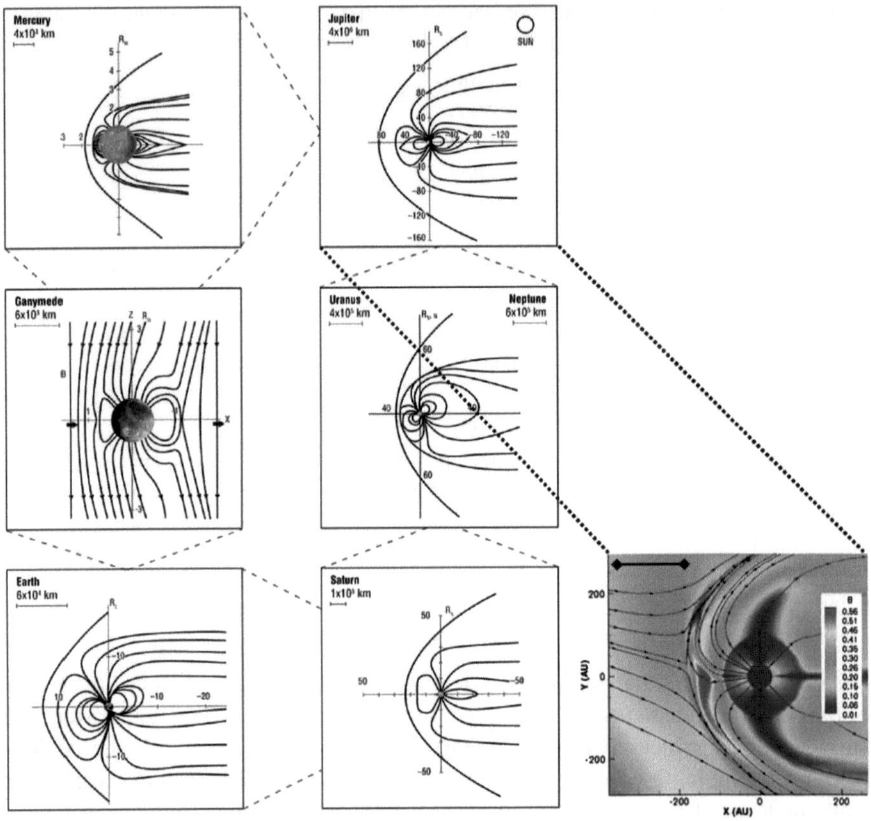

Fig. 10.1 Relative scales of the magnetospheres of the solar system adapted from Fujimoto et al. (2007) *Bars* in the *upper left* of the sketches represent scale lengths in km and are, in counterclockwise order: Mercury (4×10^3), Ganymede (6×10^3), Earth (6×10^4), Saturn (1×10^5), Uranus (4×10^5), Neptune (6×10^5), Jupiter (4×10^6). To the *right* is the heliosphere as modeled by Swisdak et al. (2010), with representing a distance of 100 AU $= 1.5 \times 10^{10}$ km

for a reconnected flux tube to drape over Jupiter's magnetosphere. Thus, even though reconnection is the principal driver of activity at Earth, it is unlikely to be so at Jupiter.

10.2 How Big Is a Magnetosphere?

Some critical parameters of the solar wind or other flowing plasma external to magnetospheres of the solar system from the smallest (Ganymede embedded in Jupiter's magnetospheric plasma) to the largest (the heliosphere embedded in the LISM) are shown in Table 10.1. These parameters are relevant to establishing the scales of the planetary magnetospheres. At all of the planets, both the Alfvénic and sonic Mach numbers are large enough to imply that magnetic and thermal pressures are small compared with dynamic pressure. By setting the ratio of the external dynamic pressure, ρu^2, to the internal pressure of a planetary dipole field at distance R_{mp} approximately to 1, one obtains an estimate of the distance to the nose of a magnetosphere, R_{mp}:

$$R_{mp}/R_o \approx \left(B_o/\rho u^2\right)^{1/6} \tag{10.1}$$

The dynamic pressure decreases rapidly with distance from the Sun (Table 10.1). Were the surface magnetic fields the same at all planets, the magnetospheres of Jupiter and Saturn would be far larger than Earth's magnetosphere simply because of the dependence of R_{mp} on dynamic pressure. Actually Saturn's surface field differs little from Earth's, but the field falls with distance from the center of the planet, r, as $(R_p/r)^3$, with R_p the planetary radius, and Saturn's radius is an order of magnitude larger than Earth's. The combined effect of these differences is to cause Saturn's nose to lie ten times further from the center of the planet than Earth's. Jupiter's magnetosphere is even larger because its surface field is more than ten times larger than Earth's.

For Ganymede and the heliosphere, the assumption that the external pressure is dominated by dynamic pressure is not valid. In these cases, a small Alfvénic Mach number implies that the confining stresses are exerted principally by magnetic pressure, and Eq. (10.1) does not provide a good estimate of the standoff distance. This is another matter that will be discussed below.

10.3 Multiple Drivers of Magnetospheric Dynamics

Dynamics differ from one magnetosphere to another. Critical to establishing the nature of magnetospheric dynamics is the relative importance of reconnection-driven convection and rotationally driven flows. For Earth's magnetosphere, it is common to consider how the two flow sources combine to govern the motion of low energy plasma, producing a plasmapause at the interface between the outer part of the magnetosphere in which reconnection-driven flow dominates and the inner part where rotation dominates (see, for example, Brice 1967; Chen 1970 and references within). The stagnation point at which rotational flow balances typical reconnection-driven convection falls well within the duskside magnetopause and well outside of Earth's surface.

The situation is far different at Mercury and Ganymede. Mercury's sidereal period is 59 days, but what is relevant is the time it takes for a point on the surface to rotate through 360° relative to the flow direction of the solar wind. This period is 176 days. Ganymede is phase-locked to Jupiter, so it does not rotate at all relative to the direction of the flow of jovian plasma (Kivelson et al. 1998). In both of these magnetospheres, rotation is largely irrelevant and the pattern of plasma flow is imposed by reconnection.

The gas giants, Jupiter and Saturn, are so large and rotate so rapidly that the stagnation point of the low energy plasma flow typically falls outside of the magnetopause (Brice and Ioannidis 1970), which makes it evident that rotational stress is important to understanding plasma properties in these systems. Yet analysis directly based on experience at Earth should be questioned. At both Jupiter and Saturn, there are sources of heavy ions (principally sulfur and oxygen) deep within the magnetosphere. In steady state, the plasma flows outward to be lost to the solar wind or down the magnetotail. As the plasma moves radially outward, conservation of angular momentum implies a reduction of angular velocity. In response to the lag, corotation enforcement current flows along field lines, coupling the ionosphere and the equatorial magnetosphere. At the equator, the field-aligned current feeds radial current in the plasma and accelerates it towards the angular velocity of the rotating ionosphere (Hill 1979). The field-aligned current system must close in the planetary ionosphere, where the effective imposition of rotation requires that the ionospheric conductance be large enough to impede significant slippage of flux tubes. At Jupiter (radius: R_J), full corotation is not imposed at distances significantly larger than 20 R_J. Furthermore, given the large spatial scale of the jovian magnetosphere, one needs to account for time delay in signal propagation. If the time delay is long compared with the plasma outflow time, the ionosphere may be unable to control the equatorial plasma (Vasyliunas 1994). Thus, signal propagation times, of minor importance at Earth, must be considered for large magnetospheric systems such as those of Jupiter and Saturn.

Additional subtleties arise as well in assessing just how significant solar wind-driven flow is in magnetospheres of the outer solar system. Dayside reconnection at Jupiter, Saturn, Uranus and Neptune may not mimic the process at Earth, where of order 10 % of the upstream electric field is effective in driving convection within the magnetosphere (Hill et al. 1983). The efficiency of reconnection has been found to depend on various properties of the plasmas within which the reconnecting fields are embedded (e.g., Borovsky 2013) and, in particular, to depend on the angle of magnetic shear across the dayside magnetopause, to decrease with increasing M_A and β (Scurry et al. 1994) and to be inhibited when the change of β across the current sheet is large (Phan et al. 2010, 2013; Swisdak et al. 2010). As both M_A and β of the solar wind increase with distance from the Sun, magnetopause reconnection is likely to be less efficient at the outer planets and more efficient at Mercury than at Earth. This speculation has been supported by spacecraft data related to magnetopause reconnection for different planetary magnetospheres, as discussed in the following sections.

10.4 Mercury and Solar Wind-Driven Dynamics

Mercury's magnetosphere is so small that the planet's surface lies less than one planetary radius below the dayside magnetopause, as illustrated schematically in Fig. 10.2. The schematics reflect an expectation that reconnection can strip off some (moderate loading) or all (extreme loading) of the dayside flux. The average solar wind Alfvénic Mach number and β at Mercury's orbit, smaller than at Earth by factors of ~0.4 and ~0.3, respectively (Table 10.1), lead one to anticipate efficient reconnection. Various observations made by the Messenger spacecraft (Slavin et al. 2010) support this expectation. The magnetosphere has been shown to experience geomagnetic activity of considerable intensity on remarkably short time scales. For example, on 29 September 2009, the lobe field magnitude in the magnetotail increased by a factor of 2–3.5 in 2–3 min. Lobe field increase is characteristic of the growth phase of a terrestrial substorm although the increase of the tail lobe field strength at Earth is typically only a factor of ~1.4 (Caan et al. 1975). The time scale for the lobe increases observed at Mercury is short compared with the ~hour time scale of a growth phase at Earth, but consistent with the expectations from the Dungey cycle for a Mercury-sized magnetosphere. Strangely no acceleration signatures of the sort observed at times of terrestrial substorms were identified in energetic particle measurements at Mercury. Furthermore, the complete stripping of dayside closed flux illustrated in Fig. 10.2c has not been observed and simulations suggest that it is the strong inductive response arising because of Mercury's large metallic core that prevents the extreme response illustrated (X. Jia, personal communication, 2013).

Direct evidence of exceptionally intense magnetopause reconnection at Mercury was reported by Slavin et al. (2012) A plot of the data acquired by Messenger near the magnetopause tailward of the southern magnetic cusp shows (Fig. 10.3) "a shower of flux-transfer events (FTEs)" that lasted 2–3 s separated in time by 8–10 s. Flux transfer events arise when reconnection between solar wind and magnetospheric flux tubes occurs over a limited region on the magnetopause (Russell and Elphic 1978), but at Earth the typical recurrence time for FTEs is 8 min. The appearance of FTEs tailward of the cusp are consistent with expectations from the

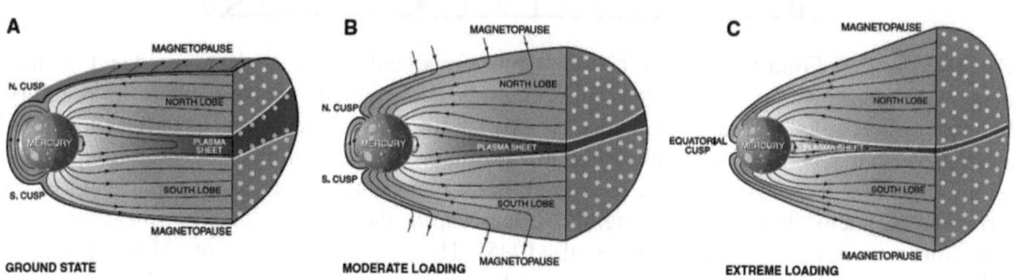

Fig. 10.2 From Slavin et al. (2010): A schematic view of Mercury's magnetosphere (**a**) in its ground state, (**b**) during moderate activity, and (**c**) during an interval of strong loading by reconnection as observed on 29 September 2009 by Messenger. Note that the stripping away of closed field lines on the dayside magnetosphere has not been observed

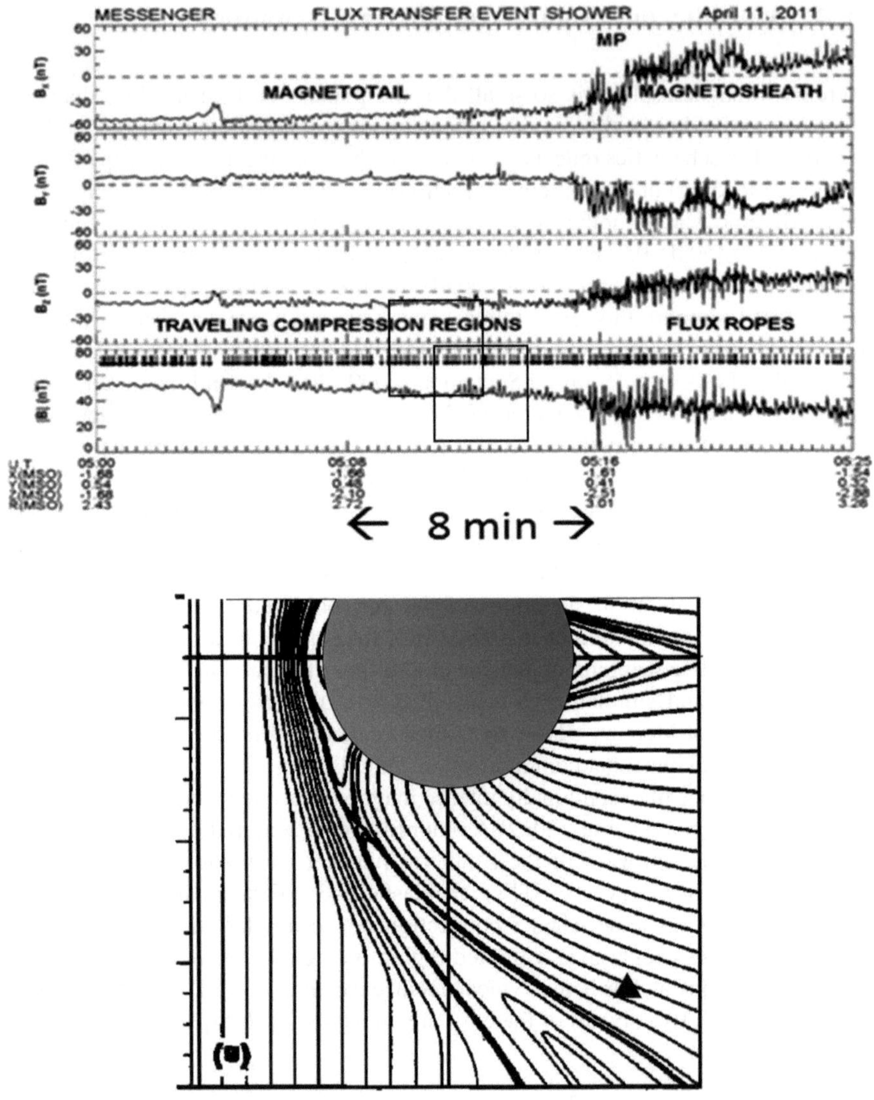

Fig. 10.3 *Above*: From Slavin et al. (2012), three components of the magnetic field and the field magnitude measured on a Messenger pass through the southern magnetosphere of Mercury. Fluctuations recorded near the magnetopause (MP), both outside and inside, have been identified as signatures of reconnection poleward of the cusp and include multiple traveling compression regions and flux ropes. *Below*: The background is a representation of the magnetic field of the terrestrial magnetosphere reconnecting above the cusp in the presence of a northward-oriented IMF from the simulation of Berchem et al. (1995). The *gray circle* represents Mercury with its radius selected so that the magnetopause standoff distance is a fraction of a Mercury radius. Messenger's approximate location at the time it acquired the data of the plot above is shown as a *black triangle*

Dungey model of reconnection for a northward IMF as illustrated in a schematic magnetosphere in Fig. 10.3 taken from a magnetohydrodynamic (MHD) simulation described by Berchem et al. (1995). The short recurrence time of FTEs is consistent with an increase of reconnection efficiency in the low Alfvénic Mach number and low beta solar wind near Mercury (Scurry et al. 1994).

10.5 Jupiter and Saturn and Rotationally-Dominated Dynamics

It has been noted above that rotational stresses imposed on the equatorial plasma of the giant planets appear to dominate the dynamics, leaving little role for reconnection-driven flows even assuming that 10% of the solar wind electric field is imposed across the magnetosphere. It is likely, however, that in making such an assumption we have overestimated the significance of magnetopause reconnection. If indeed the efficiency of reconnection varies with the Alfvénic Mach number and the β of the solar wind, the magnetic fields of Jupiter and Saturn may not reconnect efficiently with the IMF where M_A and β are ~3 times larger than at Mercury and M_A is 40 % larger than at Earth. Few reconnection signatures such as FTEs have been identified on the upstream boundary at Jupiter: on nine magnetopause encounters at Jupiter, Walker and Russell (1985) found only 14 FTE events, all transporting little magnetic flux.

Rotational forces contribute significantly to dynamics when $\rho r\, sin\theta \Omega^2$ is comparable to or large compared with other contributions to the force density. Here θ is the angle from the spin axis and Ω is the angular velocity of the planet. Rotational stresses are critical in the magnetospheres of Jupiter and Saturn because of the large spatial scales of the magnetospheres (planetary radii are more than an order of magnitude larger than Earth's radius) and the relatively rapid rotation rates (>2.2 times Earth's rotation rate) of these planets. Furthermore, plasma densities are relatively large because of the presence of sources (Io at Jupiter, a source of 10^3 kg s^{-1} of plasma and Enceladus at Saturn, a source of 100 kg s^{-1} of plasma) deep within both magnetospheres and the plasma is composed principally of sulfur and oxygen ions, so the mass per ion is, on average, ~20 proton masses. The aurora, driven at Earth by the magnetotail response to the solar wind, is found at Jupiter to be dominated by response to intense field-aligned currents that link the ionosphere to the equatorial magnetosphere where they act to drive the outflowing heavy ion plasma towards corotation (Clark et al. 2004). Thus the dynamics must be understood principally in terms of the effects of rapidly rotating plasma trapped near the centrifugal equators of the planets.

The interaction of the rotating plasma with the magnetic field modifies the magnetic configuration both through centrifugal stress and through preferential acceleration of parallel velocity of energetic particles. A plasma disc geometry develops not only on the night sides of the magnetospheres as at Earth, but on the

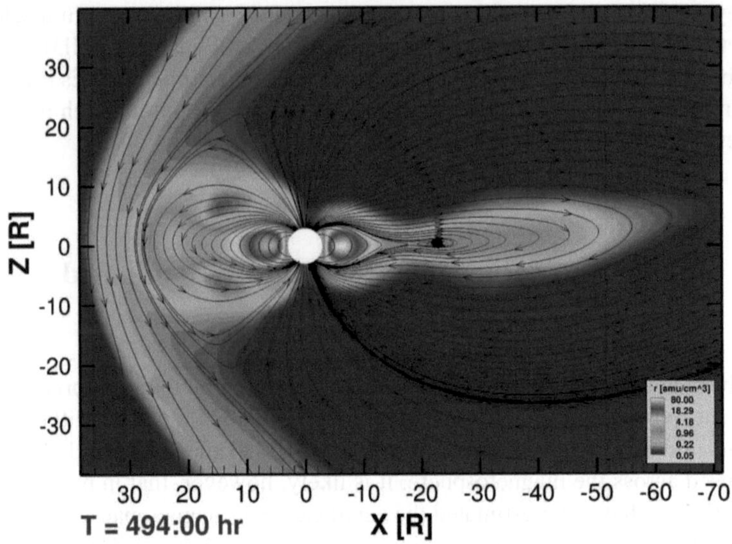

Fig. 10.4 The noon-midnight meridian of a Saturn simulation (Jia and Kivelson 2012) with a plasma source near Enceladus at 4 R_S injecting 100 kg s^{-1} water group ions. In this simulation the *lines* show the direction of the magnetic field and *color* represents density in amu cm^{-3}. The IMF is southward (i.e., parallel to the dipole field near the equator) and there is no evidence of magnetopause reconnection (Color figure online)

day sides as well. The slingshot effect of heavy ion plasma coupled with the development of pressure anisotropy ($p_{\|}/p_{\perp} > 1$) (Kivelson and Southwood 2005) stretches the field on both day and night sides. This feature is evident in a cut through the noon-midnight meridian of an MHD simulation of Saturn's magnetosphere (Jia and Kivelson 2012) in Fig. 10.4. The principal features of the magnetosphere are similar to those of Earth: an upstream shock, a compressed magnetosheath, a magnetopause separating internal and external plasmas except near the high latitude cusps, and a current sheet in the tail separating low density northern and southern lobe field lines. Plasmoids form in the tail, breaking off near midnight near 20 R_S as can be seen in the figure. The plasmoids, consistent with effects of centrifugal acceleration of outward-moving charged particles in the magnetotail, do not move radially down the tail. Instead, as they move down tail, they move towards later local times, exiting the magnetosphere on the dawn flank. In this magnetosphere, with solar wind Mach number of order 13, dynamics seem typically to be driven mainly internally, although the magnetosphere does respond to strong solar wind disturbances. Jupiter's magnetosphere also appears to be driven largely by rotational forces.

10.6 Ganymede as an Example of a Magnetosphere in an $M_A < 1$ Plasma Flow

Ganymede is the largest moon in the solar system and the only moon known to have a permanent, internally-generated magnetic field (Kivelson et al. 1996). It is embedded in Jupiter's flowing plasma for which $M_A < 1$. The small Alfvén Mach number reflects the fact that at Ganymede's orbit, deep within the jovian magnetosphere, the dynamic pressure of the plasma rotating around the planet at a speed somewhat greater than Ganymede's orbital speed is smaller than its magnetic pressure. The form of a magnetosphere changes noticeably depending on whether M_A, is bigger or smaller than 1. Planetary magnetospheres form in an $M_A > 1$ plasma flow and symmetry is dominated by the direction of the upstream flow. The dynamic pressure of the flowing external plasma causes the open flux tubes to bend strongly towards the flow direction, producing the familiar bullet shape of a planetary magnetosphere. For $M_A < 1$, the magnetic pressure dominates and symmetry is imposed by the direction of the external magnetic field. As alluded to in the introduction, the open flux tubes emerging from the polar cap form an almost cylindrical volume, bending only slightly towards the upstream field direction to form what are called "Alfvén wings" (Neubauer 1980; Southwood et al. 1980). The Alfvén wings are asymptotic to boundaries that make an angle $\alpha_A = tan^{-1}(\nu/\nu_A) = tan^{-1}(M_A)$ with the background field as illustrated in the lower panel of Fig. 10.5 from a simulation of Jia et al. (2009).

Although the field external to Ganymede's magnetosphere fluctuates little, near the magnetopause crossings the magnetic field measured on flybys of Ganymede fluctuated considerably (on time scales of 20–50 s). On the pass illustrated in the upper panel of Fig. 10.5, Galileo remained close to the upstream magnetopause and fluctuations of field and flow were present not only at the magnetopause crossings but also, at reduced intensity, throughout the pass. Initially thought to arise from Kelvin-Helmholtz waves on the boundary (Kivelson et al. 1998), the fluctuations are now interpreted as signatures of intermittent reconnection at the magnetopause.

The reinterpretation is based on analysis of the dynamics of the magnetopause in the MHD simulation of Jia et al. (2010). In the simulation (carried out for conditions upstream of Ganymede for Galileo's G8 pass of May 7, 1997 with $M_A = 0.7$), reconnection recurs at 20–50 s intervals across a wide swath of the upstream magnetopause. In the upper plot of Fig. 10.6, black traces represent B_z and $|\mathbf{B}|$ measured by the Galileo magnetometer. Colored traces in the upper plot of Fig. 10.6 are the z-component and the total magnetic field at simulation time steps separated by 2 s plotted against position along the G8 pass converted into pseudo-time. The spread at a fixed point is largest near the magnetopause crossings, implying large temporal fluctuations of both B_z and $|\mathbf{B}|$ at these positions. The velocity extracted from the simulation in the same way is plotted in Fig. 10.6 vs. pseudo-time. At any spacecraft position, the flow would vary within the spread of the colored traces, again largest at the magnetopause crossings. Reconnection as the cause of the perturbations is confirmed by the development of intense field-aligned flows that

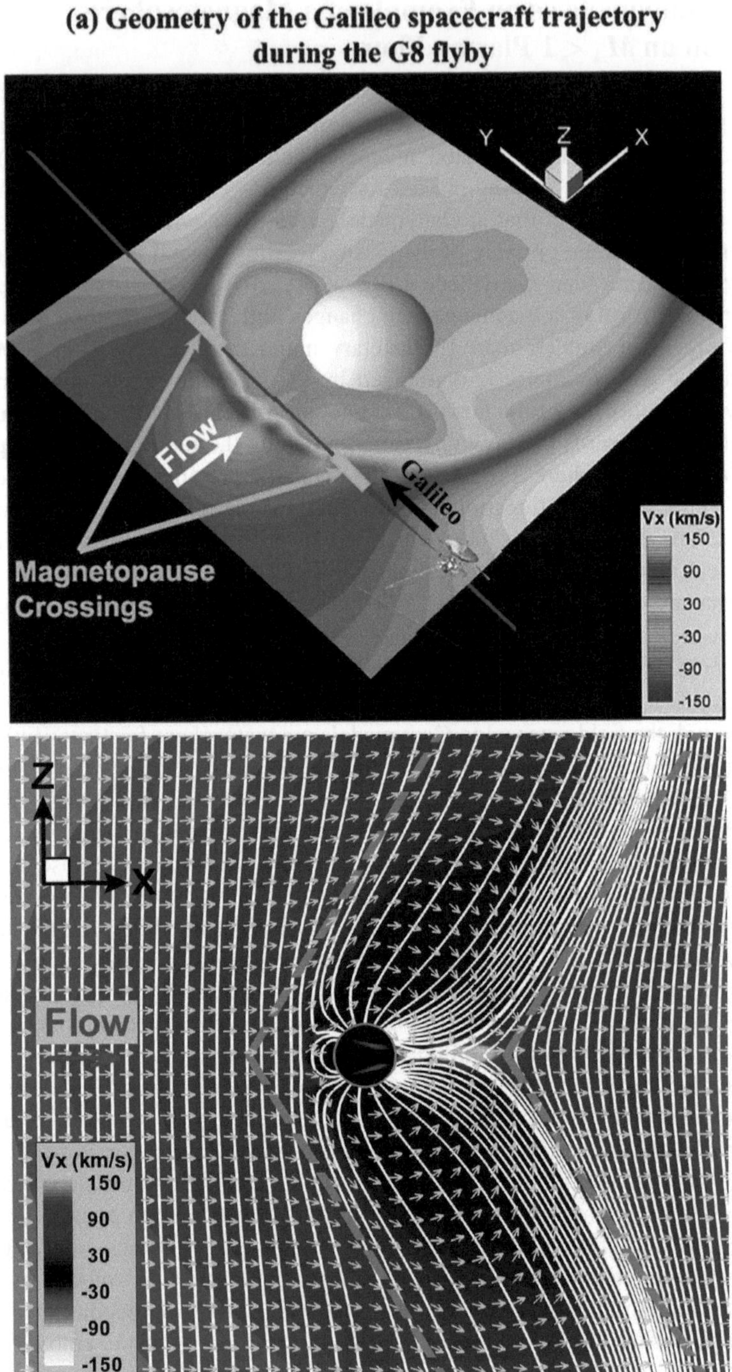

Fig. 10.5 Ganymede's magnetosphere from the MHD model of Jia et al. (2009 [*below*]; 2010 [*above*]). *Above*, a cut through the plane containing the trajectory of the Galileo spacecraft on pass G8 on May 7, 1997 near Ganymede's equator. *Color* represents the velocity along the direction (x)

appear in the simulation in conjunction with the field variations. That bursty changes of plasma flow were indeed observed by Galileo near the magnetopause crossings can be inferred from the dynamic representation of ion flux vs. energy shown in the lowest panel of Fig. 10.6. The energy associated with flow during a quiescent interval is represented by the dashed black line extracted from the simulation. Intermittent patches of energy higher than the quiescent energy are concentrated near the magnetopause crossings, with enhancements consistent with the range of changing velocity seen in the line plots from the simulation. These data, interpreted by comparison with the simulation, lead us to conclude that typical intermittency of reconnection for this magnetosphere in the small M_A regime is in the range 20–50 s, as listed in Table 10.2, comparable with the intermittency observed at Mercury.

10.7 The Biggest Magnetosphere (the Heliosphere): Another Case of $M_A < 1$ External Plasma?

The solar wind, the tenuous, flowing, magnetized plasma that flows outward from the Sun, fills the entire solar system, slowing down only as it approaches a distance at which its dynamic pressure is of order the pressure of the local interstellar medium (LISM). The Sun and solar wind create a cavity in the LISM, forming the heliosphere, the biggest of the known magnetospheres. Since their final planetary encounters, the Voyager 1 and 2 spacecraft have been heading towards the heliopause, the upstream boundary of the heliosphere. At the time of this writing, both of the Voyager spacecraft crossed the termination shock and entered a heliospheric boundary layer containing shocked solar wind, but neither had crossed the heliopause (Stone 2012). The termination shock forms because the super-magnetosonic solar wind must decelerate before it reaches the heliopause. Beyond the heliopause lies the LISM, but no in situ measurements have been made within it, so its properties have been inferred indirectly. Among those indirect measurements are those provided by the IBEX spacecraft (McComas et al. 2009a). IBEX measures the directional flux of energetic neutrals that arise from charge exchange with energized ions and are thought to come from regions beyond the termination shock (McComas et al. 2009b, 2012a, b). A remarkable feature of the observed flux of ENAs moving towards the center of the solar system is the ribbon of enhanced intensity illustrated in Fig. 10.7. If one assumes that this feature arises from processes occurring at the heliopause, its orientation and symmetry differ

Fig. 10.5 (continued) of the upstream flow. *Below*: a cut through the *x*-*z* plane, where *z* is parallel to Ganymede's spin axis. *Color* represents v_x. *Yellow arrows* are unit vectors along the flow direction. *White lines* represent the magnetic field. The *dashed red lines*, drawn at an angle M_A to the upstream unperturbed magnetic field, are seen to be close to parallel to the asymptotic orientation of open field lines (Color figure online)

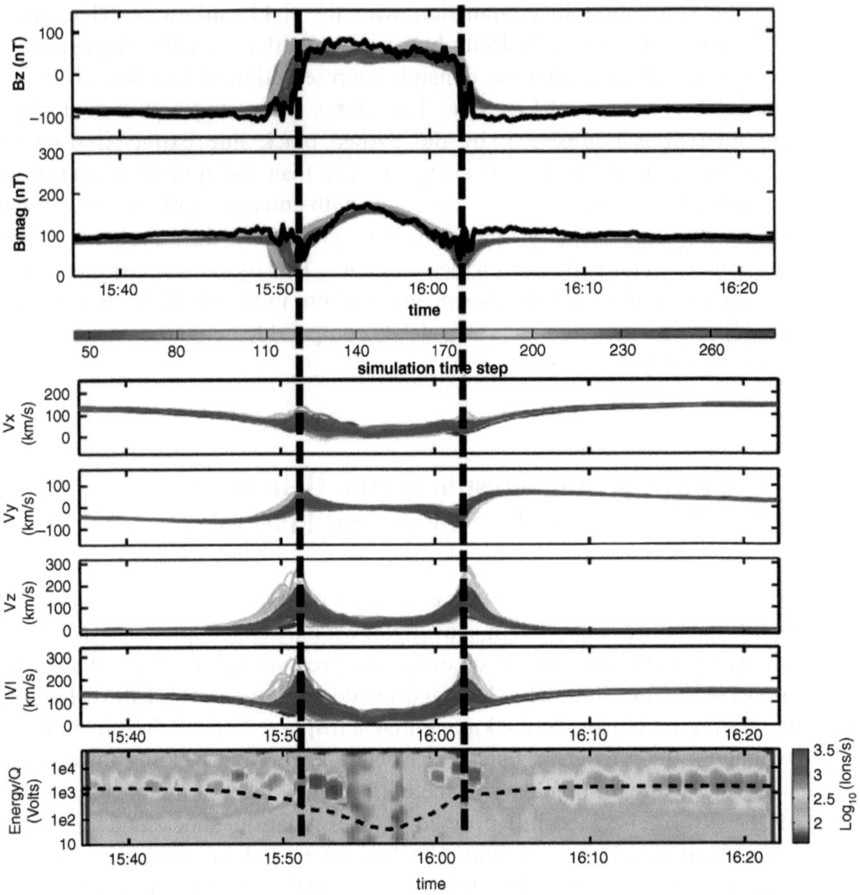

Fig. 10.6 The *upper panel* is a plot of the *z* component of the magnetic field and of the field magnitude. The *black lines* represent data from the Galileo magnetometer measurements (0.3 s cadence) plotted vs. UT on May 7, 1997. The *colored traces* are from the simulation of Jia et al. (2010) and are plotted vs. the position of Galileo relative to Ganymede at the time labeled. Traces taken from the simulation at intervals of 2 s are shown in different colors. *Heavy dashed lines* identify the magnetopause crossing from the reversal of the sign of B_z. *Below* are plotted three components of the flow velocity and its magnitude, again along the spacecraft trajectory, from different time steps of the simulation. Again *color* represents changing time (*color bar* below plot). The *lowest panel* shows the dynamic spectrum of ion number flux measured by the Galileo plasma investigation (PLS) plotted vs. UT. The color-coded ion counting rate represents the maximum response at a given E/Q during one spacecraft spin period. (The PLS data were obtained from the PPI node of the Planetary Data System.) The *black dashed line* represents the flow energy of heavy ions (with mass/charge = 16) obtained from the simulated flow speed along the Galileo trajectory extracted from a single time step of the simulation when bursty flows are absent near the magnetopause. The patchy bursts at energies above the flow energy of the background plasma are consistent with the acceleration of plasma through magnetopause reconnection (Color figure online)

Table 10.2 Parameters important for magnetopause reconnection and observed intervals between observed FTEs for selected magnetospheres

Body	M_A	β	Time scale for intermittent reconnection	Control of dynamics by magnetopause reconnection
Ganymede	<1	0.4–1.6	20–50 s	Strong
Mercury	Near 1	0.5	8–10 s	Strong
Earth	O (10)	1.7	~8 min	Intermediate
Jupiter/ Saturn	>10	1.6	Long	Weak

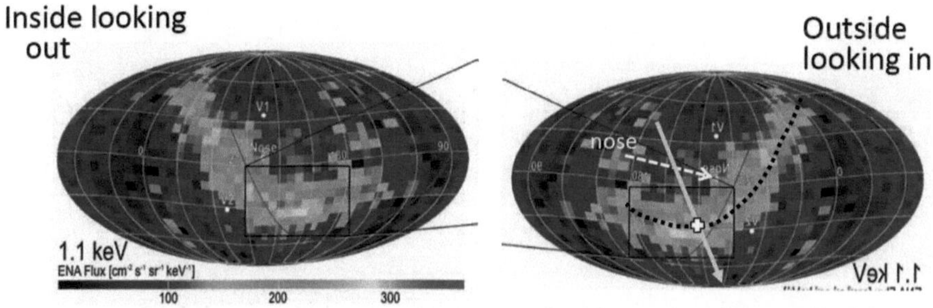

Fig. 10.7 From McComas et al. (2009b), *left*: a portion of their Fig. 10.7.1 showing all-sky maps of measured 1.1 keV ENA flux plotted vs. ecliptic coordinates (J2000). The nose of the heliosphere is marked as are the positions of Voyager 1 and 2. *Right*: the same image flipped about a *vertical* axis to represent the perspective of the ribbon from an upstream perspective. Superimposed are a *black dashed curve* that follows the center of the ribbon, a white cross marking the most intense emission, and a *yellow arrow* along the direction Kivelson and Jia (2012) propose for the field of the LISM (Color figure online)

from expectations based on early models of the interaction of the solar wind with the LISM. The symmetry center is not at the nose of the heliosphere, defined by the direction of motion of the Sun relative to the LISM.

The structure of the ribbon has been discussed (McComas et al. 2009b) in terms of processes arising from the interaction with the LISM at the heliopause, including a non-uniform distribution of pressure, a Rayleigh-Taylor instability possibly accompanied by magnetic reconnection, and localized anisotropy of ion pitch angle distributions in a region of external field compression. Kivelson and Jia (2012, 2013) suggest that the structure of the ribbon can be reproduced if the interaction with the LISM at the heliopause arises from magnetic reconnection. Their argument is based on the symmetry of the region in which magnetopause reconnection occurs at other magnetospheres. They show that for an oblique orientation of the upstream magnetic field, the shape and location of the portion of the magnetopause in which reconnection heats plasma differ for planetary magnetospheres embedded in the super-Alfvénic flow of the solar wind, and the magnetosphere of Ganymede embedded in the sub-Alfvénic flow of Jupiter's magnetospheric plasma. The different symmetry properties in sub- and super-

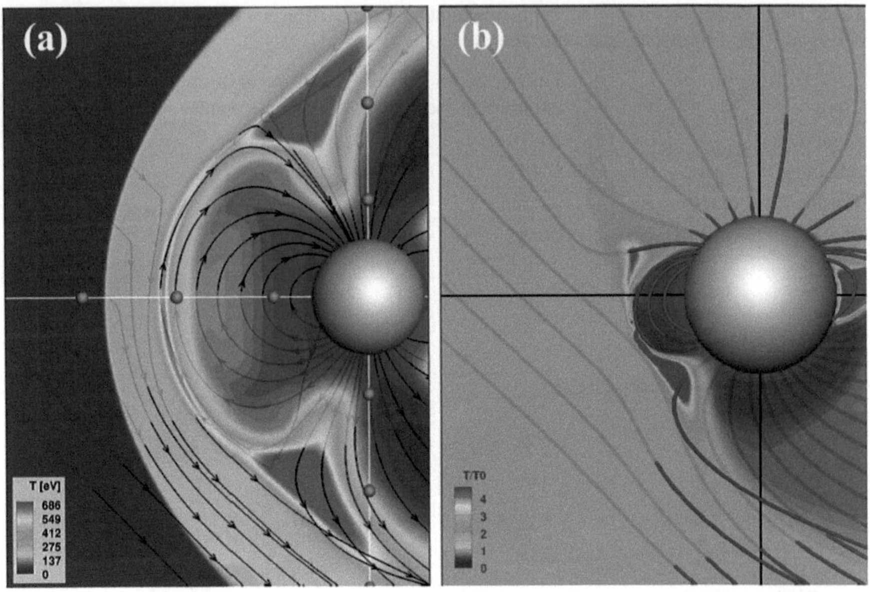

Fig. 10.8 The magnetospheres of (**a**) Earth with $M_A = 8$ and $M_F = 6$ and (**b**) Ganymede with $M_A = 0.4$, $M_F = 0.3$ in the x (flow direction) - z (dipole axis) plane. In (**a**) *black lines with arrows* (*dark* on the closer side and *grayed* on the opposite side of the x-z plane) represent magnetic field lines; *color* represents the plasma temperature (thermal pressure/density). In (**b**) field lines are *magenta. Color* represents T in the x-z plane. The simulation of the Earth was extracted from the global, 3D MHD model BATSRUS (Block Adaptive Tree Solar wind Roe-type Upwind Scheme) of the terrestrial magnetosphere (Powell et al. 1999; Gombosi et al. 2002, 2004; Ridley et al. 2004; Tóth et al. 2012) and the Ganymede simulation was run with the model of Jia et al. (2009, 2010). In both cases **B** was tilted as inferred for the heliopause (Kivelson and Jia 2012). Heating near the boundary in both images results from reconnection. Despite the tilt of the upstream field, in the regions near the equator there is little asymmetry N/S in panel (**a**) but substantial asymmetry in panel (**b**) (Color figure online)

Alfvénic regimes are clearly evident in temperature plots extracted from MHD simulations run for different upstream Mach numbers upstream conditions with the same (oblique) orientation of the upstream magnetic field relative to the dipole axis. In Fig. 10.8 are shown aspects of the field (field orientation) and plasma (effective temperature) properties in and near the x–z plane, with x along the flow, orthogonal to the z-aligned dipole axis. Figure 10.8a was run for $M_A = 8$, $M_F = 6$, typical of solar wind conditions at Earth. Figure 10.8b was run for Ganymede with $M_A = 0.4$, $M_F = 0.3$. For the high Mach number adopted for the case of Earth, the symmetry of the heated regions is to a large degree imposed by the direction of the flow, with considerable north-south symmetry of the heated regions despite the substantial tilt of the upstream magnetic field. For the low Mach number case at Ganymede, the heated region shifts southward and falls approximately where the unperturbed upstream field would be tangent to a sphere with radius of the dayside

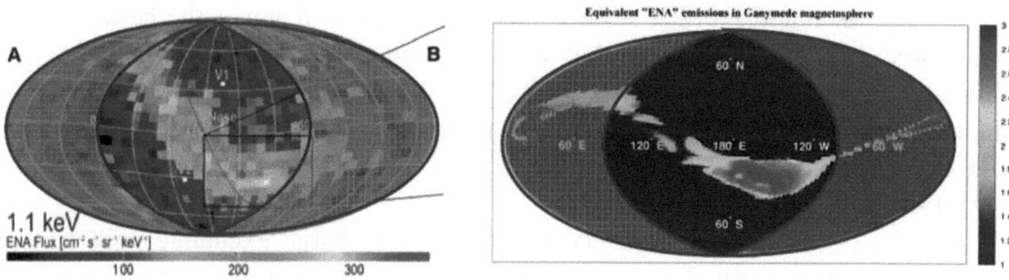

Fig. 10.9 (**a**) as on the left in Fig. 10.7 from McComas et al. (2009b). (**b**) The temperature sampled radially from the center of Ganymede in the Jia et al. simulation as a proxy for the distribution of sources of ENAs on the plane of the sky. *Shading* is used to hide the regions where properties of the upstream plasma are not likely to dominate the interaction. The offset of the hottest regions is similar in the two plots

magnetopause. The shift off the equatorial plane results because of the finite component of the upstream field along the flow direction (Fig. 10.9).

For an oblique upstream magnetic field orientation, heated plasma near Ganymede's magnetopause is found in regions shifted away from the nose of the magnetosphere and extended in the direction transverse to the field and the flow. Drawing lessons from the analogy, Kivelson and Jia (2012, 2013) propose that the flow of the LISM, like the flow upstream of Ganymede, is also sub-Alfvénic. In this case, the magnetic pressure of the LISM exceeds the dynamic pressure; the upstream field is relatively rigid and reconnection occurs preferentially along a region where the unperturbed upstream field is approximately tangent to the quasi-spherical upstream boundary of the heliosphere. The orientation of the magnetic field of the LISM shown in Fig. 10.7 can be inferred from this assumption [in agreement with the conclusions of Funsten et al. (2009) and McComas et al. (2012a)]. If the flow of the LISM onto the heliopause is sub-Alfvénic it is also sub-magnetosonic, consistent with the argument of McComas et al. (2012a) that no bow shock stands upstream of the heliopause.

Reconnection accelerates ions along the background magnetic field. The energy of the charged particles on a reconnected flux tube increases by an amount proportional to $\mathbf{v} \cdot \mathbf{v}_A$ where $\mathbf{v}$ is the initial parallel velocity. In the range of energies near 1 keV, the phase space distribution decreases with energy, so any acceleration process increases the phase space density at fixed energy. This acceleration explains why the flux of energetic ions is locally larger than the background flux where the ions have been accelerated by reconnection. An increase of the phase space density of ions near a reconnection line is consistent with a localized region of higher intensity ENA flux.

Reconnection requires a shear angle between the fields inside and outside of the heliopause. The solar wind maintains its predominantly tangential orientation even beyond the termination shock (Burlaga and Ness 2011, 2012). The orientation is not antiparallel to the proposed orientation of the upstream magnetic field. In order to account for the intensity of ENAs in both northern and southern hemispheres (see Fig. 10.7, for example), one must allow for reconnection with a large range of shear angles. Kivelson and Jia (2013) show that the change of β across the heliopause is <1, satisfying the conditions that allow reconnection for shear angles $> 50°$ (Swisdak et al. 2010; Phan et al. 2010). This suggests that for either polarity of the tangential field of the shocked solar wind in the outer heliosphere, reconnection can occur with a field oriented in the oblique manner proposed for the field of the LISM.

The ENAs that form the observed ribbon must charge exchange with an ion moving inward towards the center of the solar system if it is to be observed by IBEX. Reconnection accelerates particles along the heliopause. In order to produce inward-moving ENAs, pitch angle scattering must occur. Kinetic simulations of reconnection (Lottermoser et al. 1998) show that non-linear effects scatter the ions and the ion distribution becomes nearly isotropic within of order 200 ion inertial lengths as the plasma flows away from the reconnection site. For conditions at the heliopause, the ion inertial length is of order 1,000 km, a negligible fraction of an AU. This enables reconnection to account for inward-moving ENAs.

If the Kivelson and Jia interpretation of the form of the ribbons of enhanced intensity found on ENA maps is correct, the shape of the upstream heliopause is expected to differ markedly from the shape of a planetary magnetopause. The upstream boundary is expected to be far blunter in a sub-Alfvénic LISM flow than it would be if the flow were super-Alfvénic. The blunt shape of the upstream boundary at Ganymede is shown in Fig. 10.5, and is clearly blunter than the strongly curved upstream surface familiar from Earth's magnetosphere.

It seems likely that Voyager 1 and Voyager 2 will soon cross the heliopause and it will be of great interest to find if direct measurements confirm the importance of reconnection and the characteristics of the LISM described here.

10.8 Summary

Jim Dungey taught us how a magnetosphere works. This paper shows that concepts he developed apply to all magnetospheres and that they are central to understanding their behavior. Nonetheless, outcomes differ from one magnetosphere to another. To a considerable degree, the differences can be understood by allowing for the variation of the principal dimensionless parameters. Earth's magnetosphere is the best understood, but we now have examples of magnetospheres of different scale (Mercury, Jupiter), of different ratios of rotational to convective flow (Ganymede, Jupiter), and of different range of wave speed to flow speed in the external flow (Ganymede, Mercury). Every new parameter range makes a different kind of magnetosphere. Learning from other magnetosphere, one can speculate on the properties of plasma external to the remote upstream boundary of the solar wind, the heliopause. From the study of comparative magnetospheres, we learn new plasma science and simultaneously improve our understanding of how the terrestrial magnetosphere functions. And our appreciation of Jim's early perceptions continues to increase.

References

Berchem, J., Raeder, J., Ashour-Abdalla, M.: Magnetic flux ropes at the high-latitude magnetopause. Geophys. Res. Lett. **22**, 1189 (1995)

Borovsky, J.E.: Physical improvements to the solar wind reconnection control function for the Earth's magnetosphere. J. Geophys. Res. Space Phys **118**, 1 (2013). doi:10.1002/jgra.50110

Brice, N.M.: Bulk motion of the magnetosphere. J. Geophys. Res. **72**, 5193 (1967)

Brice, N.M., Ioannidis, G.A.: The magnetospheres of Jupiter and Earth. Icarus **13**, 173 (1970)

Burlaga, L.F., Ness, N.F.: Transition from the sector zone to the unipolar zone in the heliosheath: *Voyager 2* magnetic field observations. Astrophys. J. **737**, 35 (2011). doi:10.1088/0004-637X/737/1/35

Burlaga, L.F., Ness, N.F.: Heliosheath magnetic fields between 104 and 113 AU in a region of declining speeds and a stagnation region. Astrophys. J. **749**, 13 (2012). doi:10.1088/0004-637X/749/1/13

Caan, M.N., McPherron, R.L., Russell, C.T.: Substorm and interplanetary magnetic field effects on the geomagnetic tail lobes. J. Geophys. Res. **80**, 191 (1975)

Chen, A.J.: Penetration of low-energy protons deep into the magnetosphere. J. Geophys. Res. **75**, 2458 (1970)

Clark, J.T., Grodent, D., Cowley, S.W.H., Bunce, E.J., Zarka, P., Connerney, J.E.P., Satoh, T.: Jupiter's aurora. In: Bagenal, F., Dowling, T., McKinnon, W. (eds.) Jupiter: The Planet, Satellites and Magnetosphere. Cambridge University Press, Cambridge (2004)

D'Amicis, R., Bruno, R., Bavassano, B., Carbone, V., Sorriso-Valvo, L.: On the scaling of waiting-time distributions of the negative IMF Bz component. Ann. Geophys. **24**, 2735 (2006)

Dungey, J.W.: Interplanetary magnetic field and the auroral zones. Phys. Rev. Lett. **6**, 47 (1961)

Dungey, J.W.: Structure of the exosphere, or adventures in velocity space. In: DeWitt, C., Hieblot, J., Lebeau, A. (eds.) The Earth's Environment, pp. 505–550. Gordon and Breach, New York (1963)

Fujimoto, M., Baumjohann, W., Kabin, K., Nakamura, R., Slavin, J.A., Terada, N., Zelenyi, L.: Hermean magnetosphere-solar wind interaction. Space Sci. Rev. **132**, 2–4 (2007). doi:10.1007/s11214-007-9245

Funsten, H.O., et al.: Structures and spectral variations of the outer heliosphere in IBEX energetic neutral atom maps. Science **326**, 964 (2009)

Gombosi, T.I., Tóth, G., De Zeeuw, D.L., Hansen, K.C., Kabin, K., Powell, K.G.: Semi-relativistic magnetohydrodynamics and physics-based convergence acceleration. J. Comput. Phys. **177**, 176 (2002). doi:10.1006/jcph.2002.7009

Gombosi, T.I., et al.: Solution-adaptive magnetohydrodynamics for space plasmas: sun-to-earth simulations. Comput. Sci. Eng. **6**(2), 14 (2004). doi:10.1109/MCISE.2004.1267603

Hill, T.W.: Inertial limit on corotation. J. Geophys. Res. **84**, 6554 (1979)

Hill, T.W., Dessler, A.J., Goertz, C.K.: Magnetospheric models. In: Dessler, A.J. (ed.) Physics of the Jovian Magnetosphere, pp. 353–394. Cambridge University Press, Cambridge (1983)

Jia, X., Kivelson, M.G.: Driving Saturn's magnetospheric periodicities from the upper atmosphere/ionosphere: magnetotail response to dual sources. J. Geophys. Res. **117**, A11219 (2012). doi:10.1029/2012JA018183

Jia, X., Walker, R.J., Kivelson, M.G., Khurana, K.K., Linker, J.A.: Three-dimensional MHD simulations of Ganymede's magnetosphere. J. Geophys. Res. **113**, A06212 (2008). doi:10.1029/2007JA012748

Jia, X., Walker, R.J., Kivelson, M.G., Khurana, K.K., Linker, J.A.: Properties of Ganymede's magnetosphere inferred from improved three-dimensional MHD simulations. J. Geophys. Res. **114**, A09209 (2009). doi:10.1029/2009JA014375

Jia, X., Walker, R.J., Kivelson, M.G., Khurana, K.K., Linker, J.A.: Dynamics of Ganymede's magnetopause: intermittent reconnection under steady external conditions. J. Geophys. Res. **115**, A12202 (2010). doi:10.1029/2010JA015771

Kivelson, M.G., Jia, X.: The smallest magnetosphere as a model for the largest, AGU Fall meeting abstract SH14B-07 (2012)
Kivelson, M.G., Southwood, D.J.: Dynamical consequences of two modes of centrifugal instability in Jupiter's outer magnetosphere. J. Geophys. Res. **110**, A12209 (2005). doi:10.1029/2005JA011176
Kivelson, M.G., Jia, X.: An MHD model of Ganymede's mini-magnetosphere suggests that the heliosphere forms in a sub-Alfvénic flow, J. Geophys. Res. **118**, 1–8 (2013). doi:10.1002/2013JA019130
Kivelson, M.G., Khurana, K.K., Russell, C.T., Walker, R.J., Warnecke, J., Coroniti, F.V., Polanskey, C., Southwood, D.J., Schubert, G.: Discovery of Ganymede's magnetic field by the Galileo spacecraft. Nature **384**, 537 (1996)
Kivelson, M.G., Warnecke, J., Bennett, L., Joy, S., Khurana, K.K., Linker, J.A., Russell, C.T., Walker, R.J., Polanskey, C.: Ganymede's magnetosphere: magnetometer overview. J. Geophys. Res. **103**, 19963 (1998)
Lottermoser, R.F., Scholer, M., Matthews, A.P.: Ion kinetic effects in magnetic reconnection: hybrid simulations. J. Geophys. Res. **103**, 4547 (1998)
McComas, D.J., et al.: IBEX—interstellar boundary explorer. Space Sci. Rev. **146**, 11 (2009a). doi:10.1007/s11214-009-9499-4
McComas, D.J., et al.: Global observations of the interstellar interaction from the interstellar boundary explorer (IBEX). Science **326**, 959 (2009b). doi:10.1126/science.1180906
McComas, D.J., et al.: The heliosphere's interstellar interaction: no bow shock. Science **336**, 1291 (2012a). doi:10.1126/science.1221054
McComas, D.J., et al.: The first three years of IBEX observations and our evolving heliosphere. Astrophys. J. Suppl. **203**, 1 (2012b). doi:10.1088/0067-0049/203/1/1
Neubauer, F.M.: Nonlinear standing Alfvén wave current system at Io: theory. J. Geophys. Res. **85**, 1171 (1980)
Phan, T.D., Gosling, J.T., Paschmann, G., Pasma, C., Drake, J.F., Oieroset, M., Larson, D., Lin, R.P., Davis, M.S.: The dependence of magnetic reconnection on plasma β and magnetic shear: evidence from solar wind observations. Astrophys. J. Lett. **719**, L199 (2010)
Phan, T.D., Paschmann, G., Gosling, J.T., Øieroset, M., Fujimoto, M., Drake, J.F., Angelopoulos, V.: The dependence of magnetic reconnection on plasma β and magnetic shear: evidence from magnetopause observations. Geopys. Res. Lett. **40**, 1 (2013). doi:10.1029/2012GL054528
Powell, K.G., Roe, P.L., Linde, T.J., Gombosi, T.I., DeZeeuw, D.L.: A solution-adaptive upwind scheme for ideal magnetohydrodynamics. J. Comput. Phys. **154**, 284 (1999)
Ridley, A., Gombosi, T., Dezeeuw, D.: Ionospheric control of the magnetosphere: conductance. Ann. Geophys. **22**, 567 (2004)
Russell, C.T., Elphic, R.C.: Initial ISEE magnetometer results: magnetopause observations. Space Sci. Rev. **22**, 681 (1978)
Scurry, L.C., Russell, C.T., Gosling, J.T.: Geomagnetic activity and the beta dependence of the dayside reconnection rate. J. Geophys. Res. **99**, 14811 (1994)
Slavin, J.A., Holzer, R.E.: Solar wind flow about the terrestrial planets. 1. Modeling bow shock position and shape. J. Geophys. Res. **86**, 11401 (1981)
Slavin, J.A., et al.: MESSENGER observations of extreme loading and unloading of Mercury's magnetic tail. Science **329**, 665 (2010)
Slavin, J.A., et al.: MESSENGER observations of a flux-transfer-event shower at Mercury. J. Geophys. Res. **117**, A00M06 (2012). doi:10.1029/2012JA017926
Southwood, D.J., Kivelson, M.G., Walker, R.J., Slavin, J.A.: Io and its plasma environment. J. Geophys. Res. **85**, 5959 (1980)
Stone, E.: Voyager 1 observations of rapid changes in the heliosheath. Abstract SH13D-01. Presented at Fall Meeting, AGU, San Francisco, December 2012
Swisdak, M., Opher, M., Drake, J.F., Alouani Bibi, F.: The vector direction of the interstellar magnetic field outside the heliosphere. Astrophys. J. **710**, 1769 (2010). doi:10.1088/0004-637X/710/2/1769

Tóth, G., et al.: Adaptive numerical algorithms in space weather modeling. J. Comput. Phys. **231**, 870 (2012). doi:10.1016/j.jcp.2011.02.006

Vasyliunas, V.M.: Role of the plasma acceleration time in the dynamics of the Jovian magnetosphere. Geophys. Res. Lett. **21**, 401 (1994)

Walker, R.J., Russell, C.T.: Flux transfer events at the jovian magnetopause. J. Geophys. Res. **90**, 7397 (1985)

Chapter 11
Magnetic Reconnection in the Solar Corona: Historical Perspective and Modern Thinking

Peter Cargill

Abstract In 1953 Jim Dungey published the first paper that highlighted the key points of magnetic reconnection: in particular the change in the topology of the magnetic field and the global release of magnetic energy. The first part of this paper looks back to the 1950s and examines the context and consequences of Jim's paper. The second part presents a contemporary and personal view of how magnetic reconnection plays a central role in energy release in the solar corona, and in particular suggests that a range of reconnection events (flares, microflares and nanoflares) can account for coronal behaviour.

11.1 Introduction

Magnetic reconnection (hereafter reconnection) is recognised as an important physical process in solar, space and astrophysical plasmas. In simple terms, it involves (1) the conversion of magnetic energy into kinetic and thermal energy and, in many cases, the acceleration of charged particles and (2) a change in the magnetic topology of the system. It is now invoked to account for magnetic energy release on the Sun (see Priest and Forbes 2000 as well as Cargill et al. 2010; Cargill 2013a for non-specialist reviews) and is essential for understanding how solar and stellar winds interact with planetary magnetospheres (chapters "Dungey's Reconnection Model of the Earth's Magnetosphere: The First 40 Years", "Observing Magnetic Reconnection: The Influence of Jim Dungey", "Sun et Lumière: Solar Wind-Magnetosphere Coupling as Deduced from Ionospheric Flows and Polar Auroras" and "The Science of the Cluster Mission"). Applications to astrophysical contexts are found in Priest and Forbes (2000) and Zweibel and Yamada (2009).

P. Cargill (✉)
Space and Atmospheric Physics, The Blackett Laboratory, Imperial College, London SW7 2BW, UK

School of Mathematics and Statistics, University of St Andrews, St Andrews KY16 9SS, UK
e-mail: p.cargill@imperial.ac.uk

D. Southwood et al. (eds.), *Magnetospheric Plasma Physics: The Impact of Jim Dungey's Research*, Astrophysics and Space Science Proceedings 41,
DOI 10.1007/978-3-319-18359-6_11

Reaching our present understanding of reconnection has taken 60 years. Giovanelli (1946, 1947, 1948) was the first to address the question of energy release in solar flares, and in particular particle acceleration, in complex magnetic fields. However, as we shall discuss in Sect. 11.2, he did not make use of the principles of magnetohydrodynamics (MHD) being developed at that time. The first presentation of reconnection in the language of MHD is in Jim's 1953 paper (Dungey 1953). The application to the "open" magnetosphere followed (Dungey 1961). It is a salutary point, backed up by an examination of citations to the important papers, that it took a long time for these ideas to be widely accepted. Progress in the 1950s and early 1960s was slow, with the concept of reconnection being treated with at best a lack of comprehension and more generally ignored. Dungey (1983) and Southwood (chapter "From the Carrington Storm to the Dungey Magnetosphere") have noted that the "open magnetosphere" was finally taken seriously after the late 1970s when data from the ISEE-1 and -2 satellites became available. Even then, a substantial body of non-believers persisted.[1] Reconnection on the Sun also took some time to become established as a viable concept. While Petschek (1964) showed how fast energy release could arise, it took until the early 1970s for this to be fully assimilated into flare physics.

This chapter is split into two parts. First I examine the development of reconnection physics between the publication of Giovanelli (1946) to the NASA conference on solar flares in 1963 which contains Petschek's paper, with emphasis on Dungey (1953). The second part looks at how these ideas are reflected in present-day work on solar flares and the solar corona from a personal viewpoint. The Appendix examines citation numbers from early reconnection papers.

11.1.1 Literature

There are a number of historical articles that have been of use. The most important is Jim's own view of his career (Dungey 1994). David Stern's short interview with Jim (Stern 1986) is of great interest. Jim's "Inaugural Lecture as Professor of Physics" (Dungey 1966) at Imperial College on 3 May 1966 contains some discussion of the early days, though sadly is no longer available for half a crown. Southwood (1997) recounted his memories of Jim at Imperial College in the 1960s and 1970s. There are several relevant articles in the AGU "Pioneers of Space Physics" series: Axford (1994) especially so, though it lacks any references. Elsewhere, Cowling (1985) reminisces on his career while the book edited by Hones (1984) is an excellent "mid-term review" of reconnection in the early 1980s. As with many conferences in past times, it retains a transcript of the discussion following the talks, a tradition sadly lost. The interviews recorded as

[1] The author recalls attending a conference as late as 1985 when one speaker on the magnetotail went to great lengths to denigrate reconnection (rudely) at every opportunity.

part of the AIP oral history transcripts (http://www.aip.org/history/ohilist/transcripts.html) provide useful background, including interviews with Cowling, Hoyle, Gold etc. The Australian Academy of Science has an appreciation of Giovanelli (http://www.science.org.au/fellows/memoirs/giovanelli.html).

11.2 1946–1964: Giovanelli–Petschek via Dungey (1953)

Solar flares have been recognised as major energy releases on the Sun since at least the mid-nineteenth centuries (Carrington 1859). White-light flares such as seen by Carrington are quite rare, but with better understanding of atomic physics, and improved instrumentation, observations in transitions such as H_α became possible. Ellison (1949) presented a survey of H_α flare observations, including the large flare observed at Kodaikanal, India on 22 February 1926, shown on the left of Fig. 11.1. In present-day parlance this is a "two-ribbon flare", probably associated with a coronal mass ejection (CME). The flare ribbons broaden with time, and there are also remote brightenings. The lower right panel shows the sunspot grouping: this is a magnetically active area. In H_α, the bright regions show what is now accepted as radiation due to the collisions of energetic electrons in the chromosphere. The right panel of Fig. 11.1 shows an example of such a flare with modern instrumentation

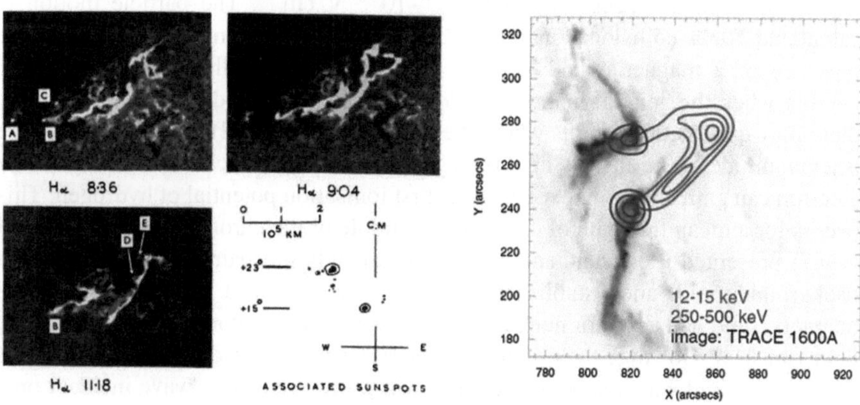

Fig. 11.1 An H_α flare observed in 1926 (adapted from Ellison 1949). The three panels show H_α emission at different times as flare progresses. Note the characteristic two ribbons of emission. The *lower right panel* shows sunspot distribution at the time of the flare. Each panel is approximately 550 Mm in the horizontal direction and 400 Mm in the vertical. The *right panel* shows a modern image of such a large flare (Krucker et al. 2008a). The image of the solar surface is obtained by the TRACE spacecraft in the UV continuum and shows the two ribbon structure. The *red* and *blue contours* are hard X-ray emission seen by the RHESSI spacecraft (in black and white the low and high energy contours are on the right and left of the figure respectively). The *blue contours* are emission due to high energy electrons in the upper chromosphere. The *red contours* are from less energetic electron in the corona, and their emission outlines the large-scale coronal structure of the flare (Color figure online)

(Krucker et al. 2008a): the ribbons are now seen in the UV continuum and there is hard X-ray emission both in the chromosphere and corona. The reader is reminded that X-ray observations of flares were not available until the early years of the space age (Tucker and Giacconi 1986), and systematic observations of coronal flares did not occur until soft X-ray imaging, particularly with the Skylab telescopes, became available. And although the corona could then be seen at eclipses, its true nature was not appreciated until the war years (1939–1945), when the detection of highly ionised states of iron proved the existence of a hot (> MK) plasma. It took even longer for the importance of the magnetic field in the corona to be fully appreciated.

11.2.1 Giovanelli, Hoyle and Cowling

This was the context in the immediate post-war years when Jim began his PhD with Fred Hoyle at Cambridge. Ron Giovanelli (1946, 1947, 1948) and Fred Hoyle (1949) both became interested in how energetic charged particles could be produced in magnetised plasmas. Both were concerned with the flare problem, Hoyle[2] also with the aurora, with of course major consequences for Jim (Dungey 1994; chapter "Dungey's Reconnection Model of the Earth's Magnetosphere: The First 40 Years").

Giovanelli (1946) argued that a dynamically evolving sunspot would induce an electric field in all space, of order 10^{-2}–10^{-3} V cm^{-1}. The particle motion is calculated for a collisional medium (Chapman and Cowling 1939), and in the presence of a magnetic field is an $\mathbf{E} \times \mathbf{B}$ drift. Giovanelli makes two points: (1) that when the general magnetic field of the Sun[3] is added to the sunspot field, there may exist neutral points where the field vanishes and (2) at such points, there is a magnitude of the electric field for a given atmospheric location above which an electron can gain energy in excess of the first ionisation potential of hydrogen. This occurs for a mean free path of 450 km, the middle of the chromosphere. Giovanelli (1947) presented more mathematical detail, including a pair of sunspots and the background field, and established that there were indeed locations where the magnetic field had a null or neutral point, i.e. all three components of the magnetic field vanished. However the principles of MHD were not used and the evolution of the sunspot field was treated as though it was a low-frequency wave incident on a conductor, a calculation from EM theory.

Giovanelli (1948, 1949) and Hoyle (1949: completed in mid-1948) addressed particle acceleration at neutral points: since these ideas have been largely superseded, we address this only briefly. The ideas behind electron runaway in the corona

[2] The largest part of Hoyle's book contains a discussion of the solar corona as being due to the accretion of interstellar material.

[3] According to Giovanelli, the existence of this general field had been questioned. Cowling (1953) makes the same point.

are introduced: for a given initial particle energy, there is a critical electric field above which collisions cannot stop the electron acceleration (Hoyle, p. 93), which pre-dates the "Dreicer field" (Dreicer 1959). They framed the discussion in terms of a "discharge", a word that came to dominate the topic for a decade or so. Hoyle introduces the corona explicitly, and argues that it is central to the flare phenomenon, with electron acceleration being easier there.

Cowling (1953: written in 1951) had several objections, including the lack of consistency of Giovanelli's approach with the requirements of MHD. More significantly, he pointed out that the current at the neutral point had to be determined from Ampère's law, which even for strong magnetic fields implied scales of order a few metres. He felt that it was unlikely that such small scales were compatible with observed flare scales. The problem persists to this day with flare scenarios that rely on a single current sheet. Cowling writes "*unless one is prepared to accept the possibility that a flare layer is only a few metres thick, curl* ***H*** *cannot have the same direction all through the layer*" so multiple current layers may instead be involved in a flare leading to "*an irregular dissipation of energy in a violently twisted field*". It would take decades for such ideas to re-emerge in the framework of multiple reconnection sites (Sect. 11.3.1). Piddington (1953) makes similar points.

11.2.2 Dungey (1953)

Jim's paper (Dungey 1953) is the birthplace of reconnection. It is couched in the language of MHD. It makes use of the freezing of magnetic field to the plasma, and conditions for its violation (Dungey 1950). When read today, it seems a very modern paper which, while perhaps short on mathematical developments, contains outstanding physical insight. Two sketches in his paper (Figures 2 and 3 there) make the key points: (1) magnetic reconnection can change the topology of the magnetic field and (2) energy release in reconnection occurs globally, and is not confined to the x-line or neutral point. I have adapted Jim's Figs. 2 and 3 in Fig. 11.2 of this paper.

To quote Jim: "*If there are two lines of force as shown in Fig. 2, the direction of the current corresponds to a field with a clockwise direction. The field therefore decays in that direction. The lines of force in Fig. 2 can now be regarded as* ***being broken and rejoined***" at the neutral point.[4] Jim goes on to note that "*the total length of the lines of force decreases...... and it follows that the energy of the field decreases*". This point is very clear in the two examples on the right of Fig. 11.2,

[4] It is worth here drawing attention to the sketch in Alfvén (1950) which has some similarities to the lower panel of Jim's Fig. 3.

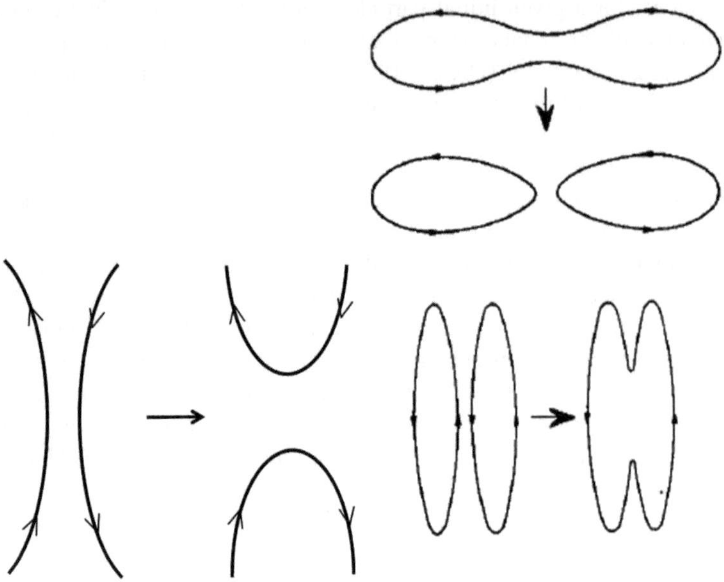

Fig. 11.2 Two sketches from Dungey (1953). The *left* shows oppositely directed field lines approaching each other and changing their connectivity. In Jim's paper this is referred to as field lines "are broken and re-joined". The *right hand figure* shows a number of other geometries with x-points (or x-lines). Note, especially in the *right hand panel*, how the shortening of the field lines gives almost all the energy release away from the x-point

and Jim concludes Sect. 11.4 by noting "*field energy is released and field energy from a relatively large region is concentrated on the particles in the vicinity of the neutral point*".

Dungey (1953) contains a third important piece of work, the collapse of an x-point (or x-line). The solid lines in Fig. 11.3 show the magnetic field lines associated with an exact x-point: $\mathbf{B} = (B_x, B_y, B_z) = -B_0(y, x, 0)$: the minus sign ensures consistency with Figure 1 of Jim's paper and B_0 is a constant, set equal to unity in Fig. 11.3. Following the analysis of Priest (1982), we perturb this: $\mathbf{B} = (B_x, B_y, B_z) = -B_0(y, \alpha^2 x, 0)$ with $\alpha > 1$: the dashed lines on Fig. 11.3 are for $\alpha = 2$. The current is then $\mathbf{J} = -(cB_0/4\pi)(\alpha^2 - 1)\hat{\mathbf{z}}$ and the Lorentz force, $\mathbf{J} \times \mathbf{B}/c = (B_0^2/4\pi)(\alpha^2 - 1)(-\alpha^2 x, y, 0)$, is denoted by the solid arrows on Fig. 11.3 (see also Fig. 1 of Jim's paper). The Lorentz force leads to the field lines on left and right being pushed towards the x-line, and those top and bottom being forced away. So the x-point collapses, like a pair of scissors closing, with the collapse eventually being terminated by formation of a current sheet with non-ideal processes becoming important.

Indeed much of the early attention to the 1953 paper focussed on x-point collapse rather than the conceptual ideas behind reconnection. Priest and Forbes (2000: Chapter 7.1) present a modern view of this topic, including the calculation of

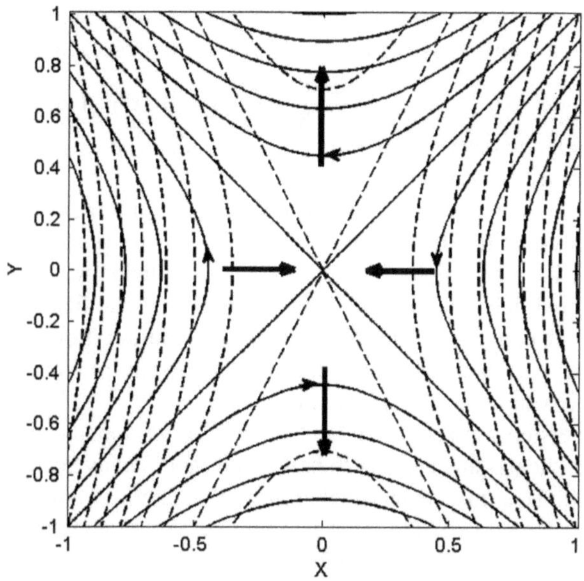

Fig. 11.3 Magnetic field during x-point collapse. The *solid lines* denote the pre-collapse field (i.e. $\alpha = 1$) and the *dashed lines* the perturbed field ($\alpha = 2$). The *thick arrows* show the direction of the Lorentz force associated with the perturbed field

Imshennik and Syrovatskii (1967) that predicts the formation of a current singularity in of order an Alfvén transit time, as well as a formal linear stability analysis of an x-point (e.g. McClymont and Craig 1993) with growth rates weakly dependant (logarithmically) on the resistivity. Numerical simulation of x-point collapse has been studied in the MHD regime (Fuentes-Fernández et al. 2011) and for collisionless plasmas (Tsiklauri and Haruki 2007), in which other recent work is discussed.

The choice of journal (Philosophical Magazine) for this important work may seem surprising. This was brought about by the rejection of the original version of the paper by Monthly Notices of the Royal Astronomical Society (MN: Dungey 1994). With the help of the RAS, I have been able to examine the MN refereeing process. Unfortunately copies of the original manuscript were not retained by the society, but one can deduce from the available documents that the paper submitted to MN was similar to the published version. A short account of this work appeared between MN rejection and final publication (Dungey 1952).

The paper was "communicated" to MN by Giovanelli (Jim was in Sydney by this time) in late June 1951. Unfortunately Jim ran into one referee (V. C. A Ferraro) who did not believe the results. MN decided to consult another referee (T. G. Cowling)[5] who, while seeming to understand the x-point analysis, came

[5] The author and editors have agreed that, over 60 years after these events, the identity of the referees be made public in the interest of historical completeness. Both referees have been deceased for well over two decades.

back with one very specific criticism about the neglect of pressure gradients: on the face of it a reasonable point since this could stop the collapse. One can calculate the required pressure as: $p = p_0 + \frac{B_0^2(\alpha^2 - 1)}{8\pi}\left(y^2 - \alpha^2 x^2\right)$. For small p_0 this is unphysical for $|y| < |\alpha x|$ (i.e. in the region either side of the x-point where the **J** × **B** force is towards the x-point), so that pressure gradients cannot prevent the collapse for a low-beta plasma. Jim did not pick up on this point in his response to Cowling. However the field increases linearly as one moves from the x-point, and this eventually becomes unphysical. Instead the local field must be matched onto a large-scale global field and perturbations of this large-scale field will feed into the vicinity of the x-point, initiating the collapse. So the question of the role of the pressure gradient is not clear-cut, but depends on how far from the x-point the local field solution is valid. Cowling hints at this aspect.

A revised version was sent to MN in late 1951. Jim responds by adding a section on pressure gradients (this appears to have been modified as Section 6 of the 1953 paper), although he argues on physical grounds that they will not change the overall picture. He explains to Ferraro what is going on. There does not appear to be a written review from Cowling, though he was encouraged to report to the RAS council (of which he was then a member) at its early 1952 meetings. [According to the RAS Council minutes, the MN editor resigned in early 1951, Council thereafter acted as editor.] Ferraro maintained his stance and claimed that "*there does not seem to be anything really new in the paper except for the x-point collapse calculation* (my wording)" which he (incorrectly) claimed was in error. So the paper was rejected. Given the seniority of both referees, it was going to be an uphill struggle for Jim to achieve any other decision. Rejection of papers is an occasional hazard, but in this particular case, given the innovation and importance of the work, one needs to ask why. From my study of the MN documents and RAS council minutes, the editorial procedures seem amateurish from a modern viewpoint. I believe that Cowling felt Jim needed to do more work on the paper, and this seems to be a straightforward difference of opinion between author and referee. Ferraro's attitude is harder to understand. And, despite subsequent publication in Phil. Mag., these shenanigans cannot have been encouraging to a young scientist.

There is one fascinating thing in the documents held at Burlington House. At the end of Jim's response to Cowling is a sketch that I have reproduced in Fig. 11.4 (the original is too damaged to copy properly). It does NOT appear in Jim's 1958 book (Dungey 1958a), though it is in his PhD thesis (see Dungey 1994): in any event, it is not the sort of thing one sketches quickly in the response to a referee. In his reply to the referee Jim says (discussing neutral point collapse): "*One can get a picture from Hoyle's theory of the aurora, and can see that the pressure gradient would not upset the flow. The flow is part of the corpuscular beam as it goes round the Earth (see sketch)*". In fact, Hoyle's theory, as described in Hoyle (1949), looks rather different from this sketch and so far as I can determine, Hoyle never published any such figure (see also Stern 1986): this figure belongs to Jim alone.

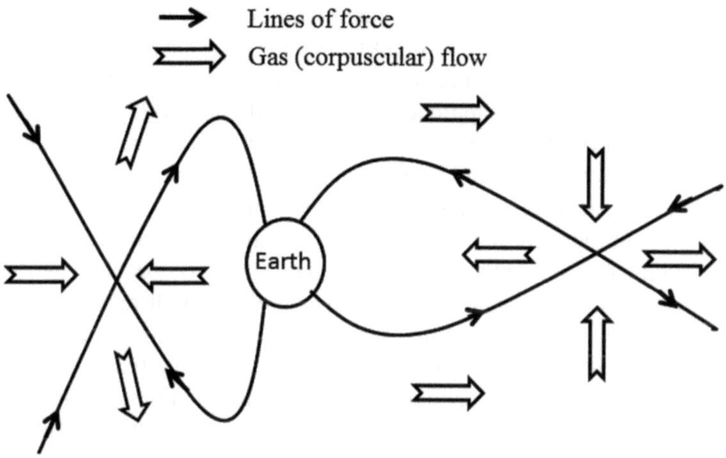

Fig. 11.4 A sketch (reproduced by the author) from Jim's response to the second referee of his 1951 submission to MN

11.2.3 Consequences: Sweet and Parker

The immediate impact of Dungey (1953) was minimal: roughly ten citations between publication and 1962.[6] Whether this was an effect of Phil. Mag. as a publication vehicle is unclear, but the opinion one comes away with from those who do cite the paper is that it was not understood (see for example, Elsasser 1954).[7] Despite this, "*whenever I was asked to give a talk, I talked about reconnection*" (Stern 1986).

In late August 1956 there was an IAU Symposium in Stockholm on "Electromagnetic Phenomena in Cosmical Physics" that Jim attended. His contribution was a paper (Dungey 1958b) entitled "The neutral point discharge theory of solar flares. A reply to Cowling's criticism", where Cowling (1953) is being replied to. One has a sense here of Jim fighting the last war, in particular with regard to Lenz's law, and indeed Cowling had never objected to a "violation" of Lenz's law: there is no such violation and Cowling understood that. Jim writes and talks about things being scientifically quite lonely in those days (Dungey 1994), and one sees someone who knows he has a brilliant idea, but can't find anyone to buy into it. His textbook (Dungey 1958a) is an outstanding exposition of space plasma physics, but there is still no sign of the ideas about the magnetosphere that came later on.

[6] Googlescholar. The usually reliable ADS does not track Phil Mag papers from the 1950s. I have carried out some cross-checks between ADS and Googlescholar on other papers published at that time and the citation numbers seem consistent.

[7] It is interesting to note that Feynman (1955) in a study of liquid helium talked of vortex "lines .... snap together and join connections in a new way" (Fig. 10 of Feynman).

The Stockholm conference involved another major step in reconnection theory. Sweet (1958a) proposed a "neutral point theory for solar flares" which gives credit to Jim's collapsing current sheet idea. This is a global model (albeit 2D), that breaks away from the classical EM approach of Giovanelli and works within the framework of MHD.[8] A quadrupolar field (two sunspot pairs) are forced together. Since the flux systems from each pair of spots cannot interpenetrate, a neutral point and extended current sheet forms between them. This introduces the idea of a remote driver (sunspot motion) for the current sheet formation, and has been very influential (see also the elegant mathematical presentation of this approach by Priest and Raadu 1975). It is though with this paper that a detour with reconnection theory began, as noted by Jim in 1994. Sweet argued that the current sheet should have a scale commensurate with the global dimension of the system of interest (L). Further, as the fields were pushed together, there must be a pressure in the current sheet that is comparable to the external magnetic pressure, which in turn must be balanced at the top and bottom of the current sheet by an outflow.

Also at the Stockholm meeting was E. N. Parker, who presented a paper on cosmic rays. Following on from Sweet's presentation, Parker (1957) published a more rigorous version of his idea. The gestation period was considerable, with the paper being submitted to JGR in June 1957, but it was still published ahead of Sweet's.[9] The part of this paper that is usually recalled today is the initial rough calculation that gives the $1/\sigma^{1/2}$ scaling of the reconnection rate (σ is the electrical conductivity), a result that was not in Sweet's paper. In contrast to Sweet, Parker attempts to find a steady state solution to the problem (see also Parker 1963). However, the assumption that the sheet length is much greater than its width is built into the model, resulting in the same slow dissipation rate. The paper though does contain two important suggestions: one is that the fields need not be anti-parallel for Sweet's process to work (see Parker's Fig. 4), nowadays known as component reconnection, and the second is to question the role that field dissipation can play in turbulence.

But the lessons from Dungey (1953) and the comments of Cowling (1953) do not seem to have been accepted. The scale of the acceleration site is still very small, and the idea of reconnection as a global process is not recognised, despite Jim's magnetospheric contributions. Perhaps those interested in the Sun were unaware of this work. Slow reconnection rates remained conventional wisdom for some time. Cowling (1962) in a general review of MHD gives a good summary of Jim's work on x-point collapse and does not mention the role of pressure gradients. He uses the phrase "*relinking of lines of force*".

[8] It appears that some workers cite Sweet (1958b) as the sole "Sweet-Parker reconnection" reference. Given the timing of the publication of Sweet (1958a), the date of the Stockholm conference in August 1956 and the content of the two papers, this is hard to understand.

[9] The long delay in publishing the proceedings of the Stockholm conference can lead to some confusion in understanding the sequence of events at that time. One must assume that the proceedings were delayed by a few late papers, not an unusual phenomenon.

An interesting point concerns the origin of the word "reconnection". It has been stated in the literature (e.g. Cargill et al. 2010; Priest and Forbes 2000) that this is due to Parker (1957). This does not appear to be the case. Dungey (1958b) says (p. 139) "*The effect of the discharge is to "reconnect" the lines of force at the neutral point, and this happens quickly*", so dating to the 1956 Stockholm conference. Parker and Krook (1956: submitted January 1956, published July 1956) also introduce the term "reconnection": they talk of "*severing and reconnecting field lines....*" in the context of flux loops arising in the dynamo problem (see especially Figures 3b and 7 of their paper). Their paper was concerned largely with solutions of the diffusion equation, and lacked the dynamics introduced by Jim. Prior to these works, the trail becomes cold. Thus the origin of "reconnection" appears to reside in two papers which have, between them, cumulatively under 30 citations (ADS).

11.2.4 October 1963 Solar Flare Conference

In the interim, Jim, having understood the three-dimensional aspect of magnetospheric reconnection, published the open magnetosphere paper (Dungey 1961). His contributions thereafter were almost exclusively in the field of magnetospheric physics, as documented elsewhere in this volume. October 1963 saw a AAS-NASA symposium on solar flares to whose proceedings Jim contributed a short paper (Dungey 1964), with his affiliation now Imperial College. There is a sense of looking forward and back: "*Petschek has demonstrated that the discharge mechanism is not seriously impeded by the pressure of the plasma as suggested by Sweet and Parker*" is stated in the Abstract. Yet Petschek (1964) did more than that. By removing the artificial constraint that the current sheet length is the global system scale, he opened the way for the development of what is now called "fast reconnection". He also returned to the implicit point of Dungey (1953) that energy release is a global process. In the compressible version of the Petschek mechanism, the energy is released at sets of slow mode shock waves that stand in the incoming plasma flow: the "diffusion region" where the magnetic topology actually changes is tiny, and the energy released there will be small. That does not make it unimportant: reconnection cannot happen without it.[10] It is also interesting to note that Sweet and Parker both acclaimed Petschek's work in the comments following his paper: one can only assume they had advance warning of what was going to be said.

[10] The subsequent paper of Levy et al. (1964: written in December 1963 shortly after the NASA/AAS meeting) took Petschek's ideas and applied them to Jim's open magnetosphere model. With fast reconnection as developed by Petschek, an understanding of how this worked in a non-symmetric system (Fig. 3 of their paper) and Jim's basic picture, this is the start of the development of a quantitative description of the open magnetosphere.

11.3 The Legacy

Modern coronal physics dates from 1973 and the Skylab observatory (Zirker 1977; Sturrock 1980; Orrall 1981). The EUV and X-ray telescopes showed the corona to be structured, dynamic and multi-thermal (Sect. 11.3.2), with flaring on many scales. Plasma and magnetic field were also seen to be ejected into the solar wind as CMEs. Many satellites have since returned data, and of relevance to what follows are: Solar Maximum Mission (SMM: 1980–1988), Yohkoh (1992–2005), Reuven Ramaty High Energy Solar Spectroscopic Imager (RHESSI: 2002–present), Hinode (2006–present) and Solar Dynamics Observatory (SDO: 2010–present). SMM, Yohkoh and RHESSI were largely directed at observing flares.

Space permits only a short discussion of the role of magnetic reconnection in the solar corona (see Priest and Forbes (2000) and Priest (2013) for more detailed reviews), so I will present my own view of aspects of solar flares and coronal heating, addressing how particle acceleration may work in flares and how coronal heating mechanism(s) could be deduced. Specific reviews on flares can be found in Vlahos et al. (2009), in the volume edited by Emslie et al. (2011), and Vilmer (2012), while the non-flaring corona is covered by Klimchuk (2006); Reale (2010) and Parnell and De Moortel (2012).

11.3.1 Solar Flares

11.3.1.1 The "Standard" Flare Model and Its Problems

Magnetic reconnection is widely argued to be able to account for the prompt energy release required in flares: in large events 10^{32} ergs in well under 1 h. Indeed in some sense little has changed since 1963: the scenario for eruptive flares is said to originate in Carmichael (1964), subsequently being developed by Sturrock (1966). An outline is shown in Fig. 11.5 (Sturrock 1974). An x-point forms in the corona above closed field lines. This is unstable, ejecting the plasma above into the interplanetary medium (the CME) and also accelerating particles that precipitate into the chromosphere. [In fact such a flare geometry has an even longer history: see Dungey (1958a: reproduced as Fig. 11.6 here), or Fig. 3 of Hoyle and Wickramasinghe (1961).] The modern scenario, to a great degree motivated by results from the Yohkoh spacecraft (e.g. Shibata and Magara 2011; Fig. 11.7; Magara et al. 1997) involves a prominence eruption, with the outward motion of the material driving fast reconnection at a coronal current sheet. The causality of CME and flare has been debated extensively (e.g. Gosling 1993; Hudson et al. 1995), but it appears that an overall instability or loss of equilibrium in the coronal magnetic field occurs, possibly initiated by the emergence of new magnetic flux through the photosphere. Such a model for eruptive flares is now referred to in

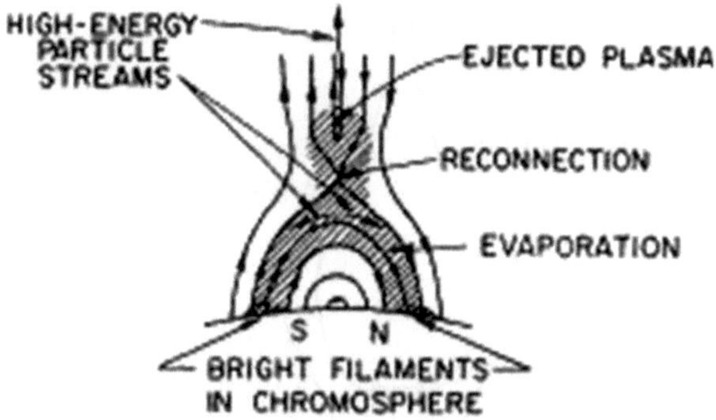

Fig. 11.5 A sketch due to Sturrock (1974) of a scenario for large, eruptive flares. There is a single coronal x-line at which reconnection proceeds. Above this, plasma is ejected into the solar wind (note that this sketch was prepared prior to the recognition of the origin of CMEs). Accelerated particles from the x-line stream towards the chromosphere, producing both the hard X-ray emission and the chromospheric flare. The coronal field lines fill with plasma emitting in soft X-rays

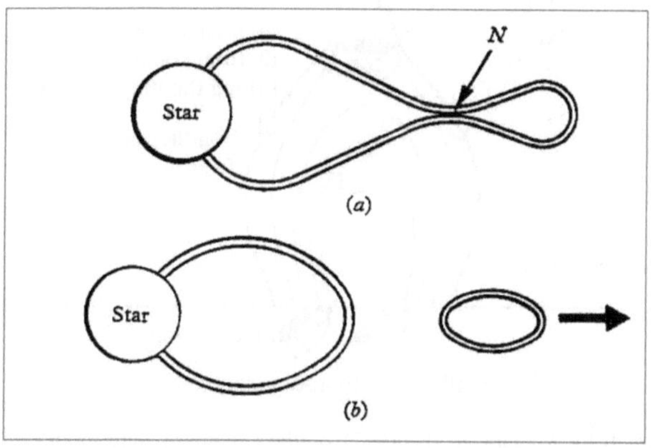

Fig. 11.6 The ejection of magnetic flux from a star's corona (Dungey 1958a)

some quarters as a "standard flare model" (Shibata and Magara 2011) and has been the subject of numerous MHD numerical simulations and cartoons.[11]

However, there are difficulties. Peterson and Winckler (1959) detected high energy radiation from the Sun associated with 0.5–1 MeV electrons, and over the next decade it gradually became apparent from hard X-ray (HXR) observations that copious numbers of electrons above 20 keV were being produced, with the radiation being due to Bremsstrahlung as the electrons collided with the chromosphere. Brown (1971) developed a "thick target model" for this emission, and was able to

[11] The web site http://www.solarmuri.ssl.berkeley.edu/~hhudson/cartoons/ collated by Hugh Hudson is a source of both cartoons and amusement.

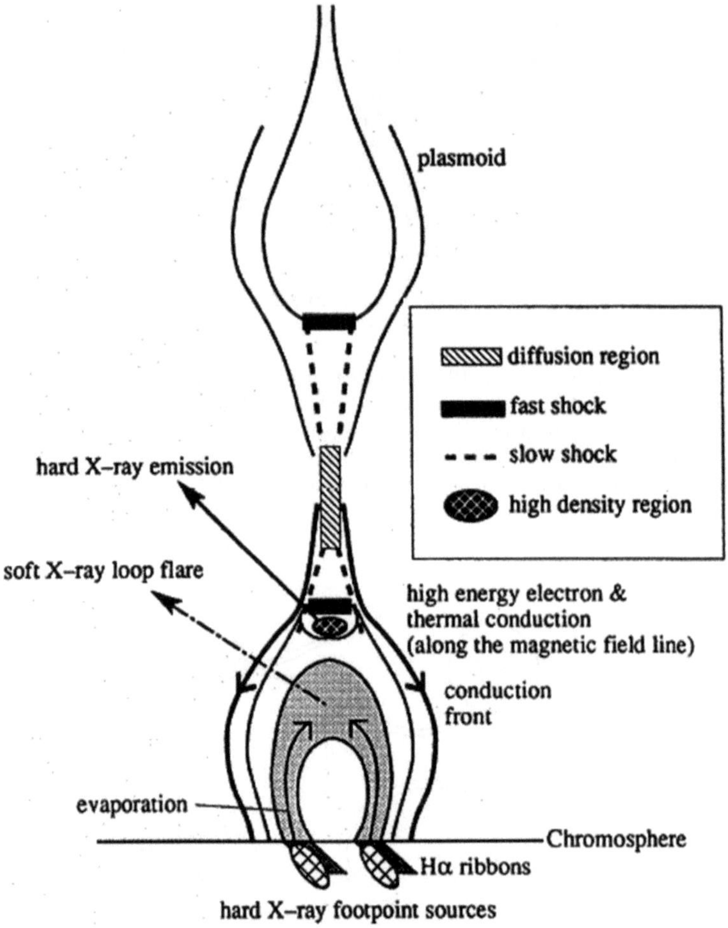

Fig. 11.7 A sketch by Magara et al. (1997) of the "standard flare model". Note the overall similarity to Sturrock's sketch, but now with the ejection of a "plasmoid" which would be part of the CME. The basic components of Petschek reconnection are present, as well as a termination shock where the reconnection jet runs into closed field lines not involved in the flare/eruption

deduce the electron spectrum from the photon one in a simple way, permitting an estimate of the energy in these accelerated electrons. This raised two fundamental problems. One is that, due to the low efficiency of the Bremsstrahlung process that produces the detected HXR photons, the number of electrons required to produce the HXR emission is large: perhaps 10^{35} per second. Almost all the energy in the accelerated electrons goes into plasma heating, producing the chromospheric (and occasionally white light) flare, as well as coronal soft X-rays. The second problem is that, given observations of the total flare energy obtained over all wavelengths, the fraction of the energy release that ends up in $>$10 keV electrons is large, up to tens of percent in large flares (see papers in Emslie et al. 2011). The energy in sub-MeV ions is unknown, but there seems to be no reason why it cannot be large. Thus the problem posed from Giovanelli onwards, namely fast energy release in flares, becomes a great deal harder to solve. To quantify, if one considers "reasonable" pre-flare quantities: a volume of 10^{27} cm^3 and a density of 10^9 cm^{-3}, one has to accelerate the entire coronal density in a few seconds, Cowling's (1953) concerns about current sheet thickness not withstanding!

MHD models for flares cannot account for this particle acceleration, but it is widely argued that in eruptive flares the magnetic geometry they imply (large monolithic current sheets or collapsing magnetic traps) are suitable facilitators. This may be complacent. A difficulty concerns charge and current quasi-neutrality. If for example only electrons are accelerated, and subsequently leave the corona (not always the case: Krucker et al. 2008b) then one must (a) replenish the corona to maintain quasi-neutrality and (b) deal with the magnetic field induced by the fleeing electrons by means of a return current. These points have been debated for several decades with no particularly strong conclusions (but see Sect. 11.3.1.3).

11.3.1.2 Another Way to Look at Solar Flares

The large number of flares seen over the lifetime of SMM permitted a statistical survey using the Hard X-ray Burst Spectrometer (HXRBS) instrument. Crosby et al. (1993: also Crosby 2011) analysed the HXRBS data, in particular the peak HXR flux, and an estimate of the energy in >20 keV electrons, the latter using the thick target model. They showed that the distribution of flares as a function of flare energy was a power law $N(E) \sim E^{-1.5}$ over three decades. The power law distribution is suggestive of a similarity in the physical process in these flares, and this is unlikely to be driven by an eruption. When we note that for every flare that releases 10^{32} ergs there are many that release 10^{29} ergs, the inadequacy of the term "standard model" for flares becomes even clearer.

Drawing on the work on avalanche models of Bak et al. (1987: see also Bak 1999); Lu and Hamilton (1991) proposed that a flare could be studied in terms of a driven system which underwent relaxation when certain conditions were met between neighbouring points. In reality, this meant that the neighbouring fields were sheared by a critical amount, a current sheet formed and dissipated. Lu and Hamilton showed that the distribution of events as a function of energy released was $E^{-1.4}$. This precise scaling should not be taken too seriously since their original model was quite simple. More sophisticated versions have confirmed the overall idea that driven systems typical of the solar corona do involve energy release over a wide range of event sizes. What this suggests is that large flares involve many reconnection sites operating in harmony, triggered within a few Alfvén transit times, and small flares may involve fewer dissipation sites, as Cowling suggested as a way around the small volume associated with current sheets.

Multiple reconnection sites have implications for flare particle acceleration. An early study was due to Anastasiadis and Vlahos (1991) who considered a system with multiple quasi-perpendicular shock waves at which particles underwent drift acceleration. At a single shock drift acceleration is limited by both the energy gain and the number of particles accelerated before escape: the largest energy gains occur to very few particles. However, if particles interact with second and subsequent shocks, greater energy gains are possible. This approach can be extended to multiple current sheets within Lu and Hamilton type models (e.g. Vlahos et al. 2004) and also in MHD models where the accelerated particles are "test particles" (Turkmani et al. 2006; Cargill et al. 2006; Gordovskyy and Browning 2011).

These examples still have particles interacting with a very narrow current sheet. More recent advances have been made with self-consistent models using hybrid and particle codes, though of necessity simpler fields are used than are likely to be present in the corona. The work of Fermo et al. (2010) and Drake et al. (2013) is especially illuminating. They study the coalescence of many magnetic islands in a scenario that involves multiple current sheets undergoing reconnection. Figure 11.8 (from Fermo et al.) sketches the procedure. From the viewpoint of particle acceleration, the important thing is the shortening of the field lines as reconnection occurs. Provided the electrons or protons have energy in excess of a critical threshold (V_A for protons, $>> V_A$ for electrons), they can then be accelerated very efficiently. The particles conserve adiabatic invariants and, as the field lines snap closed, their parallel energy must increase. The acceleration at a single reconnection event is fairly modest, but in a "turbulent" medium with multiple x- and o-points, systematic and significant energy gain arises (Drake et al. 2013). Note

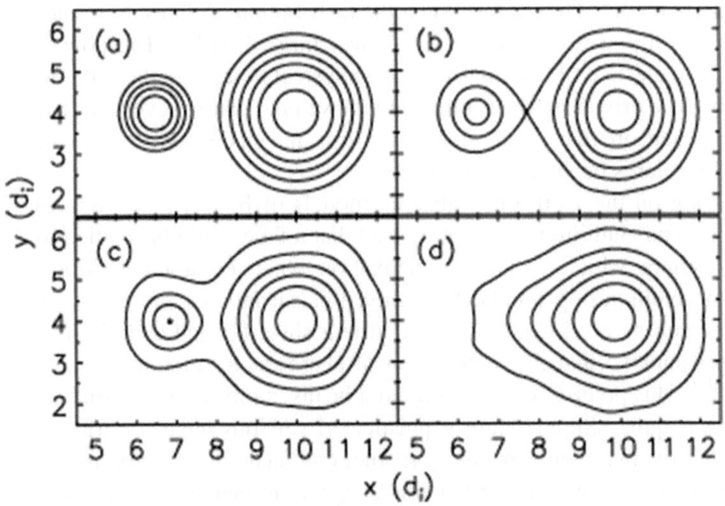

Fig. 11.8 The interaction of two magnetic islands from a particle in cell simulation (Fermo et al. 2010). Time increases from panel (**a**) to panel (**d**). Magnetic field lines in the plane are shown. An x-line forms (panel **b**), and the smaller island is eaten by the larger. Note the shrinkage of the reconnected field lines in panel (**d**): this is an example of the globality of magnetic energy release in reconnection

that the particle acceleration is "global" in that it occurs as the magnetic islands contract: again we see the foresight of Jim's argument on this point.

Such models can also provide an idea of what terminates the acceleration process: test particle models can be alarmingly efficient, and observations also suggest that limiting occurs (Krucker et al. 2010) at a rough equipartition between particle and magnetic energy. Drake et al. note that their acceleration process increases only the parallel energy, which contributes in an MHD description to a parallel pressure due to energetic particles. When this reaches a critical value, the firehose instability sets in, the condition being: $p_{\|} > p_{\perp} + \frac{B^2}{4\pi}$, and non-linearly must convert parallel to perpendicular energy, So the condition is just that a significant fraction of the available field energy is converted to energetic particles. In a driven system there will be a state of marginal stability around the firehose instability condition.

11.3.1.3 Flares: Past and Future

We began this chapter by describing Giovanelli's work that tried to account for chromospheric flare emission. In some regards little has changed. Despite manifest objections, there is still a viewpoint that there is a magic neutral point or x-point or x-line or current sheet that can account for the flare properties. While it is clear that major topological rearrangement of the magnetic field is required in large eruptive flares, and this may indeed involve some form of neutral point, it seems very difficult to understand how the acceleration requirements of the entire flare can be met with such a model, or indeed with any model relying on what are effectively laminar fields.

The fact that multiple accelerators, either distributed over a fraction of the solar corona, or present within an evolved turbulent current sheet, permit ready particle acceleration is now established and I contend that such models are the only sensible framework in which to consider flare particle acceleration. These scenarios can accelerate particles to the required energies, can account for acceleration of selective ions, and can accelerate a considerable number of particles.

We still do not understand the full electrodynamic consequences of massive acceleration in the corona, nor how charge neutrality is maintained. The proposal to "move" the acceleration site to the upper chromosphere (Brown et al. 2009; Fletcher and Hudson 2008) is one option. While this does not alter greatly the fraction of the flare energy in the accelerated particles, it does alleviate greatly the number of particles needed, since particles are re-accelerated after collisions. We noted elsewhere (Cargill et al. 2012c) that the inability to satisfy global electrodynamic considerations in the corona may force particle acceleration elsewhere. On the onset of magnetic reconnection in the corona, there are three channels available for magnetic energy: flows, heating and particle acceleration. The latter is the fastest. If the acceleration processes is terminated because the electrodynamic constraints (e.g. return current) cannot be satisfied, then energy must flow into the other channels, with the addition of a Poynting flux (large-amplitude Alfvén wave), out of the corona. Progress on this scenario is being made (Fletcher and Hudson 2008; Russell and Fletcher 2013).

11.3.2 The Non-flaring Solar Corona

It has taken many years for magnetic reconnection to become established as a mechanism for heating the non-flaring magnetically closed corona, in particular the bright active regions (ARs) that are our main focus. Heating by Alfvén waves has been discussed for several decades despite the well-known problems of their dissipation (see Wentzel 1974, 1976 and Ionson 1978 for possible resolutions). Yet despite a vast literature, few papers have actually addressed the problem in a way of interest for active regions (e.g. Ofman et al. 1998; van Ballegooijen et al. 2011). Wave heating of open field regions (coronal holes) is another matter (Cranmer 2009) but will not be discussed here. It should also be stressed that wave heating can arise as a result of the possible damping of waves generated during reconnection.

11.3.2.1 Early Reconnection Models for Coronal Heating

The origins of reconnection as a heating mechanism go back to the October 1963 symposium on solar flares. Gold (1964), following Gold and Hoyle (1960), argued that the magnetic field in the corona becomes highly tangled due to photospheric motions, as shown in Fig. 11.9, and subsequently dissipates.[12] This work was ignored until Tucker (1973) published the first "modern" paper on coronal heating by reconnection (though he did not actually mention "reconnection"). While the plasma physics invoked to dissipate the coronal currents (the ion acoustic instability) may not be relevant, the intent is clear: magnetic energy is fed into the corona through slow (1 km s^{-1}) photospheric motions, builds up, and dissipates such that a quasi-steady state is maintained. Levine (1974a, b) then proposed that the corona was "heated" at multiple current sheets. He argued that particle acceleration would be the principal mode of magnetic energy release, and subsequent thermalisation would lead to heating. Also in this epoch Parker (1972) argued that current sheets would be formed spontaneously in the corona, though the methodology leading to the result remained controversial for a long time. But the seed was sown.

[12] H. Hudson states that this cartoon is due to Jack Piddington. There is no evidence for this in the published literature, although Gold offers no references in his article. Also Maxwell's equation in the solar interior should not be taken literally.

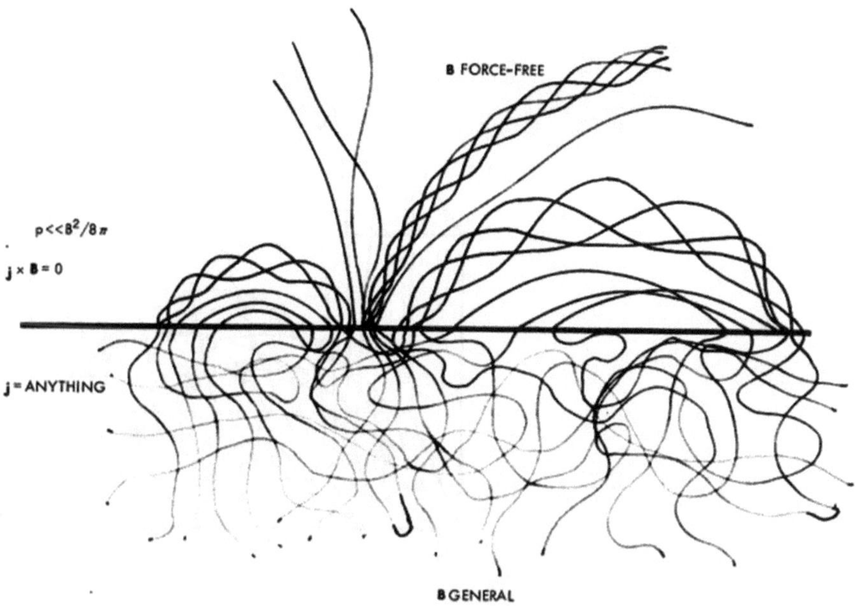

Fig. 11.9 A sketch of the tangled magnetic field in corona (Gold 1964). Magnetic energy is injected at the photosphere, forming both large-scale organised structures, later called coronal loops, and more random field patterns. Gold argued that, averaged over time, the energy injected will be dissipated

11.3.2.2 From Steady to Time-Dependent Coronal Heating: Nanoflares

The simplest approach to understanding the corona is in terms of a steady heating mechanism (e.g. papers in Orrall 1981). It was argued that Sklyab coronal measurements could be fit "adequately" with such models and this viewpoint persisted widely until at least the mid-1990s, and in some quarters to this day. However, evidence for time-dependent heating started to come from other sources. Lin et al. (1984) detected a new class of HXR brightenings around tens of keV, lasting up to a minute and with an energy release of 10^{28} ergs. These are microflares, and this revealed that the principles behind flares, fast energy dissipation and particle acceleration, extend to much lower energy events. Subsequent measurements from RHESSI (Christe et al. 2008; Hannah et al. 2008) show that microflares are ubiquitous, and it seems feasible that the fraction of energy that goes to accelerated particles is smaller than in flares. In turn, this raises the possibility that impulsive heating through magnetic reconnection could extend over an even wider range of energy.

What size of heating event can be expected? Motions at the photosphere lead to a Poynting flux into the corona: $S \approx V_t B_t B_a / 4\pi$, where subscript "$t$" ("$a$") denotes a quantity in the plane of (vertical to) the solar surface. For a coronal loop of length L, a heating event releases $Q \approx L B_t^2 / 8\pi \,\text{ergs}\,\text{cm}^{-2}$ if all the injected energy is

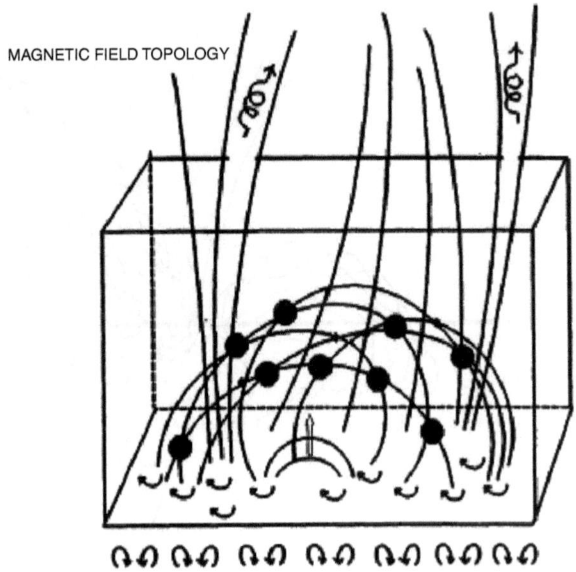

Fig. 11.10 A sketch of a nanoflare-heated corona (Vlahos 1994). Note the interacting field lines and the *large dots* representing energy release sites

dissipated.[13] For a given threshold for the event to occur (B_t/B_a must exceed some value), the time needed to build up the magnetic energy is $t \approx (B_t/B_a)(L/2V_t)$ s. Typical values for ARs are $V_t = 0.5$ km s^{-1}, $B_a = 100$ G, $B_t = 30$ G, $L = 50$ Mm and a cross-section of 10^{14} cm^2. Then $S \sim 10^7$ ergs cm^{-2} s^{-1}, $Q \sim 10^{25}$ ergs and $t = 15{,}000$ s. This calculation was presented by Parker (1988) who termed these heating events "nanoflares" in view of the magnitude of Q. Note also that S is in agreement with observed energy requirements for ARs (Withbroe and Noyes 1977). The sketch of Vlahos (1994) gives an idea of what a nanoflare-heated corona looks like (Fig. 11.10).

The value of B_t/B_a at which dissipation begins is important since magnetic energy must be built up before being released in the nanoflare. If the threshold is too low (e.g. $B_t/B_a < 0.1$), then neither the observed level of radiation from an AR nor the temperature of maximum emission can be accounted for. Despite this, numerical models insist that such a low threshold (a) exists and (b) can account for coronal heating (e.g. Rappazzo et al. 2008, 2010; Dahlburg et al. 2012; Wilmot-Smith et al. 2011). This cannot be the case and one should also note that use of reduced MHD (RMHD) in some of these models is inappropriate since it assumes $B_t << B_a$ and that B_a is constant. Thus equilibria that may be possible in full MHD do not exist in RMHD, so that spurious dynamical evolution may occur. [Evidence for the attainment of higher B_t/B_a in full MHD models is unclear because computational constraints mean they cannot be run for realistic parameters. However

[13] The form of reconnection envisioned is component reconnection, as introduced by Parker (1957). Two neighbouring thin flux bundles with oppositely directed components B_t will reconnect if B_t is large enough.

Dahlburg et al. (2005, 2009) and Bowness et al. (2013) show that highly stressed states may be attained.]

We noted above the continuous distribution of events from flares to microflares, and so one must ask whether even smaller reconnection-type heating events with lower energies can be detected. The simple answer is that they can (see Aschwanden and Parnell 2002), but issues arise in the interpretation of such observations (Parnell 2004; Parenti et al. 2006), so that no conclusion can be drawn at this time of their energetic significance (Hudson 1991). It should also be made clear that, despite their name, nanoflares are not to be viewed as mini-flares. Present evidence (Hannah 2012; private communication) suggests that there is a minimal quantity of accelerated particles in active regions, though data from the FOXSI and NuStar spacecraft will clarify this. If nanoflares are responsible for AR heating, then there is a fundamental change in the way particle acceleration occurs in such small events when compared with flares. We return to this later.

How can one then deduce anything about magnetic reconnection in the corona? New observations from Hinode and SDO suggest a possible path, but we first need to address how the solar corona responds to impulsive (nanoflare) heating. Figure 11.11 shows the coronal temperature and density as a function of time in a coronal loop of length 150 Mm subject to nanoflare heating. Much of the physics is governed by the radiation from the thin transition region (TR) that lies between chromosphere and corona.[14] The coronal temperature and density are governed by (a) the energy input (flare, nanoflare), (b) conductive losses to the TR, (c) optically thin radiation to space and (d) enthalpy gain or loss to and from the TR (see Cargill et al. 1995; Klimchuk et al. 2008; Bradshaw and Cargill 2010; Cargill et al. 2012a, b). As shown in Fig. 11.11, the response to heating follows three phases: (1) an initial rapid temperature increase, (2) conductive cooling, but since the TR cannot radiate away the heat flux, there is an upward enthalpy flux, increasing the coronal density, (3) as the corona cools and the density rises, the coronal heat flux falls, and eventually is not strong enough to power the TR. However, the TR still has to radiate, and to do this there is an enthalpy flux from the corona so that the coronal density now falls.

The key to interpreting the observations lies in the emission measure, $EM(T) = \int n_e^2 dh \text{ cm}^{-5}$, where dh is along the line of sight. When the observations have good temperature coverage in the range 1–10 MK, EM(T) can be inferred from the emission from a range of spectral lines, subject to knowledge of the relevant atomic physics and instrument performance. It can also be generated easily from theoretical models: EM(T) from Fig. 11.11 is shown in Fig. 11.12. For this case of a single loop being heated and allowed to cool, we see (a) there is plasma between the peak and below 1 MK with a characteristic slope $EM(T) \sim T^2$ and (b) there is a hot component above the maximum extending to 10 MK. The lower

[14] The transition region width in a closed static magnetic loop is of order a few % of the total loop length, yet the radiation from the TR exceeds that from the corona by a factor of two.

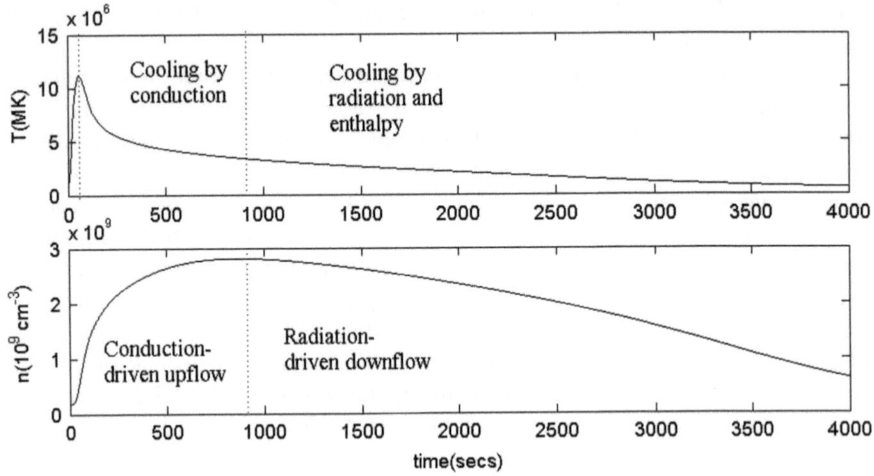

Fig. 11.11 The evolution of the temperature and density in a coronal loop of length 80 mm. A triangular heating pulse of width 100 s and peak amplitude 0.1 ergs cm^{-3} s^{-1} is applied at $t = 0$. The *first vertical dotted line* shows the peak of the heating. The *second dotted lines* show the time of maximum density and change in the energetics of the loop. For a loop sub-element of dimension 100 km^2, the total energy released is 4×10^{24} ergs

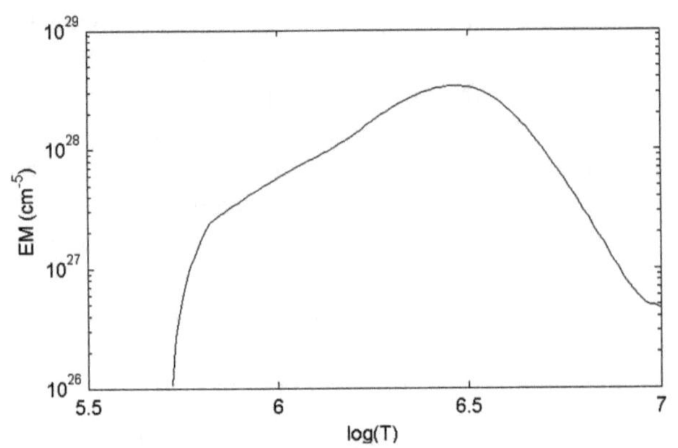

Fig. 11.12 The time-integrated emission measure as a function of temperature corresponding to the heated loop shown in this figure

temperature slope is determined by the radiative/enthalpy cooling phase, and can also be derived analytically (Cargill 1994; Cargill and Klimchuk 2004). This is a central *prediction* of nanoflare heating models.

Analysis of active regions using data from SDO and Hinode has been carried out by Warren et al. (2011, 2012); Tripathi et al. (2011); Schmelz and Pathak (2012); Bradshaw et al. (2012) and Reep et al. (2013). Warren et al. (2011) showed that, for a single AR, the EM slope was steeper than shown in Fig. 11.10, implying a deficit of plasma at cooler temperatures. The interpretation was that the loop was reheated before it could cool below 1–2 MK, contrary to the earlier requirements.

Subsequent studies of almost 20 ARs showed that in some cases slopes comparable to Fig. 11.11 were found, but in others the slopes were in the range 3–5.

What are the implications for coronal reconnection? We have carried out an in-depth study of EM slopes for a range of nanoflare heating models (Cargill 2014). For the ARs with shallow slopes, the picture of stressing field lines to high shear over 2–3 h prior to the nanoflare remains valid. The other results pose more of a problem. If the nanoflares repeat every few hundred seconds, then the Poynting flux is inadequate to power the corona except if a highly stressed field relaxes by only a small amount. This seems to be at variance with current MHD models, though that may be due to their (still) over-simplistic nature.

11.3.2.3 Coronal Heating by Reconnection: Where Next

The above discussion demonstrates the difficulty in analysing the role of reconnection in a remotely-sensed magnetically complex system. Yet, the work of Warren et al. (2012) represents a huge step forward in analysing the corona, even though the preliminary conclusions may be difficult to understand. Despite many concerns about the data analysis (and most of the concerns are things one has to live with), such multi-temperature observations permit quantitative science. The key data does not come from the spectacular images of the Sun, but from EUV spectroscopy.[15]

The future must involve further in-depth studies to reveal the true parameter dependence of the corona. In addition, various spectroscopic techniques that we discussed in Cargill (2013a) need to be followed up, as well as a study of high temperature non-flaring plasma, given that this may be the "smoking gun" of nanoflare heating (Cargill 1995; Testa et al. 2011; Testa and Reale 2012). One exciting glimpse of the future was revealed by the Hi-C rocket flight (Cirtain et al. 2013; Cargill 2013b; Testa et al. 2013) which provided spatial resolution of the EUV corona on scales of <200 km, a factor five better than what had gone before. In particular, Cirtain et al. claimed to see evidence for "magnetic braiding", the reconnection process due to Parker (1972) described earlier, Cargill suggested instead that small-scale kink instabilities were evident, while Testa et al. detected signatures of nanoflares with energy $>10^{23}$ ergs. With a five minute rocket flight, what was deemed as "adequate" spatial resolution was redefined.

Thus a future mission would require <200 km spatial resolution, high time resolution, but with broad temperature coverage as well as EUV spectroscopy with a temperature coverage comparable to that flown on Hinode, but with much improved spatial and temporal resolution. This poses formidable data handling

[15] Despite this, neither the STEREO nor SDO missions flew an imaging spectrometer. It is strange that the best temperature coverage of a corona is from the EUVE spacecraft which constructed EM (T) profiles of stellar coronae between 1 and 20 MK by using every Fe line between Fe IX and Fe XXIV (e.g. Sanz-Forcada et al. 2003; Cargill and Klimchuk 2006).

requirements, both with downlink, distribution and analysis, so the spacecraft will be in near-Earth orbit. Hopefully the forthcoming JAXA-led Solar-C mission will meet these requirements.

Finally the contrasting role of the level of particle acceleration in flares, microflares and nanoflares needs to be considered. It would be an important result if smaller events are associated with a smaller percentage of the total released energy residing in energetic particles, and lower maximum accelerated particle energies. A way forward may lie in the relative complexities of the magnetic field associated with different sized events. In a flare, a particle has the opportunity to interact with many separate accelerators: individual current sheets, or multiple magnetic island merging, whereas in a nanoflare few interactions seem likely. Further, if each accelerator can initially accelerate (inject) N particles, and those N particles can each interact with M other accelerators, the multiplicative effects will be considerable. Calculations (e.g. Vlahos et al. 2004; Drake et al. 2013) examine $M >> 1$, but it would be of interest to investigate intermediate regimes.

11.4 Conclusions

Sixty years after Jim's paper was published in Phil Mag, magnetic reconnection is viewed as the dominant mechanism for the dissipation of magnetic energy in the outer solar atmosphere. Jim's name is commonly associated with his pioneering work on the role of magnetic reconnection in the magnetosphere, and he is now widely recognised as the initiator of modern magnetospheric physics. The importance of his early work to solar coronal physics is perhaps not so widely understood.[16] Yet the important points in that paper, the change in field topology and the global nature of energy release in reconnection are central to understanding the role it plays on the Sun.

If one looks back to the reconnection papers from the 1940s and 1950s, the main subject of interest was the acceleration of particles. Sixty years on, collisionless reconnection[17] and particle acceleration still pose major problems in our understanding of reconnection. If we pose the question: how has attempting to understand the solar corona influenced the direction of reconnection physics, there are a number of answers. Firstly, it was solar flares (along with the aurora) which

[16] I gave an "encore" of my Dungey lecture in Feb 2013 to a team studying the solar corona at the International Space Science Institute (ISSI). I asked the audience if they knew who Jim Dungey was: some of the older members did, but younger ones, or those with a sheltered upbringing, did not. I then asked if they knew what he had done in the 1950s: very few did.

[17] When I first got to know Jim at Imperial in the mid-1990s, he was very interested in the nature of collisionless reconnection, and indeed the final Section of Dungey (1994) looks at the status at that time. The advances in collisionless reconnection in the past two decades have been a triumph of plasma physics, space plasma physics, and analysis of spacecraft data. Eastwood (this volume) discusses this further.

instigated the study of dissipation and acceleration in dynamic magnetic fields. Beyond that simplistic answer, there seem to be two (related) points. One is that the particle acceleration requirements for solar flares defy understanding if one considers simple reconnection pictures. This in turn has forced workers to construct scenarios where very efficient particle acceleration can take place and has led to the ideas discussed here of multiple, interacting reconnection sites. Another consequence is the realisation that a stressed magnetic field can relax through reconnection with a wide range of changes in the value of the magnetic energy. Perhaps the solar corona is unique in this aspect, but it seems likely that any astrophysical object with a lively dynamo could exhibit similar behaviour. It does seem that while these problems are not solved, the right questions are now being asked.

Dungey (1953) marks the start of modern reconnection physics. Had that paper not been published, as could easily have been the case, what would the outcome have been? Jim writes himself that things were tough in the 1950s. Magnetic reconnection did not take a terribly promising path in subsequent years, yet even the detours acknowledge the importance of what Jim did. How would our understanding of magnetic energy release have changed, or in particular have been delayed, without his early thoughts? Thankfully Jim did persevere in 1952/1953 and founded the topic of reconnection. For that everyone must be grateful.

Acknowledgements I thank the Royal Astronomical Society for access to the documents pertaining to the submission of Dungey (1953) to Monthly Notices, and to Jenny Higham, the Society librarian, for her invaluable help in locating them, and for directing me to the minutes of RAS Council in 1951–1953. I am grateful to David Southwood for his organisation of JimFest on January 10, 2013, and as president of the RAS for overseeing the establishment of the Dungey lecture, the first of which it was my privilege to deliver. He also made two important suggestions that improved this paper. I have had very useful discussions with Hugh Hudson, as well as enjoying his cartoon collection. I thank him, Jonathan Eastwood and Jim Klimchuk for their comments on the paper. Much of the work in Sect. 11.3.2 was carried out in an ISSI team led by Steve Bradshaw and Helen Mason.

Appendix. Citations of Key Papers

The below table shows the cumulative citations (all sources) for some of the more important papers discussed in this paper using three sources. Unfortunately ADS does not cover Phil Mag in 1953. The numbers of citations in this Table were complied in early 2013 using the data available at that time.The ADS numbers are total citations with refereed publications as defined by ADS in brackets.

Paper	ADS	ISI	Googlescholar	Present annual rate (approx.: ADS)	Year of changes in citations
Giovanelli (1946)	85 (77)	109	203	3–9	
Giovanelli (1947)	87 (84)	83	177	1–3	

(continued)

Paper	ADS	ISI	Googlescholar	Present annual rate (approx.: ADS)	Year of changes in citations
Hoyle (1949)	–	57	111		
Dungey (1953)	–	216	304		
Parker (1957)	591 (546)	620	782	40 (Level)	1984, 2002
Sweet (1958a)	464 (413)	428	420	30 (Level)	1971, 1999
Dungey (1961)	1484 (1,430)	1,805	2,526	65 (Increasing)	1967, 2004
Parker (1963)	374 (356)	413	545	~12	
Petschek (1964)	728 (669)	833[18]	1,025	36 (Level)	1972, 1992

The variation of numbers for each data base should be noted. Googlescholar adopts a "kitchen sink" approach with a wide range of reports etc. included. ADS has a more limited database. ISI only counts citations post-1970.

Also of interest are "breaks" or rapid changes (increases in every case) in citation rates, and the present annual rate: the "break" years should be considered to be ± 2. All numbers are only refereed papers to avoid the disease of conference proceedings. It is straightforward to do this with ADS and I have done this for five papers shown in the table. For example, ADS indicated that Petschek's paper had roughly 20 **total** citations between 1964 and 1972, and settled down at an average of roughly 8 per year to 1992 since when the average number of annual citations has increased to of order 35. [ISI shows Dungey (1953) receiving between one and ten citations a year since 1970, with no obvious trends.]

What is one to make of this? Jim's 1961 paper is in some ways the most surprising in that while once the idea permeated the community, there were enough believers from the mid-1960s to keep the "Dungey magnetosphere" to the fore, there was no rapid increase around 1980 when the first ISEE results confirmed the open magnetosphere idea. The jump in 2004/2005 can be attributed to the first wave of Cluster results which established beyond doubt the presence of both sub-solar reconnection and lobe reconnection (Dungey 1963). One can attribute the changes to both Sweet and Petschek in the early 1970s to the realisation that the corona was highly dynamic, and the jump in both Sweet and Parker's citations around 1999/2000 to the huge upsurge in work on collisionless reconnection at that time. Of course one would need to de-trend these results to reflect the much larger communities today, and the greater number of citations that seem to be de rigueur in every paper, but on the other hand most papers citation numbers decreases with the time from publication.

[18] Due to "double counting" there are difficulties in the ISI estimate.

References

Alfvén, H.: Discussion of the origin of the terrestrial and solar magnetic fields. Tellus **2**, 74 (1950)

Anastasiadis, A., Vlahos, L.: Particle acceleration inside a gas of shock waves. Astron. Astrophys. **245**, 271 (1991)

Aschwanden, M.J., Parnell, C.E.: Nanoflare statistics from first principles: fractal geometry and temperature synthesis. Astrophys. J. **572**, 1048 (2002)

Axford, W.I.: The good old days. J. Geophys. Res. **99**, 19199 (1994)

Bak, P.: How Nature Works. CUP, Cambridge (1999)

Bak, P., Tang, C., Wiesenfeld, K.: Self-organized criticality: an explanation of 1/f noise. Phys. Rev. Lett. **59**, 381 (1987)

Bowness, R., Hood, A.W., Parnell, C.E.: Coronal heating and nanoflares: current sheet formation and heating. Astron. Astrophys. **560**, 89 (2013)

Bradshaw, S.J., Cargill, P.J.: The cooling of coronal plasmas. III. Enthalpy Transfer as a Mechanism for Energy Loss. Astrophys. J. **717**, 163 (2010)

Bradshaw, S.J., Klimchuk, J.A., Reep, J.W.: Diagnosing the time-dependence of active region core heating from the emission measure. I. low-frequency nanoflares. Astrophys. J. **758**, 53 (2012)

Brown, J.C.: The deduction of energy spectra of non-thermal electrons in flares from the observed dynamic spectra of hard x-ray bursts. Sol. Phys. **18**, 489 (1971)

Brown, J.C., Turkmani, R., Kontar, E.P., MacKinnon, A.L., Vlahos, L.: Local re-acceleration and a modified thick target model of solar flare electrons. Astron. Astrophys. **508**, 993 (2009)

Cargill, P.J.: Some implications of the nanoflare concept. Astrophys. J. **422**, 381 (1994)

Cargill, P.J.: Diagnostics of coronal heating. In: Kuhn, J., Penn, M. (eds.) Infrared Tools for Solar Astrophysics: What's Next? p. 17. World Scientific Publishing, Singapore (1995)

Cargill, P.J.: From flares to nanoflares: magnetic reconnection at work on the Sun. Astron. Geophys. **54**, 3.12 (2013a)

Cargill, P.J.: Towards even smaller length scales. Nature **493**, 485 (2013b)

Cargill, P.J.: Active region emission measure distributions and implications for nanoflare heating. Astrophys. J. **784**, 49 (2014)

Cargill, P.J., Klimchuk, J.A.: Nanoflare heating of the corona revisited. Astrophys. J. **605**, 911 (2004)

Cargill, P.J., Klimchuk, J.A.: On the temperature—emission measure distribution in stellar corona. Astrophys. J. **643**, 300 (2006)

Cargill, P.J., Mariska, J.T., Antiochos, S.K.: The cooling of solar flare plasmas: 1. Theoretical considerations. Astrophys. J. **439**, 1034 (1995)

Cargill, P.J., Vlahos, L., Turkmani, R., Galsgaard, K., Isliker, H.: Particle acceleration in solar flares: the role of a turbulent global coronal magnetic field. Space Sci. Rev. **124**, 249 (2006)

Cargill, P.J., Parnell, C.E., Browning, P.K., de Moortel, I., Hood, A.W.: Magnetic reconnection in the solar atmosphere: from proposal to paradigm. Astron. Geophys. **51**, 3.31 (2010)

Cargill, P.J., Bradshaw, S.J., Klimchuk, J.A.: Enthalpy-based thermal evolution of loops. II. Improvements to the model. Astrophys. J. **752**, 161 (2012a)

Cargill, P.J., Bradshaw, S.J., Klimchuk, J.A.: Enthalpy-based thermal evolution of loops. III. Comparison of zero-dimensional models. Astrophys. J. **758**, 5 (2012b)

Cargill, P.J., Vlahos, L., Baumann, G., Drake, J.F., Nordlund, A.: Particle acceleration in solar flares. Space Sci. Rev. **173**, 223 (2012c)

Carmichael, H.: A process for flares. In: NASA/AAS Symposium on Solar Flares, vol. SP-50, p. 451 (1964)

Carrington, R.C.: Description of a singular appearance seen on the Sun on Sept 1, 1859. Mon. Not. Roy. Astron. Soc. **20**, 13 (1859)

Chapman, S., Cowling, T.G.: The Mathematical Theory of Non-uniform Gases. CUP, Cambridge (1939)

Christe, S., Hannah, I.G., Krucker, S., McTiernan, J., Lin, R.P.: RHESSI microflare statistics. 1. Flare-finding and frequency distributions. Astrophys. J. **677**, 1385 (2008)

Cirtain, J.W., et al.: Energy release in the solar corona from spatially resolved magnetic braids. Nature **493**, 501 (2013)
Cowling, T.G.: Solar electrodynamics. In: Kuiper, U. (ed.) The Sun, p. 532. Chicago Press, Chicago (1953)
Cowling, T.G.: Magnetohydrodynamics. Rep. Prog. Phys. **25**, 244 (1962)
Cowling, T.G.: Astronomer by accident. Annu. Rev. Astron. Astrophys. **23**, 1 (1985)
Cranmer, S.R.: Coronal holes. Living Rev. Sol. Phys. **6**, 3 (2009)
Crosby, N.B.: Frequency distributions: from the sun to the earth. Nonlinear Process. Geophys. **18**, 791 (2011)
Crosby, N.B., Aschwanden, M.J., Dennis, B.R.: Frequency distributions and correlations of solar flare x-ray parameters. Sol. Phys **143**, 275 (1993)
Dahlburg, R.B., Klimchuk, J.A., Antiochos, S.K.: An explanation for the "switch-on" nature of magnetic energy release and its application to coronal heating. Astrophys. J. **622**, 1191 (2005)
Dahlburg, R.B., Liu, J.-H., Klimchuk, J.A., Nigro, G.: Explosive instability and coronal heating. Astrophys. J. **704**, 1059 (2009)
Dahlburg, R.B., Einaudi, G., Rappazzo, A.F., Velli, M.: Turbulent coronal heating mechanisms: coupling of dynamics and thermodynamics. Astron. Astrophys. **544**, L20 (2012)
Drake, J.F., Swisdak, M., Fermo, R.: The power-law spectra of energetic particles during multi-island magnetic reconnection. Astrophys. J. Lett. **763**, L5 (2013)
Dreicer, H.: Electron and ion runaway in a fully ionized gas: I. Phys. Rev. **115**, 238 (1959)
Dungey, J.W.: A note on magnetic fields in conducting materials. Math. Proc. Cambridge Philos. Soc. **46**, 651 (1950)
Dungey, J.W.: Theory of magnetic storms and auroras. Nature **170**, 795 (1952)
Dungey, J.W.: Conditions for the occurrence of electrical discharges in astrophysical systems. Philos. Mag. **44**, 725 (1953)
Dungey, J.W.: Cosmic electrodynamics. CUP, London (1958a)
Dungey, J.W.: The neutral point discharge theory of solar flares: a reply to Cowling's criticism. IAU Symp. **6**, 135 (1958b)
Dungey, J.W.: Interplanetary magnetic field and the auroral zones. Phys. Rev. Lett. **6**, 47 (1961)
Dungey, J.W.: The magnetosphere: Imperial College Inaugural lecture. Imperial College of Science and Technology, London (1966)
Dungey, J.W.: Citation classic. Curr. Contents **49**, 20 (1983)
Dungey, J.W.: Memories, maxims, and motives. J. Geophys. Res. **99**, 19189 (1994)
Dungey, J.W.: The structure of the exosphere or adventures in velocity space. In: Geophysics, the Earth's Environment. Gordon and Breach, New York (1963)
Dungey, J.W.: Remarks on the discharge theory of flares. In: NASA/AAS Symposium on Solar Flares, vol. SP-50, p. 415 (1964)
Ellison, M.A.: Characteristic properties of chromospheric flares. Mon. Not. Roy. Astron. Soc. **109**, 3 (1949)
Elsasser, W.M.: Dimensional relations in magnetohydrodynamics. Phys. Rev. **95**, 1 (1954)
Emslie, A.G., Dennis, B.R., Hudson, H., Lin, R.P.: Preface. Space Sci. Revs. **159**, 1 (2011)
Fermo, R.L., Drake, J.F., Swisdak, M.: A statistical model of magnetic islands in a current layer. Phys. Plasmas **17**, 01702 (2010)
Feynman, R.P.: Applications of quantum mechanics to liquid helium. In: Progress in Low Temperature Physics, Chapter 2. North Holland (1955)
Fletcher, L., Hudson, H.S.: Impulsive phase flare energy transport by large-scale Alfvén waves and the electron acceleration problem. Astrophys. J. **675**, 1645 (2008)
Fuentes-Fernández, J., Parnell, C., Hood, A.: Magnetohydrodynamic dynamical relaxation of coronal magnetic fields. II. 2D magnetic X-points. Astron. Astrophys. **536**, 32 (2011)
Giovanelli, R.G.: A theory of chromospheric flares. Nature **158**, 81 (1946)
Giovanelli, R.G.: Magnetic and electric phenomena in the Sun's atmosphere associated with sunspots. Mon. Not. Roy. Astron. Soc. **107**, 338 (1947)
Giovanelli, R.G.: Chromospheric flares. Mon. Not. Roy. Astron. Soc. **108**, 163 (1948)

Giovanelli, R.G.: The emission of radiation from flares. Mon. Not. Roy. Astron. Soc. **109**, 337 (1949)

Gold, T., Hoyle, F.: On the origin of solar flares. Mon. Not. Roy. Astron. Soc. **120**, 89 (1960)

Gold, T.: Magnetic energy shredding in the solar atmosphere. In: NASA/AAS Symposium on Solar Flares, vol. SP-50, p. 389 (1964)

Gordovskyy, M., Browning, P.K.: Particle acceleration by magnetic reconnection in a twisted coronal loop. Astrophys. J. **729**, 101 (2011)

Gosling, J.T.: The solar flare myth. J. Geophys. Res. **98**, 18937 (1993)

Hannah, I.G., Christe, S., Krucker, S., Hurford, G.J., Hudson, H.S., Lin, R.P.: RHESSI microflare statistics. II. x-ray imaging, spectroscopy, and energy distributions. Astrophys. J. **677**, 704 (2008)

Hones, E.W.: Magnetic reconnection. Geophys. Monogr. **34**, 4–14 (1984)

Hoyle, F.: Recent researches in solar physics. CUP, Cambridge (1949)

Hoyle, F., Wickramasinghe, N.C.: A note on the origin of the Sun's polar field. Mon. Not. Roy. Astron. Soc. **123**, 51 (1961)

Hudson, H.S.: Solar flares, microflares, nanoflares, and coronal heating. Sol. Phys. **133**, 357 (1991)

Hudson, H., Haisch, B., Strong, K.T.: Comment on 'The solar flare myth' by J. T. Gosling. J. Geophys. Res. **100**, 3473 (1995)

Imshennik, V.S., Syrovatskii, S.I.: Two-dimensional flow of an ideally conducting gas in the vicinity of the zero line of a magnetic field, Sov. Phys.—JETP **25**, 656 (1967)

Ionson, J.A.: Resonant absorption of Alfvénic surface waves and the heating of solar coronal loops. Astrophys. J. **226**, 650 (1978)

Klimchuk, J.A.: On solving the coronal heating problem. Sol. Phys. **234**, 41 (2006)

Klimchuk, J.A., Patsourakis, S., Cargill, P.J.: Highly efficient modelling of dynamic coronal loops. Astrophys. J. **682**, 1351 (2008)

Krucker, S., Hurford, G.J., MacKinnon, A.L., Shih, A.Y., Lin, R.P.: Coronal γ-ray bremsstrahlung from solar flare-accelerated electrons. Astrophys. J. Lett. **678**, L63 (2008a)

Krucker, S., Battaglia, M., Cargill, P.J., et al.: Hard X-ray emission from the solar corona. Astron. Astrophys. Rev. **16**, 155 (2008b)

Krucker, S., Hudson, H.S., Glesener, L., White, S., Masuda, S., Wuelser, J.-P., Lin, R.P.: Measurements of the coronal acceleration region of a solar flare. Astrophys. J. **714**, 1108 (2010)

Levine, R.H.: Acceleration of thermal particles in collapsing magnetic regions. Astrophys. J. **190**, 447 (1974a)

Levine, R.H.: A new theory of coronal heating. Astrophys. J. **190**, 457 (1974b)

Levy, R.H., Petschek, H.E., Siscoe, G.L.: Aerodynamic aspects of the magnetospheric flow. AIAA J. **2**, 2065 (1964)

Lin, R.P., Schwartz, R.A., Kane, S.R., Pelling, R.M., Hurley, K.C.: Solar hard X-ray microflares. Astrophys. J. **283**, 421 (1984)

Lu, E.T., Hamilton, R.J.: Avalanches and the distribution of solar flares. Astrophys. J. Lett. **380**, L89 (1991)

Magara, T., Shibata, K., Yokoyama, T.: Evolution of eruptive flares. I. plasmoid dynamics in eruptive flares. Astrophys. J. **487**, 437 (1997)

McClymont, A.H., Craig, I.J.D.: Linear theory of fast reconnection at an X-type neutral point. Astrophys. J. **405**, 207 (1993)

Ofman, L., Klimchuk, J.A., Davila, J.M.: A self-consistent model for the resonant heating of coronal loops: the effects of coupling with the chromosphere. Astrophys. J. **493**, 474 (1998)

Orrall, F.Q.: Active regions. In: Proceedings of 3rd Skylab Workshop, Colorado Associated University Press, Boulder (1981)

Parenti, S., Buchlin, E., Cargill, P.J., Galtier, S., Vial, J.-C.: Modeling the radiative signatures of turbulent heating in coronal loops. Astrophys. J. **651**, 1219 (2006)

Parker, E.N.: Sweet's mechanism for merging magnetic fields in conducting fluids. J. Geophys. Res. **62**, 509 (1957)

Parker, E.N.: The solar flare phenomenon and the theory of reconnection and annihilation of magnetic fields. Astrophys. J. Suppl. **8**, 177 (1963)
Parker, E.N.: Topological dissipation and the small-scale fields in turbulent gases. Astrophys. J. **174**, 499 (1972)
Parker, E.N.: Nanoflares and the solar X-ray corona. Astrophys. J. **330**, 474 (1988)
Parker, E.N., Krook, M.: Diffusion and severing of magnetic lines of force. Astrophys. J. **124**, 214 (1956)
Parnell, C.E.: The role of dynamic brightenings in coronal heating. ESA **SP-575**, 227 (2004)
Parnell, C.E., De Moortel, I.: A contemporary view of coronal heating. Phil. Trans. R. Soc. A **370**, 321 (2012)
Peterson, L.E., Winckler, J.R.: A gamma ray burst from a flare. J. Geophys. Res. **64**, 697 (1959)
Petschek, H.E.: Magnetic field annihilation. In: NASA/AAS Symposium on Solar Flares, vol. SP-50, p. 425 (1964)
Piddington, J.H.: Theories of solar phenomenon depending on sunspot fields moving on the chromosphere and corona. Mon. Not. Roy. Astron. Soc. **113**, 188 (1953)
Priest, E.R.: Solar Magnetohydrodynamics. Reidel, Dordrecht (1982)
Priest, E.R.: Magnetohydrodynamics of the Sun. CUP, Cambridge (2013)
Priest, E.R., Forbes, T.G.: Magnetic Reconnection. CUP, Cambridge (2000)
Priest, E.R., Raadu, M.A.: Preflare current sheets in the solar atmosphere. Sol. Phys. **43**, 177 (1975)
Rappazzo, A.F., Velli, M., Einaudi, G., Dahlburg, R.B.: Nonlinear dynamics of the parker scenario for coronal heating. Astrophys. J. **677**, 1348 (2008)
Rappazzo, A.F., Velli, M., Einaudi, G.: Shear photospheric forcing and the origin of turbulence in coronal loops. Astrophys. J. **722**, 65 (2010)
Reale, F.: Coronal loops: observations and modeling of confined plasma. Living Rev. Sol. Phys. **7**, 5 (2010)
Reep, J.W., Bradshaw, S.J., Klimchuk, J.A.: Diagnosing the time dependence of active region core heating from the emission measure. II. nanoflare trains. Astrophys. J. **764**, 193 (2013)
Russell, A.J.B., Fletcher, L.: Propagation of Alfvénic waves from corona to chromosphere and consequences for solar flares. Astrophys. J. **765**, 81 (2013)
Sanz-Forcada, J., Brickhouse, N.S., Dupree, A.K.: The structure of stellar coronae in active binary systems. Astrophys. J. Suppl. **145**, 147 (2003)
Schmelz, J.T., Pathak, S.: The cold shoulder: emission measure distributions of active region cores. Astrophys. J. **756**, 126 (2012)
Shibata, K., Magara, T.: Solar flares: magnetohydrodynamic processes. Living Rev. Sol. Phys. **8**, 6 (2011)
Southwood, D.J.: An education in space physics. In: Gillmor, S. (ed.) Discovery of the Magnetosphere; History of Geophysics Series, vol. 7, p. 185. American Geophysical Union, Washington (1997)
Stern, D.P.: A conversation with Jim Dungey. Eos **67**(51), 1986 (1986)
Sturrock, P.A.: Model for the high energy phase of solar flares. Nature **211**, 695 (1966)
Sturrock, P.A.: Particle acceleration in solar flares. IAU Symp. **57**, 437 (1974)
Sturrock, P.A.: Solar flares. In: Proceedings of 2nd Skylab Workshop, Colorado Associated University Press, Boulder (1980)
Sweet, P.A.: The neutral point theory of solar flares. IAU Symp. **6**, 123 (1958a)
Sweet, P.A.: The production of high energy particles in solar flares. Nuovo Cimento Suppl. **8** (Ser 10), 188 (1958b)
Testa, P., Reale, F.: Hinode/EIS spectroscopic validation of very hot plasma imaged with the solar dynamics observatory in non-flaring active region cores. Astrophys. J. Lett. **750**, 10 (2012)
Testa, P., Reale, F., Landi, E., DeLuca, E.E., Kashyap, V.: Temperature distribution of a non-flaring active region from simultaneous Hinode XRT and EIS observations. Astrophys. J. **728**, 30 (2011)

Testa, P., et al.: Observing coronal nanoflares in active region moss. Astrophys. J. Lett. **770**, 1 (2013)
Tripathi, D., Klimchuk, J.A., Mason, H.E.: Emission measure distribution and heating of two active region cores. Astrophys. J. **740**, 111 (2011)
Tsiklauri, D., Haruki, T.: Magnetic reconnection during collisionless, stressed, X-point collapse using particle-in-cell simulation. Phys. Plasmas **14**, 119205 (2007)
Tucker, W.H.: Heating of solar active regions by magnetic energy dissipation: the steady state case. Astrophys. J. **186**, 285 (1973)
Tucker, W.H., Giacconi, R.: The X-ray universe. Harvard, Cambridge, MA (1986)
Turkmani, R., Cargill, P.J., Galsgaard, K., Vlahos, L., Isliker, H.: Particle acceleration in stochastic current sheets in stressed coronal active regions. Astron. Astrophys. **449**, 749 (2006)
van Ballegooijen, A.A., Asgari-Targhi, M., Cranmer, S.R., DeLuca, E.E.: Heating of the solar chromosphere and corona by Alfvén wave turbulence. Astrophys. J. **736**, 3 (2011)
Vilmer, N.: Solar flares and energetic particles. Phil. Trans. R. Soc. A. **370**, 3241 (2012)
Vlahos, L.: Theory of fragmented energy release on the Sun. Space Sci. Rev. **68**, 39 (1994)
Vlahos, L., Isliker, H., Lepreti, F.: Particle acceleration in an evolving network of unstable current sheets. Astrophys. J. **608**, 540 (2004)
Vlahos, L., Krucker, S., Cargill, P.J.: Solar flares: a strongly turbulent particle accelerator. In: Vlahos, L., Cargill, P.J. (eds.) Turbulence in Space Plasmas. Springer, Berlin (2009)
Warren, H.P.D., Brooks, H., Winebarger, A.R.: Constraints on the heating of high-temperature active region loops: observations from hinode and the solar dynamics observatory. Astrophys. J. **734**, 90 (2011)
Warren, H.P., Winebarger, A.R., Brooks, D.H.: A systematic survey of high-temperature emission in solar active regions. Astrophys. J. **759**, 141 (2012)
Wentzel, D.G.: Coronal heating by Alfvén waves. Sol. Phys. **39**, 129 (1974)
Wentzel, D.G.: Coronal heating by Alfvén waves. II. Sol. Phys. **50**, 343 (1976)
Wilmot-Smith, A.L., Pontin, D.I., Yeates, A.R., Hornig, G.: Heating of braided coronal loops. Astron. Astrophys. **536**, 67 (2011)
Withbroe, G.L., Noyes, R.W.: Mass and energy flow in the solar chromosphere and corona. Annu. Rev. Astron. Astrophys. **15**, 363 (1977)
Zirker, J.: Coronal holes. In: Proceedings of 1st Skylab Workshop, Colorado Associated University Press, Boulder (1977)
Zweibel, E.G., Yamada, M.: Magnetic reconnection in astrophysical and laboratory plasmas. Annu. Rev. Astron. Astrophys. **47**, 291 (2009)

Chapter 12
From the Carrington Storm to the Dungey Magnetosphere

David Southwood

Abstract A short history is presented of the evolution of our understanding of direct physical coupling between the solar and terrestrial environments. We start with the questions raised with the great magnetic storm of 1859. We then trace the discoveries, theories, personalities and controversies over the next 100 years. The story ends with Dungey's publication of his simple synthesis of the open or reconnection model of the magnetosphere in 1961 and its general acceptance about 20 years later.

12.1 Introduction

In 1961, Jim Dungey published a very dense two-page paper in Physical Review Letters (Dungey 1961a) entitled "Interplanetary Magnetic Field and the Auroral Zones" which now one can see resolved a century's work and controversy over how the outer regions of the Sun and Earth were physically coupled. The paper was based on an idea contained in his Ph.D. thesis in 1950 but, having had much difficulty in publishing the central work of his thesis in the early 1950s, he only published the paper after 11 years gestation and once he realised that there was experimental evidence for the idea. In classical manner enlightenment came suddenly from Parnassus. In Jim's case he was in Montparnasse, Paris, stirring a cup of coffee in a café when, looking at the pattern of the motion in the cup, he suddenly realised that his model explained the long known pattern of polar ionospheric disturbance currents and that the pattern was directly driven by the solar wind flow past the Earth (Dungey 1983).

Jim's theory linking the location of the aurora, a solar source of electrical energy, and the magnetic disturbances seen during geomagnetically disturbed times gave the basis for the solution of a problem of 100 years standing but also brought together the approaches of two antagonistic schools with different views on how

D. Southwood (✉)
Blackett Laboratory, Imperial College, London SW7 2BZ, UK
e-mail: d.southwood@imperial.ac.uk

D. Southwood et al. (eds.), *Magnetospheric Plasma Physics: The Impact of Jim Dungey's Research*, Astrophysics and Space Science Proceedings 41,
DOI 10.1007/978-3-319-18359-6_12

Sun and Earth were linked. The antagonism did not melt overnight and controversy remained. Even with the tool to thought represented by Jim's paper, there were still 20 more years to go until Jim's open or reconnection model of the magnetosphere became the standard model (Dungey 1983).

What science background did Jim's final answer need? Stan Cowley (chapter "Dungey's Reconnection Model of the Earth's Magnetosphere: The First 40 Years") and Steve Milan (chapter "Sun et Lumière: Solar Wind-Magnetosphere Coupling as Deduced from Ionospheric Flows and Polar Auroras") elucidate further the reconnection or open magnetosphere model. However, for now, one may note that Jim's model brought together apparently contradictory features in the ideas of the competing schools. Jim's model showed that the geomagnetic currents in the ionosphere were driven by an emf or voltage that was ultimately provided by solar material ejected from the Sun, a possibility long ignored by Jim's compatriots. Field aligned currents, such as the founder of the Scandinavian school, Birkeland (1908) predicted, were a natural feature of the model (Fig. 12.1). These had been excluded by the way the British codified magnetic data in terms of equivalent source currents although Goldie (1937) was a British exception. The model even allowed for a charged particle to follow a field line from the Sun and end up in the Earth's polar cap as the Scandinavian school had usually assumed. However, in contrast, it also

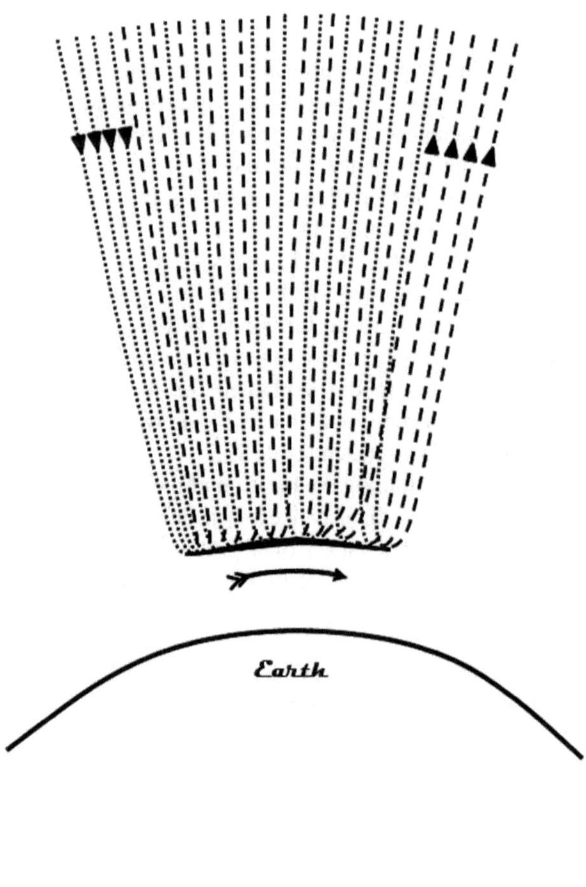

Fig. 12.1 Reproduction of the sketch on p. 105 of Birkeland (1908). The author envisages neutral streams of charged particles flowing down the magnetic field into the auroral ionosphere. It is not clear in the text, but the *dotted* and *dashed lines* presumably represent the streaming charges of opposite sign. The incident electrons (or cathode rays) cause the auroral light display. On the flanks current flows in or out of the upper atmosphere (ionosphere). Within the ionosphere there is a horizontal current in the direction of the *arrow*

allowed for the creation of magnetospheric cavity through the Earth's field pressure holding off the solar wind to create a cavity around the Earth as the leader of the British school Chapman (Chapman and Ferraro 1930) had suggested. However by indicating how the solar wind voltage could enter the terrestrial cavity, it also gave a basis for understanding how the storm magnetic depression built up as a ring current developed around the Earth, which had remained a great frustration for Chapman and Ferraro (1933). The development of magnetohydrodynamics by Alfvén, leader of the Scandinavian school in the 1940s, was an enormous step as was Jim's Ph.D. supervisor Fred Hoyle's audacious suggestion (Hoyle 1949) of allowing for a significant interplanetary field (see chapter "Dungey's Reconnection Model of the Earth's Magnetosphere: The First 40 Years"). Jim's model only worked if one recognised that the frozen flux theorem at the heart of MHD broke down and reconnection occurred. This fact meant that, at times, Jim was subjected to strong scepticism from both sides. Even publishing his thesis was delayed for several years within the British system. At the same time, as Stan Cowley reports in chapter "Dungey's Reconnection Model of the Earth's Magnetosphere: The First 40 Years", diehard supporters of the Scandinavian school would vow to resist the idea of reconnection. Nowadays this is all a curious history. Despite that hindsight, it should not be forgotten as it does serve to emphasise the importance of Jim's achievement.

12.2 The 1859 Storm

It is well-known that from August 28 through to September 3, 1859 there was a great magnetic storm, now called the Carrington storm (Carrington 1860). Moreover it is generally agreed that the observations of white light flares made by Carrington and the subsequent measurements of large global terrestrial magnetic disturbances founded the field of solar-terrestrial science, i.e. the study of the direct material (as opposed to radiative) connection between Sun and Earth.

The global structure and dynamics of the 1859 storm itself remains a matter of present research (Tsurutani et al. 2003; Clauer and Siscoe 2006; Li et al. 2006) because of its scale, with aurorae seen in the tropics and enormous deflections of the Earth's field, and the hazards such a disturbance would present today to our technologically "wired" society. Now the context of the environmental discussions is clear and the major issues are the response to and possible side effects of such a large disruption of the system.

12.3 Human Fallibility

From 1859 to 1961 and then to around 1980 when the idea achieved general acceptance, is a long time. The development of the science of magnetohydrodynamics in the 1940s was probably a necessary precursor. In my view, however, only partly can one explain the long delays by the speed of scientific development. Part of the blame is that scientists are human and subject to human emotions. At times they shy from confronting received ideas of the day, from opening up to ideas that challenge those ideas and also there are issues of personality that preclude effective communication. Scientists, even the greatest scientists, are not always the dispassionate processors of hard facts that perhaps one might hope or assume.

12.4 The Nineteenth Century

There is a certain irony in the attribution of title of originator of solar-terrestrial science to Carrington. At least as important, were the magnetic observations of Balfour Stewart at Kew (Stewart 1860) and indeed the world-wide records of magnetic disturbance from the stations set up a few years before by Edward Sabine. Indeed as noted in Sabine's obituary (Anon. 1883) as early as 1852, Sabine (1852) had noted from analysis of the records from his Toronto station, that magnetic variations could be separated into a repeatable diurnal cycle and an irregular part. However he recorded that the irregularity correlated very closely with fluctuations in the number of sunspots.

It also seems true that Balfour Stewart indeed deserves much credit for keeping the idea that there might be a material connection between Sun and Earth following Carrington's report. Carrington himself remained publicly sceptical. In the report of the November 1859 meeting of the Royal Astronomical Society, one reads: "*Mr. Carrington exhibited at the 1859 November Meeting of the Society a complete diagram of the disk of the sun at the time, and copies of the photographic records of the variations of the three magnetic elements, as obtained at Kew, ...*".

When challenged from the floor as to whether he believed that there could be a direct material connection, he replied cautiously ".... *While the contemporary occurrence may deserve noting, he would not have it supposed that he even leans towards hastily connecting them. One swallow does not make a summer.*" (Carrington 1860). Caution was appropriate. It should be borne in mind that in that era, Kelvin and Rayleigh ruled the British scientific roost and neither was inclined to accept that material could move directly through space from Sun to Earth. Kelvin made his view clear in his presidential address to the Royal Society in 1892 (Kelvin 1892) whilst also referring directly to Sabine's (1852) proposal as well as the events of 1859.

12.5 Kristian Birkeland

Balfour Stewart does seem to have passed the idea of a direct connection to his student (and successor as professor of physics at Manchester University) Arthur Schuster but it is a long time before anyone started to suggest what exactly could constitute the connection. The first concrete suggestion was by Kristian Birkeland, whose specific interest was not so much geomagnetism but to establish the physical nature of the auroral displays regularly seen in the polar regions. Around the turn of the nineteenth century, he led successive expeditions to Finnmark in northern Norway to observe the aurora and to record associated magnetic and atmospheric disturbances. He also deployed stations to the north, east and west in Spitsbergen, Novaya Zemlya (Russia) and Iceland.

Birkeland himself was a remarkable man and his life story is recorded well in two books (Jago 2001; Egeland and Burke 2005). Although he made scientific achievements in many areas, including helping establish what became the largest company in Norway, it is the aurora that seemed his first driving interest. Indeed as both books make clear, he often used the more lucrative aspects of his work (such as fixing nitrogen from air to make fertiliser or producing an electromagnetic gun) to fund more work on the aurora. He showed (Birkeland 1901, 1908, 1913) by triangulation that the light displays were not a meteorological phenomenon. They occurred at much too high an altitude (~100 km). Moreover, he measured magnetic deflections below the auroral displays that show that they are associated with horizontal electrical currents flowing in the atmosphere. He then proposed that the displays are the result of bombardment of the atmosphere from above by electrons. This proposal, correct in fact, was all the more remarkable that it was made only a matter of years after J. J. Thompson's discovery of the electron in 1897. However, Birkeland's knowledge of the experimental behaviour of electrons (or cathode rays) was already great. In parallel with his auroral expeditions, he had set up in Oslo (named Kristiania at the time) an experimental apparatus consisting of a magnetised sphere in a vacuum chamber in which he saw that the electrons were guided towards the poles. This type of experiment, which he called a terrella, was taken up and worked on in various sites around the world until the 1970s. The term terrella is sometimes even attributed to Birkeland. However, Stern (1989) points out that he took the term from Gilbert's sixteenth century work De Magnete (Gilbert 1600).

The bombardment idea and the idea that the electrons originated from the Sun were part of the 800 page report (Birkeland 1908, 1913) of his polar expeditions copies of which were widely circulated (Jago 2001) even to the crowned heads of Europe. As described by Jago, the impact on the scientific world was good except in the United Kingdom. Kelvin's long held view that there could not be a material connection between Sun and Earth was influential. Indeed in 1911, Schuster saw fit to specifically refute Birkeland's proposal of a stream of electrons from the Sun (Schuster 1911). Schuster was probably not biased against there being a material connection between Sun and Earth. However, he pointed out an obvious flaw in Birkeland's argument as he understood it. A charged stream of electrons from the Sun would quench itself as the accumulating negative charge on Earth would repel incoming charges. Birkeland however had already in the second auroral expedition report (Birkeland 1908) suggested that it was a charge neutral stream of positively and negatively charged particles that emanates from the Sun and finds access to the Earth's upper atmosphere in the polar regions by flowing along the magnetic field.

Indeed he proposed that the horizontal currents he detected are closed by field aligned currents flowing between interplanetary space and the upper atmosphere. Part of Birkeland's figure (from p. 105, Birkeland 1908) is redrawn in Fig. 12.1. It seems clear that, by this time, Birkeland had in mind a stream of positively and negatively charged particles guided by the field in a thin sheet with the horizontal upper atmospheric (i.e. ionospheric) current driven by downward and upward currents on the flanks.

In the close of his paper, Schuster (1911) seems to refer to a later theory of Birkeland without expanding more. He proposes that the option for a material link lies in magnetic guidance as Birkeland has proposed and the ionising effects of the radiation (changing the conductivity of the atmosphere) to be the basis for the currents. Where Schuster goes wrong is in concluding that the emf (or voltage) driving the current must be associated with the Earth's rotation. In practice, it is the Sun that provides this but it would be 30 years before such a proposition would be made (Alfvén 1939, 1940).

The basic significance of there being field-aligned currents flowing into and out of the upper atmosphere was thenceforward largely ignored by the large majority of solar terrestrial scientists for about half a century until the discovery of magnetic perturbations from such currents in spacecraft measurements over the auroral zone in 1966 (Zmuda et al. 1966). Even then the idea was only slowly absorbed by the community as a whole.

Unfortunately, serious damage was done by Birkeland not being taken seriously in the British science community which was dominant world-wide at the time. There is something deeper perhaps. Jago (2001) describes Birkeland as taking the British attitude personally. She regards the absence of positive reaction in the British community to the 1908 report as the point of separation of the community into two schools, British and Scandinavian. Furthermore, Jago indicates that the schools were antagonistic and also the separation was based on rather personal antipathy as well as philosophical differences over what was the right basic approach. Arriving in the field about a half a century later, I have to admit that it looks so.

Birkeland himself did not give up the idea of an electrically neutral stream guided by the terrestrial field, the electrons of which impact the upper atmosphere to generate the auroral display. He encouraged Carl Størmer to calculate the possible orbits for charged particles coming from infinity into a dipolar magnetic field (see e.g. Stoermer 1917a, b). This work underpinned much early work on the penetration of cosmic rays, once it was realised that they were energetic charged particles. Its contribution to auroral physics was limited simply because, for charged particles coming from the Sun, the Earth did not present a simple dipole magnetic field. In fact, the presence of a field in interplanetary space substantially larger than that predicted by either the dipole of the Earth or the Sun would turn out a crucial feature of the ultimate answer. It would be through this that the Sun could impose an electromotive force on the Earth's upper atmosphere.

12.6 Sidney Chapman

The basis for seeing the difference in philosophy between British and Scandinavians is seen best by looking at the work of Sidney Chapman.

In 1918, Sidney Chapman, the person who was to become the leader of the British school, published his first important paper (Chapman 1918). It was a statistical study of geomagnetic storm disturbance fields with a theory appended. The theory put forward by Chapman, almost certainly in ignorance of at least the detail of Birkeland's proposals in the previous two decades, proposed that a stream of charged particles largely of one sign was the cause of the storm. Lindemann (1919) raised the same objections as Schuster had towards Birkeland's idea more than a decade before. Chapman made no further theoretical proposal until the next decade. In a series of papers Chapman and Ferraro (1930, 1931a, b, 1932, 1933) produced a new theory of geomagnetic storms. By statistical work using data from around the globe and removing systematically the quiet day variation, Chapman had shown that a geomagnetic storm began with a compression of the magnetic field world-wide. For the initial phase, Chapman and Ferraro postulated a neutral stream of charged particles coming from the Sun whose perfect conductivity caused the confinement of the terrestrial field in a cavity. The compression of the field envisaged by Chapman and Ferraro during the storm commencement is shown in Figs. 12.2 and 12.3, based on sketches in Chapman and Bartels (1940).

Chapman and Ferraro's approach was not a complete solution. They modelled the confinement of the terrestrial field by using an image dipole to achieve the exclusion of the Earth's field from the stream. The boundary then is, by symmetry, planar as shown in the left hand sketch in Fig. 12.2. The full three-dimensional solution for the formation of the cavity cannot be done analytically. In fact, Jim Dungey published an analytic two-dimensional solution of the Chapman-Ferraro problem in 1961 (Dungey 1961b) and three-dimensional numerical solutions appeared only later in the 1960s (Mead and Beard 1964). The points marked Q are where the field goes to zero. Particles could enter directly near these points and, traveling along the interior field, would arrive in the atmosphere at high latitudes but not on the night side. Figure 12.3 shows a sketch of what was envisaged in the equatorial plane.

Although the first papers describing the cavity formation were clearly on the right track in explaining the initial compression, the model did not explain well why subsequently the terrestrial field is depressed as the storm main phase develops. The depression was attributed to a ring current developing about the Earth in the equatorial plane but, having excluded the solar material from the cavity around the Earth, Chapman and Ferraro had difficulty in explaining how the ionised material would enter the cavity and form a closed ring around the Earth. The vague arrows in the sketch on the right of Fig. 12.3 indicate the idea that somehow charged particles would leak into the vacuum field behind the magnetic barrier on the flanks.

Figures 12.1 and 12.3 illustrate the fundamental difference between the notions of Birkeland and Chapman and Ferraro. Birkeland assumed that somehow the

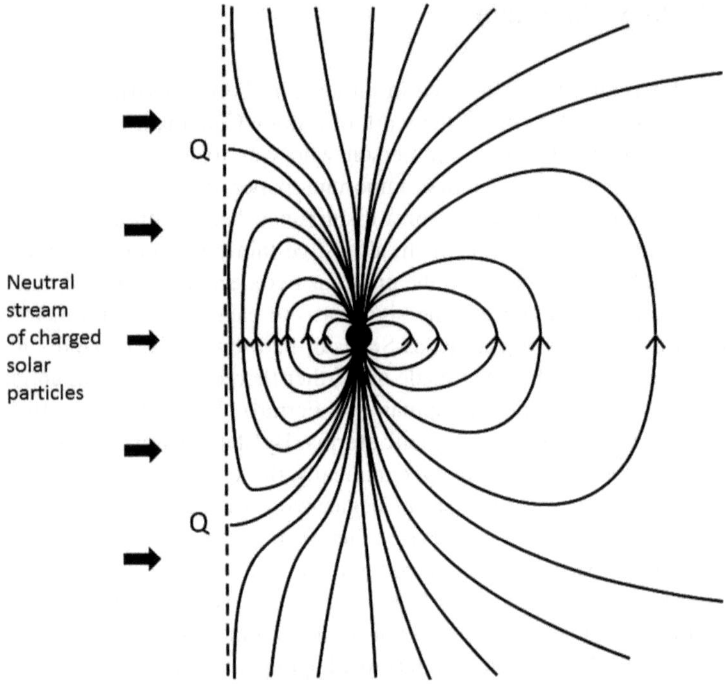

Fig. 12.2 A reproduction of Chapman and Ferraro's picture for the noon-midnight meridian of the enclosure of the terrestrial magnetic field by an advancing stream of solar charged particles

auroral particles can access directly the terrestrial magnetic field and travel along it. Chapman and Ferraro drew attention to the fact that a stream of material incident more or less at right angles to the terrestrial field would be expected to deflect and exclude solar material.

The boundary between the solar and terrestrial regimes was predicted to be thin. The incident negatively charged electrons are turned through 180° within the current layer as are the protons (Ferraro 1952). An electrostatic field keeps the protons (which start with higher inertia) from penetrating further. With the advent of the space age and the discovery of the solar wind, the magnetopause, a thin boundary between the solar and terrestrial plasma was identified as the Chapman-Ferraro boundary (Cahill and Amazeen 1963). Of course, by that time, the solar wind was known to be continuously present and the boundary a permanent feature of the solar-terrestrial system.

Chapman and Ferraro were aware there were severe limitations in their model. Their 1933 work closes with the following paragraph:

> In concluding this section we wish again to state that we are conscious of many unsatisfactory features in our present treatment which pose rather than overcome the theoretical difficulties of the problem. We intend however, to make further efforts after a solution along any avenues that may seem open.

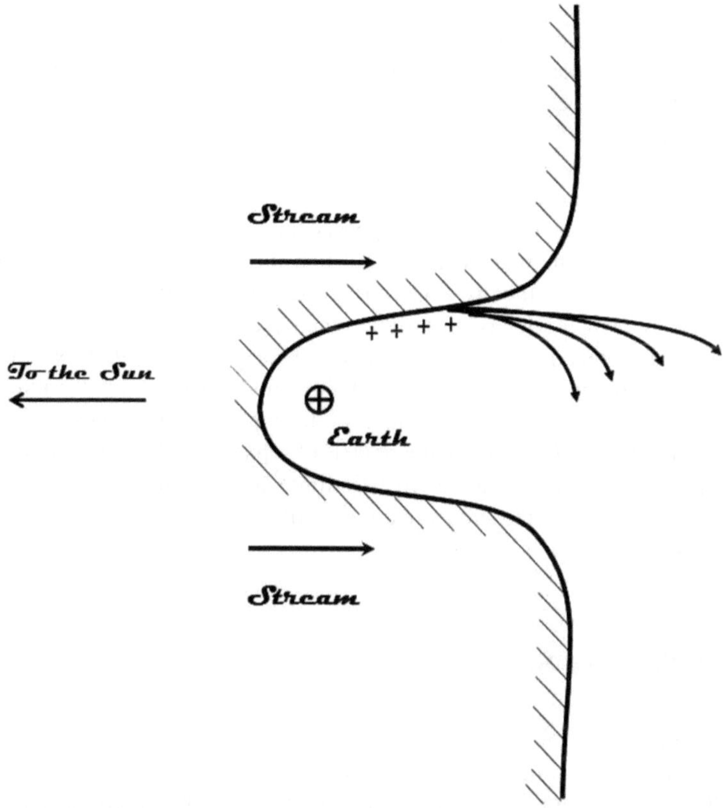

Fig. 12.3 The equatorial view of the Chapman-Ferraro picture. The *curved lines* on the flanks are intended to represent the possibility of entry of solar material there, which would then contribute to the main phase ring current

12.7 Hannes Alfvén

Birkeland died in 1917. There was no mention of his work and ideas in the works of Chapman and Ferraro. Indeed, he was ignored from then on fairly uniformly within the British school although Chapman's major work on geomagnetism (Chapman and Bartels 1940) does refer to Birkeland's work and not wholly negatively. Nonetheless, it fell to the Swedish scientist, Hannes Alfvén, to keep his ideas alive. Almost 10 years after the Chapman-Ferraro model, Alfvén developed a very different model of geomagnetic storms (Alfvén 1939, 1940). There is no attempt to accommodate the Chapman-Ferraro model in Alfvén's model. His model focuses on two things, namely to explain the ring current and the access of charged solar particles to the top of the atmosphere to provide the aurora. Alfvén himself was a great admirer of Birkeland and the model does resemble a development of his ideas. However it is wrong. Importantly it assumes that the incident charged particle stream will be magnetised and simply flow onto the Earth field. In other words there is no allowance for a magnetopause. However the model introduces the idea of an electric field being produced in the terrestrial environment by the solar interaction, effectively the introduction of a voltage (or emf in the old nomenclature) imposed from outside on the terrestrial system. In practice, Alfvén's electric field is in the opposite direction to the one that actually exists but its introduction is an important step.

A vigorous refutation of Alfvén's proposal came from Cowling (1942). Cowling, a former student of Chapman's, points out that the theory ignores the Chapman-Ferraro model and specifically does not explain the impulsive start of storms. Alfvén's paper uses the ideas of particle mean drift to describe charged particle motion that later became known as adiabatic theory (Alfvén 1950; Northrop and Teller 1960). Although conceding that Alfvén's model does tackle the main phase of storms and so the possibility of a ring current, Cowling's paper is a strong refutation of Alfvén's. Once again the issue is of separation of positive and negative charges and the corresponding electrostatic force and also the lack of treatment of the diamagnetic effect of the solar stream (effectively as illustrated in the Chapman-Ferraro theory).

12.8 A Personal Story

It is clear in retrospect that Alfvén did not accept the refutation. Moreover, he probably felt that Chapman had suggested Cowling write a critique of Alfvén's model. There is no point here in going further into the individual rights and wrongs in the old battle. A schism was now clear between what might be called the British school and the Scandinavian.

In 1942, there was a world war under way. Alfvén living in neutral Sweden had to wait until the end of hostilities to confront Chapman with his storm model. His opportunity came in 1946. Chapman had just moved from Imperial College to Oxford University. Alfvén arranged to make a 1 day visit to Chapman in Oxford. The story that now follows was told to me directly by Alfvén in the late 1970s. I was visiting the University of California San Diego where he then was. Whether the story was elaborated for effect, I don't know. What I do know is that 30 years on Alfvén still resented his treatment at the hands of Chapman.

Alfvén described coming down by train from London to Oxford to find Chapman awaiting him at the station. Politeness itself, Chapman asked if it was Alfvén's first visit to the city of dreaming spires. It was and so Chapman proposes that they must undertake some essential tourism. Alfvén says that he wished to discuss his storm theory but Chapman demurs and starts a walk around the tourist sites of Oxford. After the Ashmolean and St. Mary's church and a couple of colleges, Alfvén asks if they can discuss his theory. "Oh, but it is time for lunch at my college, so later" replies Chapman. Lunch is followed by Chapman insisting on a stroll along the banks of the Isis and then tea. As Alfvén told it, he kept raising his theory and being politely told that later they would have the discussion. The day ends with Alfvén never having got Chapman to a blackboard or even a blank sheet of paper to outline his ideas and confront their differences. Instead, Alfvén finds himself being put on the train by an ever-polite Chapman. His final words were, "but when will discuss my theory?" The response was, "Maybe, next time". The story is perhaps too good to be true but precisely true or not, I am sure that Alfvén felt it was how he had been treated. The schism was complete. Chapman and his school dismissed the Scandinavians and Alfvén was determined that Chapman was wrong. Of course both were a bit right and still quite a lot wrong but the absence of communication certainly did not help progress.

12.9 The Birkeland Symposium 1967

The antagonism between Alfvén and Chapman probably led to one of the most remarkable displays of the gap between the Chapman's school and the followers of Birkeland (Borowitz 2008). Leaping a little in time, let us go to 1967, the centenary of Birkeland's birth. As noted, Birkeland's work and ideas were ignored by much of the world geophysical community in the 1920s, 1930s and 1940s. However, that was not so in his homeland. No doubt in part because he had had a part in founding the enormous industrial enterprise Norsk Hydro, as well as because of the memory of his original expeditions to the north to measure the aurora, he had remained well-known in Scandinavia. Even today, Birkeland appears on the Norwegian 200 Kr note and even has his portrait featuring on the tail of an Air Norwegian Boeing 737-800 airliner (call sign LN-NOQ).

In 1967, the Norwegian Academy of science decided to mark his centenary with a major symposium (Egeland and Holtet, 1968). Ironically, the Birkeland Symposium was to occur in the same year that Cummings and Dessler (1967) were to point out that the magnetic fields recently detected by a US Air Force spacecraft passing over the auroral zone (reported by Zmuda et al. 1966) were likely to be due to the field-aligned currents that he had predicted in the first decade of the century. However, of course, no one knew that at the planning stage of the celebration. It was nevertheless intended to mark seriously the centenary of the great scientist. What could be more fitting for celebration of a nationally respected scientist than to invite as a keynote speaker, Sydney Chapman, at that time arguably the best known of solar terrestrial scientists? Chapman had had by 1967 a very distinguished scientific career spanning 50 years. He had made major contributions in many fields of aeronomy, geomagnetism and solar physics. Indeed, he had been not only one of the originators of the idea of the International Geophysical Year but with Lloyd Berkner, he had steered the global project through to its successful culmination in 1957 with the dawn of the space age.

I was a raw postgraduate student at the time and did not attend the Birkeland symposium. Nonetheless, Chapman's keynote talk was published in the conference proceedings (Chapman 1968) and so there is little doubt about what he intended to say and was said. A phrase from the written paper that expresses the dismissive overall tone of his paper is: "*Though Birkeland was certainly interested in the aurora and devoted a great effort to organization and support to expeditions to increase our knowledge of it, it must be confessed that his direct observational contributions were slight...*" According to eyewitnesses, in particular Alex Dessler (now of Texas A&G University) and Gordon Rostoker (University of Alberta), Chapman's oral patronising style exceeded that of the published paper and even the non-specialists in the audience were stunned by the put-down of Birkeland.

Why did this event occur? Chapman's dismissal of Birkeland's observational contributions is remarkable and straightforwardly wrong. Coming as it did in the year where his prediction of field-aligned currents was to be proven by space observations, the dismissal of his ideas is unfortunate and looks foolish 40 or so years later. Anyone who met Sydney Chapman would have been struck by his polite rather genteel nature. On the few occasions when I, a junior member of the community, met or was introduced to him, he always treated me with respect and interest.

The answer to the conundrum of his behaviour could well lie in a story told by Chapman's student, Vincent Ferraro (Ferraro 1969) given at a meeting to mark Sydney Chapman's 80th birthday, "*In 1918, Chapman added at the end of his paper on the average characteristics of magnetic storms an atmospheric-type theory of the origin of such storms, which he ascribed to the action of a stream of charged particles from the sun, mainly of one sign.*" In other words, more than a decade and a half later, Chapman (1918) had made the same mistake as Birkeland, in proposing that the aurora and allied disturbances originated from streams of electrons from the Sun. Chapman went into print (Chapman 1918). It was not to be Schuster who refuted Chapman but Lindemann (1919), who pointed out, just as Schuster had in 1911, that the theory could not work because of the electrostatic forces that would quench the flow. In Ferraro's words again: "*Lindemann (i.e. the future Lord Cherwell—friend of Churchill) in 1919 criticized Chapman's numerical development of the theory chiefly on the grounds that it involved an accumulation of charge on the earth's atmosphere which would, by electrostatic repulsion, prevent the supposed continued entry of further charges.*" Chapman's mistake is basic and it is hard to believe that two great scientists made the same error. As Ferraro indicates, since there had been a numerical development of the theory (and so likely a fair amount of effort devoted) Chapman's feelings must have been made even more acute.

For me, this story from 50 years before is probably at the root of Chapman's undoubted lapse of respect for protocol in 1967. Chapman certainly referred to his 1918 theory as phony (Akasofu 1995). However distaste for reopening a wound may also explain other aspects of what happened in intervening years. The personalities of the two men differed greatly and the way they did science matched that. Chapman was a mathematician by nature. Alfvén had started his career as an electrical engineer. Chapman was reported to refer to him as that Swedish engineer (Borowitz 2008). Ironically, Chapman had done a degree in engineering in Manchester before going to Cambridge to study mathematics. There was a deep difference in scientific approach. Chapman from the earliest days, i.e. his 1918 work on geomagnetic storms, preferred a statistical approach.

In 2010, I received a paper that was eventually published as Akasofu (2011). One paragraph in the original draft text quoted Chapman on the field aligned current issue. Akasofu withdrew the remarks from the published version. It is probably worth giving Chapman (through Akasofu) the final word: "*Chapman mentioned in his letter to me on 13 April 1969, "the history of studies of geomagnetic disturbances is a tangled skein," and he continued "---but I did overlook something (a three-dimensional current system, the author's insertion) to which I was blind and they (Birkeland and Alfvén, the author's insertion) saw. Perhaps people listened too much to me ---.*" This sounds to me like Chapman speaking.

12.10 Dungey's Model

As already noted, the final illumination concerning the basic solar terrestrial coupling mechanism needed the medium of a cup of coffee in a Parisian café. A citation classic resulted and Dungey (1961a) continues to be cited regularly to this day. Cowley (chapter "Dungey's Reconnection Model of the Earth's Magnetosphere: The First 40 Years") and Milan (chapter "Sun et Lumière: Solar Wind-Magnetosphere Coupling as Deduced from Ionospheric Flows and Polar Auroras") go through many of the issues of the model. Stan Cowley shows Jim's simple hand drawn figure, which, once correctly interpreted, gives the key. Figure 12.4 shows the ideas of the sketch with a little filling out. The sketch is a projection in the noon-midnight meridian and is certainly not supposed to be drawn to scale. The solar wind is assumed to be present and to be carrying with it an extended solar field (much larger than the solar dipole would provide at the Earth's distance). It was this interplanetary field which Hoyle (1949) proposed to Dungey that is the key. Figure 12.4 is shown with a southward interplanetary field component, the simplest case. The terrestrial polar cap field can connect through the magnetopause marked by the dashed line and thus to the interplanetary field. (In fact, no magnetopause was shown in Jim's original sketch.) Magnetic reconnection (a major topic of Jim Dungey's Ph.D. thesis 11 years earlier) in the vicinity of neutral points on the dayside and nightside of the Earth (marked X on the sketch) allows the connection of magnetic flux. Indeed the process need not balance day and night and so the amount of magnetic flux in the polar caps that connects to the interplanetary field can vary with the consequences that Cowley and Milan describe in chapters "Dungey's Reconnection Model of the Earth's Magnetosphere: The First 40 Years" and "Sun et Lumière: Solar Wind-Magnetosphere Coupling as Deduced from Ionospheric Flows and Polar Auroras", respectively. However, most importantly, everywhere except in the immediate vicinity of regions where the field is very small, magnetohydrodynamics holds sway and one may picture the voltage associated with the motional electric field in the solar material flowing outward from the Sun projected down to the ionosphere. In Fig. 12.4 the open arrows

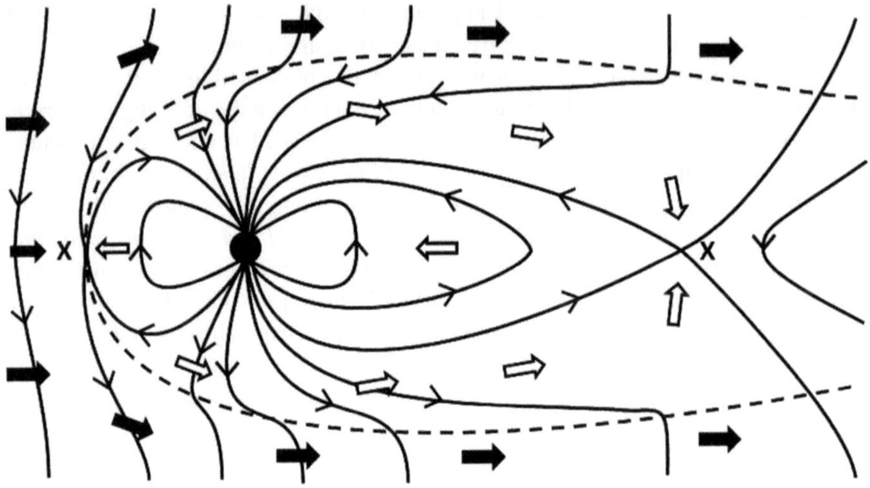

Fig. 12.4 An illustrative sketch of Dungey's open magnetosphere. The sketch is drawn in the noon-midnight meridian. The *dashed curve* represents the magnetopause. The *dark arrows* represent the solar wind flow deflected around the Earth. The magnetic field lines thread the magnetopause and the solar wind electric field is projected into the polar cap. A secondary flow is thereby driven (*open arrows*) inside the magnetosphere. On the dayside and nightside magnetic reconnection occurs at the locations indicated by X. The simplest case of purely southward external field is assumed

indicate that the transfer of momentum sets up an internal flow in the terrestrial system. This is due to the solar voltage (or emf) imposed on the terrestrial system whose absence Schuster (1911) had noted and Alfvén (1940) had tried to provide. If a voltage is being imposed across a conductor there must be electrical currents not only in the conducting medium itself but also connecting ultimately to whatever is driving the system. Figure 12.5 shows the dawn-dusk meridian of the system (again for the simplest purely southward field case). The external electric field penetrates along the field lines which in steady state are equipotentials. The dawn-dusk electric field is imposed across the dissipative conducting ionosphere. Current must flow in the polar cap ionosphere in the electric field direction from dawn to dusk. Thus in this simplest of pictures, field aligned (Birkeland) currents are expected to flow in on the dawn side and out on the dusk side. In fact, the possibility of even considering the locations of vertical current flow had been ruled out in the pre-space age era by Chapman's school of thought through their insistence on taking ground-based magnetic data and interpreting the effect in the ionosphere as an *equivalent* current system. It is a mathematical fact that attributing a set of magnetic measurements made on the surface of a sphere to a set of electrical currents flowing in a spherical layer above that surface has a unique solution only if the currents are entirely confined within that layer. Indeed that was the basis of the idea of equivalent current system. Chapman, despite his original engineering degree was clearly by nature a mathematician. However, the Dungey model makes it unavoidable that there are field aligned currents as Birkeland had surmised.

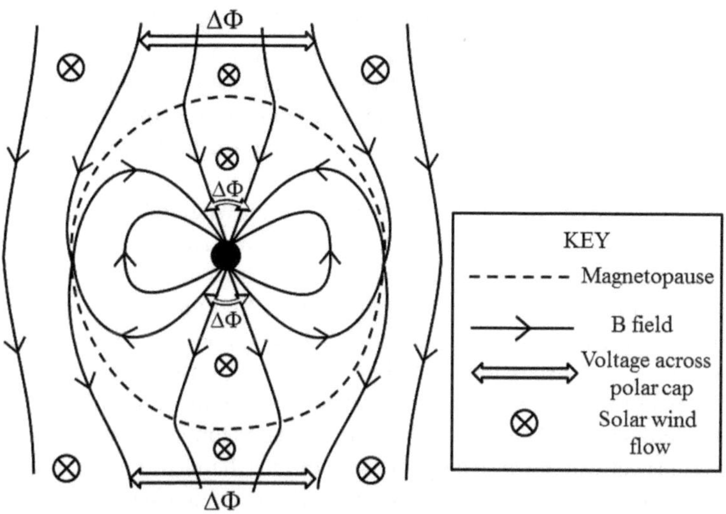

Fig. 12.5 Dawn-dusk meridian view of the Dungey magnetosphere. The solar wind flows into the page. The electric field voltage is projected down the (equipotential) field lines of the polar cap to the polar ionosphere. The simplest case of purely southward external field is assumed

Any utility in using equivalent horizontal current systems is undermined if field aligned current is a fundamental feature of the actual system (Fukushima 1969, 1976, 1994).

12.11 The First 20 Years of the Reconnection Model

Happily in the end, Dungey's (1961a) insights squared the circle, solved the problem and led to today's sophisticated understanding of the system. However the philosophical divisions in the community certainly delayed the acceptance of the Dungey picture. Here I have my direct personal experience to rely on in saying that an element of the delay was associated with departures from the breadth of thought and dispassionate thinking that one associates with the ideal scientist. The community divisions created in previous years had left their mark.

Experimental evidence of the southward interplanetary magnetic field's significance in the scale of geomagnetic activity emerged fairly soon in the 1960s. As a new postgraduate student, I found a paper by Fairfield and Cahill (1966) that fitted rather well with the ideas of Jim (and not with anyone else). This was an immense relief as he was my Ph.D. supervisor. As so many of the succinctly expressed ideas in Jim's 1961 paper allowed one then to deduce the basis of so much else that was emerging from spacecraft in the solar terrestrial system, I felt from that time onward that I had a head start on most others in the field (Southwood 1997).

However controversy raged. The schisms now were at least threefold. There were those who believed in a purely MHD magnetosphere (i.e. the Chapman-Ferraro idea in which the currents on the boundary separating the terrestrial and solar regimes were effectively diamagnetic. Then there was the Dungey school for whom the electric field of the magnetised solar medium could penetrate the boundary thereby rendering not only the boundary currents resistive but also allowing the projection into the ionosphere of a solar wind induced voltage. Lastly, there was a Scandinavian school whose leader, Hannes Alfvén, did not believe that Cahill and Amazeen's (1963) detection of a thin boundary or current sheet (the magnetopause) between the terrestrial and external regime represented a triumph for the Chapman-Ferraro theory of 30 years before. I found this latter fact baffling at the time and in my discussions with him in the 1970s asked him why it was not so. His response was that Chapman had never predicted there would be a significant magnetic field outside. That was in fact true. I pressed him about the presence of the thin current sheet which was surely a feature of the Chapman-Ferraro model. "That was where Dungey was clever" was his response to my utter surprise. Alfvén's own working on the thin sheet aspect was published in a short paper in 1968 (Alfvén 1968). The simple symmetry used is more appropriate to tail reconnection than the magnetopause. The paper was greeted with great interest in the Dungey group and an extension of the work formed the basis of Stan Cowley's doctoral thesis (Cowley 1971).

Even use of the word "reconnection" became difficult. In 1970, Aubry et al. (1970) published what seemed to me "hanging evidence" for the Dungey model by detecting with the OGO5 spacecraft in situ the inward motion of the magnetopause in the presence of an external southward field. The paper does not reference Jim nor was the word reconnection used. One of the authors, Chris Russell, wrote to me with his characteristic, even naïve, candour about this lacuna recently. His note says, "*It is very clear to me that Dungey was not part of the club in the 1960's and 70's. When we found the inward moving magnetopause when the IMF turned southward we did not use the term reconnection for it and I think we did not reference Jim.**" A footnote to his note adds "* *this was not due to ignorance but to wanting to get the paper published.*"

As the case for a reconnection magnetosphere became increasingly unanswerable, some changed school. Ian Axford was the most notable (Axford et al. 1965). He had previously propounded an MHD model of the magnetosphere across whose boundary there would be a viscous transfer of stress (Axford and Hines 1961). Elsewhere, new terms were introduced to escape using the term reconnection, such as magnetic merging or magnetic annihilation. The final triumph of Jim's view came around 1980. The pivotal observations of plasma and magnetic fields by the ISEE (International Sun Earth Explorer) 1 and 2 spacecraft in a region where reconnection was heating and accelerating plasma in the magnetopause was reported by Paschmann et al. (1979). By 1983, Jim was able to feel that his view had prevailed (Dungey 1983).

The rest is a happier history as is shown by the accompanying chapters by Stan Cowley and Steve Milan (chapters "Dungey's Reconnection Model of the Earth's

Magnetosphere: The First 40 Years" and "Sun et Lumière: Solar Wind-Magnetosphere Coupling as Deduced from Ionospheric Flows and Polar Auroras").

Acknowledgements I am very happy to thank Stan Cowley for a careful reading of the text and for pointing out a couple of sources of which I was not aware.

References

Akasofu, S.-I. (1995), A Note on the Chapman-Ferraro Theory, in Physics of the Magnetopause edited by Song, P., B. U. O. Sonnerup, and M. F. Thomsen, (Geophysical Monograph Series) AGU, Washington pp 5–10

Akasofu, S.-I.: The scientific legacy of Sydney Chapman. Eos. Trans. AGU **92**(34), 281 (2011)

Alfvén, H.: Some properties of magnetospheric neutral surfaces. J. Geophys. Res. **73**, 4379–4381 (1968)

Alfvén, H.: Cosmical Electrodynamics, 1st edn., p. 237. Oxford University Press, New York (1950)

Alfvén, H.: A theory of magnetic storms and of the aurorae. K. Sven. Vetenskapakad. Handl. Ser. 3, **18**(3) (1939) Reprinted in part with commentary by A.J. Dessler and J. Wilcox: Eos, **51**, 180–194 (1970)

Alfvén, H.: A theory of magnetic storms and of the aurorae, II, The aurorae; III, The magnetic disturbances. K. Sven. Vetenskapakad. Handl. Ser. 3, **18**(9) (1940)

Anon: Obituary—Sir Edward Sabine. Observatory **6**, 232–233 (1883)

Aubry, M.P., Russell, C.T., Kivelson, M.G.: Inward motion of the magnetopause before a substorm. J. Geophys. Res. **75**, 7018–7031 (1970). doi:10.1029/JA075i034p07018

Axford, W.I., and C.O. Hines (1961), A unifying theory of high-latitude geophysical phenomena and geomagnetic storms, Can. J. Phys., 39, 1433–1464

Axford, W.I., Petschek, H.E., Siscoe, G.L.: Tail of the magnetosphere. J. Geophys. Res. **70**, 1231–1236 (1965)

Birkeland, K.: Expedition Norvegienne de 1899–1900 pour l'etude des aurores boréales, Skr. Nor. Vidensk. Akad. KI. 1 Mat. Naturvidensk. Kl., No. I. (1901)

Birkeland, K.: The Norwegian Aurora Polaris Expedition, 1902–1903, vol. I, section 1. H. Aschehoug, Oslo (1908)

Birkeland, K.: The Norwegian Aurora Polaris Expedition, 1902–1903, vol. I, section 2. H. Aschehoug, Oslo (1913)

Borowitz, S.: The Norwegian and the Englishman. Phys. Perspect. **10**, 287–294 (2008). doi:10.1007/s00016-007-0372-0

Cahill, L.J., Amazeen, P.G.: The boundary of the geomagnetic field. J. Geophys. Res. **68**, 1835–1843 (1963)

Carrington, R.C.: Description of a singular appearance seen in the Sun on September I, 1959. Mon. Not. R. Astron. Soc. **20**, 13–15 (1860)

Chapman, S.: An outline of a theory of magnetic storms. Proc. R. Soc. Ser. A **97**, 61–83 (1918)

Chapman, S.: Historical introduction to aurora and magnetic storms. In: Egeland, A., Holtet, J.A. (eds.) The Birkeland Symposium on Aurora and Magnetic Storms, pp. 21–29. Centre National de la Recherche Scientifique, Paris (1968)

Chapman, S., Bartels, J.: Geomagnetism. Oxford University Press, Oxford (1940)

Chapman, S., Ferraro, V.C.A.: A new theory of magnetic storms. Nature **126**, 129–130 (1930)

Chapman, S., Ferraro, V.C.A.: A new theory of magnetic storms, Part I. The initial phase. Terr. Magn. Atmos. Electr. **36**, 77–97 (1931)

Chapman, S., Ferraro, V.C.A.: A new theory of magnetic storms, I. The initial phase. Terr. Magn. Atmos. Electr. (now, J. Geophys. Res.) **36**, 77–97, 171–186 (1931b)

Chapman, S., Ferraro, V.C.A.: A new theory of magnetic storms, I. The initial phase (continued). Terr. Magn. Atmos. Electr. (now, J. Geophys. Res.) **37**, 147–156, 421–429 (1932)

Chapman, S., Ferraro, V.C.A.: A new theory of magnetic storms, II. The main phase. Terr. Magn. Atmos. Electr. (now, J. Geophys. Res.) **38**, 79–96 (1933)

Clauer, C.R., Siscoe, G.L.: The great historical geomagnetic storm of 1859: a modern look. Adv. Space Res. **38**, 117–118 (2006)

Cowley, S.W.H.: The adiabatic flow model of a neutral sheet. Cosmic Electrodyn. **2**, 90 (1971)

Cowling, T.G.: On Alfvén's theory of magnetic storms and of the aurorae. Terr. Magn. Atmos. Electr. (now, J. Geophys. Res.) **47**, 209–214 (1942)

Cummings, W.D., Dessler, A.J.: Field-aligned currents in the magnetosphere. J. Geophys. Res. **72** (3), 1007–1013 (1967). doi:10.1029/JZ072i003p01007

Dungey, J.W.: Interplanetary magnetic field and the auroral zones. Phys. Rev. Lett. **6**, 47 (1961a)

Dungey, J.W.: The steady state of the Chapman Ferraro problem in two dimensions. J. Geophys. Res. **35**, 1043 (1961b)

Dungey, J.W: http://garfield.library.upenn.edu/classics1983/A1983RQ98100001.pdf (1983)

Egeland, A., Burke, W.J.: Kristian Birkeland, the First Space Scientist, Astrophysics and Space Science Library. Springer, Dordrecht (2005)

Egeland, A., Holtet, J.: The Birkeland Symposium on Aurora and Magnetic Storms. Centre National de la Recherche Scientifique, Paris (1968)

Fairfield, D.H., Cahill Jr., L.J.: Transition region magnetic field and polar magnetic disturbances. J. Geophys. Res. **71**(1), 155–169 (1966). doi:10.1029/JZ071i001p00155

Ferraro, V.C.A.: On the theory of the first phase of a geomagnetic storm: new illustrative calculation based on an idealised (plane not cylindrical) model field distribution. J. Geophys. Res. **57**, 15–49 (1952). doi:10.1029/JZ057i001p00015

Ferraro, V.C.A.: The birth of a theory. In: Akasofu, S.-I., Fogle, B., Haurwitz, B. (eds.) Sidney Chapman, Eighty, pp. 14–18. University of Colorado Press, Boulder (1969)

Fukushima, N.: Equivalence in ground geomagnetic effect of Chapman-Vestine's and Birkeland-Alfvén's current systems for polar magnetic storms. Rep. Ionos. Space Res. Jpn. **23**, 219–227 (1969)

Fukushima, N.: Equivalence in ground geomagnetic effect of vertical currents connected with Pedersen currents in the uniform conductivity ionosphere. Rep. Ionos. Space Res. Jpn. **30**, 35–40 (1976)

Fukushima, N.: Some topics and historical episodes in geomagnetism and aeronomy. J. Geophys. Res. **99**(A10), 19113–19142 (1994)

Gilbert, W.: De Magnete (On the Magnet) translated by Silvanus P. Thompson, reprinted from the 1900 edition published by Basic Books, New York, 1958. Online text: http://ebooks.adelaide.edu.au/g/gilbert/william/on-the-magnet/book1.2.html (1600)

Goldie, A.H.R.: The electric current-systems of magnetic storms. Terr. Magn. Atmos. Electr. (now J. Geophys. Res.) **42**(2), 105–107. doi: 10.1029/TE042i002p00105 (1937)

Hoyle, F.: Some Recent Researches in Solar Physics. Cambridge University Press, Cambridge (1949)

Jago, L.: The Northern lights: how one man sacrificed love, happiness and sanity to unlock the secrets of space. Hamish Hamilton, London (2001)

Kelvin, Lord W. T. (1892), Royal Society Presidential Address, Nature, 47, 107–110

Li, X., Temerin, M., Tsurutani, B.T., Alex, S.: Modeling of 1–2 September 1859 super magnetic storm. Adv. Space Res. **38**, 273–279 (2006)

Lindemann, F.A.: Note on the theory of magnetic storms. Philos. Mag. **38**, 669–684 (1919)

Mead, G.D., Beard, D.B.: Shape of the geomagnetic field solar wind boundary. J. Geophys. Res. **69** (7), 1169–1179 (1964). doi:10.1029/JZ069i007p01169

Northrop, T. G. and E. Teller (1960), Stability of the Adiabatic Motion of Charged Particles in the Earth's Field, Phys. Rev. 117, 215–225

Paschmann, G., Sonnerup, B.U.Ö., Papamastorakis, I., Sckopke, N., Haerendel, G., Bame, S.J., Asbridge, J.R., Gosling, J.T., Russell, C.T., Elphic, R.C.: Plasma acceleration at the Earth's magnetopause: evidence for reconnection. Nature **282**, 243–246 (1979). doi:10.1038/282243a0

Sabine, E.: On periodical laws discoverable in the mean effects of the larger magnetic disturbances. No. II. Philos. Trans. R. Soc. Lond. **142**, 103–124 (1852). doi:10.1098/rstl.1852.0009

Schuster, A.: The origin of magnetic storms. Proc. R. Soc. Ser. A **85**(575), 44–50 (1911)

Southwood, D.J.: An education in space physics. In: Gillmor, S. (ed.) Discovery of the Magnetosphere, History of Geophysics, vol. 7, p. 185. American Geophysics Unit, Washington (1997)

Stern, D.P.: A brief history of magnetospheric physics before the spaceflight era. Rev. Geophys. **27**, 103–114 (1989)

Stewart, B.: On the great magnetic disturbance of August 28 to September 7, 1859, as recorded by photography at the Kew observatory. Proc. R. Soc. Lond. **11**, 407–410 (1860)

Stoermer, C.: Corpuscular theory of the aurora borealis. Terr. Magn. Atmos. Electr. (now J. Geophys. Res.) **22**, 23–34 (1917a)

Stoermer, C.: Corpuscular theory of the aurora borealis. Terr. Magn. Atmos. Electr. (now J. Geophys. Res.) **22**, 97–112 (1917b)

Tsurutani, B.T., Gonzalez, W.D., Lakhina, G.S., Alex, S.: The extreme magnetic storm of 1–2 September 1859. J. Geophys. Res. **108**(A7), 1268 (2003). doi:10.1029/2002JA009504

Zmuda, A.J., Martin, J.H., Heuring, F.T.: Transverse hydromagnetic disturbances at 1100 km in the auroral region. J. Geophys. Res. **71**, 5033–5045 (1966)

MIX
Papier aus verantwortungsvollen Quellen
Paper from responsible sources
FSC® C105338

If you have any concerns about our products,
you can contact us on
ProductSafety@springernature.com

In case Publisher is established outside the EU,
the EU authorized representative is:
Springer Nature Customer Service Center GmbH
Europaplatz 3, 69115 Heidelberg, Germany

Printed by Libri Plureos GmbH
in Hamburg, Germany